Progress in Colloid & Polymer Science · Vol. 68

Progress in Colloid & Polymer Science

Editors: H.-G. Kilian (Ulm) and A. Weiss (Munich)

Frontiers in Colloid Science
In Memoriam
Professor Dr. Bun-ichi Tamamushi

guest editors:
M. Nakagaki (Kyoto), K. Shinoda (Yokohama),
and E. Matijević (Potsdam N. Y.)

Springer-Verlag Berlin Heidelberg GmbH 1983

ISBN 978-3-662-16072-5   ISBN 978-3-7985-1703-5 (eBook)
DOI 10.1007/978-3-7985-1703-5
ISSN 0340-255 X

© Springer-Verlag Berlin Heidelberg 1983 – Production: H. Frey
Originally published by Dr. Dietrich Steinkopff Verlag GmbH & Co. KG, Darmstadt in 1983
Softcover reprint of the hardcover 1st edition 1983

Type-Setting and Printing: Hans Meister KG, Druck- und Verlagshaus, Kassel

Contents

For the Memory of Professor Dr. B. Tamamushi

While a "Festschrift" in honor of Professor Bunichi Tamamushi was being prepared the sad new reached us that he died at the age of 84. In him our discipline has lost a pioneer who made contributions to colloid and surface science for half a century.

Professor Tamamushi was born on October 18, 1898. He graduated from the Tokyo University in 1922 where he also received his D. Sci. degree in 1935. During 1927–1929 he studied at the Kaiser-Wilhelm-Institut in Berlin with Professor Freundlich. Professor Tamamushi spent all of his life in academic positions. After graduation he worked as assistant at the Rikagaku Research Institute which was followed by 25 years of teaching at the Musashi High School. From 1949 till 1959 he was appointed Professor of the Tokyo University. After his retirement he continued to work at Tokyo Women's College (1959–1969), Musashi University (1969–1975) and as Honorary Director of the Neza Chemical Institute till his death.

Tamamushi's research dealt with many problems of colloid and surface science, including properties of clays, surfactants, liquid crystals, colloid stability, adsorption phenomena, two dimensional equations of state, rheology and thixotropy to mention a few. Although he spent a good part of his earlier professional life teaching at a high school, Professor Tamamushi was able to continuously carry out research.

Professor Tamamushi was also active in professional societies. He was Regional Editor for Asia of this Journal from 1970 till 1981 and he was on the Editorial Board of "Biorheology". As a member of the Division of Colloid Chemistry of the Chemical Society of Japan he participated from its beginning in the meetings and other affairs of this organization. For his outstanding work he was honored with the prestigious "Wolfgang-Ostwald-Preis" of the Kolloid-Gesellschaft in 1975 and was the first recipient of

the Chemical Education Prize of the Chemical Society of Japan in 1976.

Personally, Professor Tamamushi was one of the kindest men who treasured beauty and friendship. He spoke fluently English and German and he enjoyed greatly the company of colleagues from his country and from abroad. Professor Tamamushi kept extensive contacts with scientists in all parts of the world and regularly attended international meetings at which his discussions and words of wisdom were always appreciated.

He came from a distinguished family and it may be of interest to mention that Professor Tamamushi's grandfather was assigned to the first Japanese Embassy in the USA, where he wrote a rather extensive diary of his experiences in the host country.

In Professor Tamamushi many of us have lost a dear friend and the colloid science one of its distinguished members.

Kozo Shinoda
Egon Matijević

Solution behavior of surfactants: The importance of surfactant phase and the continuous change in HLB of surfactant*)

K. Shinoda**)

Department of Applied Chemistry, Faculty of Engineering, Yokohama National University, Yokohama, Japan

Abstract: The pseudo-phase dispersion model gave the basis for the invention of ionic surfactants with a bivalent counterion applicable in hard water on one hand, and for the discovery of infinitely aggregated micelle, i.e., surfactant phase on the other hand. In earlier days, surfactants soluble in water or oil were the focus of interest, but presently, a surfactant balanced with hydrophile lipophile properties is the center of attention. With such balance the surfactant is practically insoluble in water and hydrocarbon. But, the solubility of water or oil in surfactant phase becomes infinitely large and normal or inversed micelles are formed, provided the HLB of the surfactant is slightly changed to hydrophilic or lipophilic.

Micellar dispersion, surfactant phase separation and reversed micellar dispersion are inclusively discussed based on the concept of the relative dissolution of water and oil in surfactant phase. Solubilization of oil (or water) in micelle (or reversed micelle) and surfactant phase is also inclusively understood as the solubility in pseudo-surfactant phase and surfactant phase. The oil-water interfacial tension passes through minimum when the hydrophile lipophile property of surfactant just balances for a given oil and water. The relations among the interfacial tensions, the types of oils, the amount of cosurfactant, temperature etc. are elucidated.

Key words: Surfactant phase, micellar dispersion, reversed micelle, interfacial tension minimum, continuous change of HLB, solubilization at optimum HLB.

Introduction

It is a great pleasure and honor to participate the International Symposium on surfactant in solution by delivering the opening overview address on "Solution Behavior of Surfactants."

The important and urgent topics that scientists have to solve are deficiency of energy and resources, environmental preservation, etc. Many of these problems will be solved by the use of surfactants, i.e., tertiary oil recovery by surfactant flooding, COM fuel, mineral concentration by floatation, recycle of materials, water purification, emulsion paints, etc.

On the other hand, biologically important substances such as vitamins, hormones, antibiotics, bile salts, lecithin, etc. are surface active substances. Hence, it is my belief to devise effective and powerful surfactants from the studies of solution behavior of surfactants one of the most urgent and important research regions of today.

Pseudo-phase (micellar) dispersion in surfactant solution and separation of surfactant phase

Surfactant dissolves in water in singly dispersed state up to the saturation concentration above which usually micelles are formed at temperatures above the melting point (Krafft point) of hydrated surfactant, and hydrated solid surfactant phase separates below the melting point [1]. The aggregation number of surfactant per micelle is finite and we designate it pseudo-phase [1]. Thermodynamic functions such as the activity and the partial molal quantities stay nearly constant with the surfactant concentration above the critical micelle concentration (cmc). The micellar solution is a transparent phase because the size of micelles is usually small compared with the wave length of light.

Ordinary ionic surfactants are salted out and cannot be used in hard water, because the calcium salts of surfactant anions is solid and precipitated in hard water.

The melting point of hydrated solid soap is instantly raised more than 100 °C in the presence of a very small amount of calcium ions and calcium soap

---

*) Dedicated to Professor Dr. B. Tamamushi.

**) Overview lecture at the International Symposium on Surfactants in Solution, Lund, Sweden, June 27–July 2, 1982.

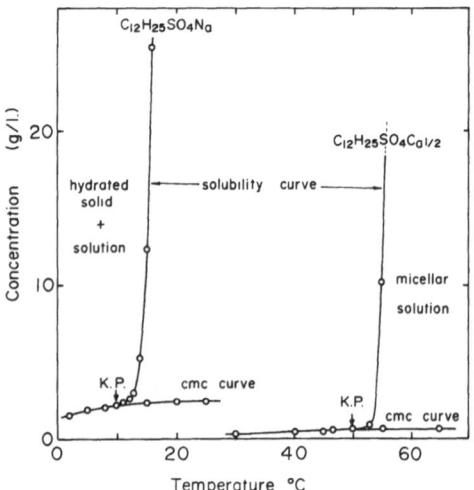

Fig. 1. Phase diagram of calcium and sodium dodecyl sulfates in water close to the Krafft points (= K.P.)

point based on the recognition of the pseudo-phase dispersion resulted in ionic surfactants applicable in hard water. Since the counterion of $R_{12}(OCH_2CH_2)_nSO_4Ca_{1/2}$ is bivalent, the electrical potential on micelle surface is smaller than sodium salt and the hydrophilic property is less. By the addition of a large amount of $CaCl_2$ or $MgCl_2$ the aggregation number increased and finally surfactant phase was separated [3, 5, 6]. Namely, the aggregation number became infinite due to the addition of a large amount of bivalent anions. In other words, the hydrophile/lipophile balance of these ionic surfactants shifted towards lipophilic so that the solubility of water in surfactant phase became finite.

Surfactant phase separation is so often observed in solutions of nonionic surfactants above the cloud point, because the hydrophile/lipophile property of the nonionics shift to more lipophilic with a temperature rise.

precipitates. It is not raised so much if the ionic group is sulfate, sulfonate, etc. Hence, the Krafft point is depressed below the room temperature by various devices, among which the introduction of alkylene oxides between polar group and hydrocarbon chain is useful. The Krafft point is raised with the hydrocarbon chain length of surfactant, so that the same devices are equally effective to use longer chain surfactants.

The Krafft points and the cmc values of such surfactants are listed in table 1 [2–5].

The Krafft point is markedly depressed by the introduction of alkylene oxide groups. Hence, an interpretation of the physical meaning of the Krafft

Importance of insolubility of surfactant in solvent

Efficient surfactant molecule has to be composed of strong lipophilic (hydrophobic) and hydrophilic (lipophobic) groups. Because of this the saturation concentrations of singly dispersed species in water and hydrocarbon are both very small, and surfactant phase is in equilibrium with very dilute surfactant solution of water and oil. This is a necessary condition to be highly adsorptive at the interface from oil as well as from water, to efficiently depress the interfacial tension, to dissolve (or solubilize) a larger amount of oil and water in surfactant phase.

Although the devices to increase the hydrophilic and lipophilic groups of surfactant are important, there is a certain limit because the melting point is raised with the molecular size. The surfactant has to be in a liquid state in the solution. Otherwise, dissolution of solvent in surfactant phase or micellar dispersion in solvent does not occur [6]. Hence, the devices to depress the Krafft point of surfactant is important [2–9].

Importance of solubility of solvent in surfactant

Although the solubility of efficient surfactant in solvent is small, the solubility of water (or hydrocarbon) in surfactant is large, because solvent is usually a small molecule and the structure of surfactant phase may be perceived as multiple bi-molecular leaflets. If the solubility of water (or oil) in surfactants phase is infinite, the surfactant disperses in water (or oil) forming micelles. Infinite solubility of solvent in

Table 1. Krafft points and cmc values of ionic surfactant applicable in hard water

Surfactants	cmc (m mole/l) at 25 °C	Krafft point °C	References
$C_{12}H_{25}SO_4Na$	8.1	9	2, 3
$C_{12}H_{25}SO_4Ca_{1/2}$	2.4	50	3
$C_{12}H_{25}OCH_2CH_2SO_4Ca_{1/2}$	0.94	15	2, 3
$C_{12}H_{25}(OCH_2CH_2)_2SO_4Ca_{1/2}$	0.71	<0	2, 3
$C_{12}H_{25}OCH_2CH_2SO_4Mg_{1/2}$	0.96	<0	3
$R_{12}OCH_2CH_2SO_4Na$	4.2	5	2
$R_{12}(OCH_2CHCH_3)_1SO_4Na$	2.7 ~ 3.0	<0	4, 5
$R_{12}(OCH_2CHCH_3)_2SO_4Na$	1.5	<0	4
$R_{12}(OCH_2CHCH_3)_3SO_4Na$	1.1	<0	5
$R_{16}SO_4Na$	0.68 (at 50 °C)	44	4, 5
$R_{16}(OCH_2CH_2)_3SO_4Na$	–	35	5
$R_{16}(OCH_2CHCH_3)_3SO_4Na$	–	28	5

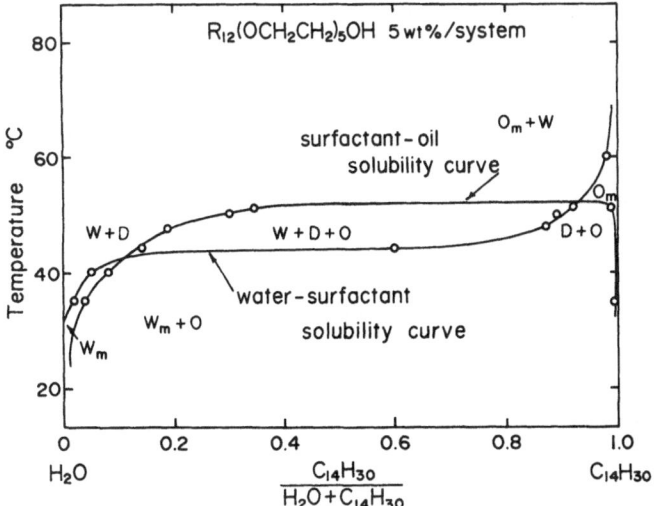

Fig. 2. Phase diagram of H_2O-$C_{14}H_{30}$ containing 5 wt%/system of $R_{12}(OCH_2CH_2)_5OH$ as a function of temperature. Abscissa is the weight fraction of $C_{14}H_{30}$ in solvents. D = surfactant

surfactant is a necessary condition for micellar dispersion, i.e., soluble surfactants [6]. Surfactants soluble in water or oil have earlier been the center of attention and surfactants neither soluble in water nor in oil were usually discarded. When the hydrophile/lipophile property of surfactant is just balanced; however, the solubilities of water and oil in surfactant are both finite and a surfactant phase is obtained.

a) Continuous change of HLB (= hydrophile/lipophile balance) in nonionic surfactant

These situations are clearly shown in the phase diagram of water-nonionic surfactant-oil system as a function of temperature in figure 2. Water dissolves in nonionic surfactant phase infinitely at lower temperature and aqueous micellar solution, W_m, is obtained on one hand and hydrocarbon will dissolve infinitely at higher temperature and reversed micellar solution, O_m, is obtained on the other hand, because nonionic surfactant is hydrophilic at lower temperature and lipophilic at higher temperature. Below the surfactant-oil solubility curve in figure 2 surfactant and oil phases separate and above the water-surfactant solubility curve water and surfactant phases separate. So that 3 phase region consisted of water, surfactant and oil phases is observerd at the intermediate temperature at which the hydrophile/lipophile property of surfactant just balances. Realms W_m and O_m are one phase, because oil and water dissolve completely in water and oil phase, respectively.

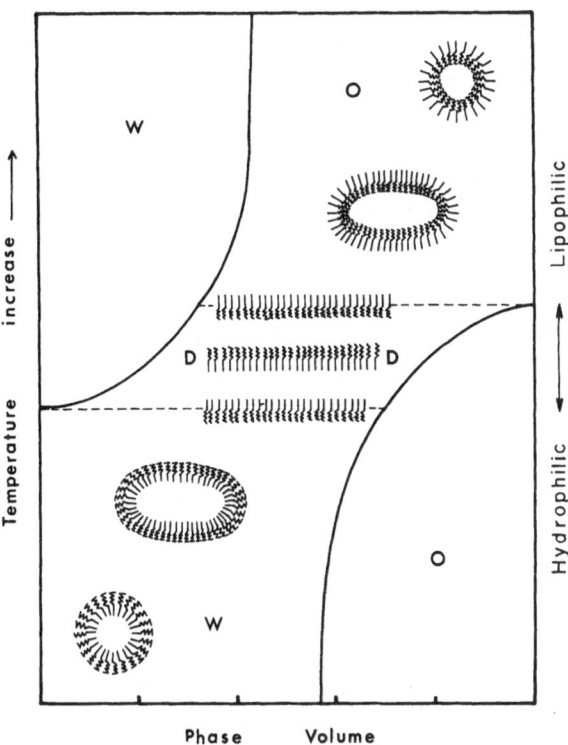

Fig. 3. Schematic illustration of the change of solution behavior of surfactant with the hydrophile lipophile balance in water-surfactant-hydrocarbon system

The change of the phase volumes and solution behavior with temperature in 1 : 1 mixture of oil and water containing surfactant are illustrated schematically in figure 3.

Phase diagram similar to figure 2 is obtained by decreasing the hydrophilic chain length instead of temperature raise as shown in figure 4, [10]. This fact

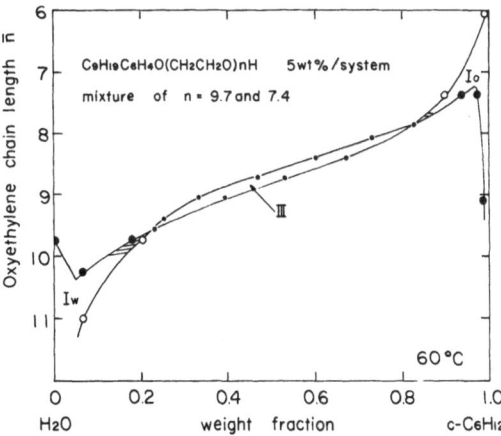

Fig. 4. The effect of the average oxyethylene chain length of nonionics on the phase diagram of water/cyclohexane/nonionic surfactant system (reference 10)

is important, because the variable changed from temperature to hydrophile/lipophile balance of nonionic surfactant.

b) Continuous change of HLB in ionic surfactant

Temperature change usually does not affect much to the HLB of ionic surfactant. The HLB was changed by mixing ionic surfactant with co-surfactant in the presence of salts. Phase diagram of $R_{12}SO_4Na/R_8(OCH_2CH_2)_2H/3$ wt% NaCl aqueous solution/ $C_{10}H_{22}$ at 25 °C is shown in figure 5, [11]. The pattern of the diagram is similar to figure 4.

In this case the composition of hydrophilic ionic surfactant and lipophilic cosurfactant was changed instead of ethyleneoxide chain length. The continuous change from aqueous micellar solution, surfactant phase to non-aqueous micellar solution is observed with the composition change and the solubilization of water or oil was very large. Hence, the former conclusion is equally applicable to ionic and nonionic surfactant, provided the HLB of two surfactants are reasonably close. Unlike nonionics, phase equilibria are insensitive to temperature change and very useful

for many practical purposes. If salt is not added, the solubilization of oil in aqueous micellar solution is small as shown in the left hand side of figure 5. Similarly if we used a too lipophilic cosurfactant or a too hydrophilic ionic surfactant without salt, solution behavior of ionic surfactants would be different from nonionic ones and the two could not be treated uniformly.

Importance of surfactant phase

The importance of surfactant phase in nonionic surfactant-water-hydrocarbon system was revealed in 1968 [12]. Ever since many scientists studied surfactant phase and found various important facts [13–15]. The types of emulsions invert around the temperature at which surfactant phase appears [12]. The interfacial tension between oil and water is minimum at three phase region [13, 16–18]. The term "Middle phase" so frequently used by tertiary oil recovery scientists is nothing else but surfactant phase [16–18]. The amount of surfactant required to dissolve a mixture of water and oil is minimum in three phase region [12, 14]. Surfactant, water and oil phases are perceived as mostly continuous in surfactant phase, because the self diffusion of respective components is high [15]. Probable structure of surfactant phase is illustrated in figure 6.

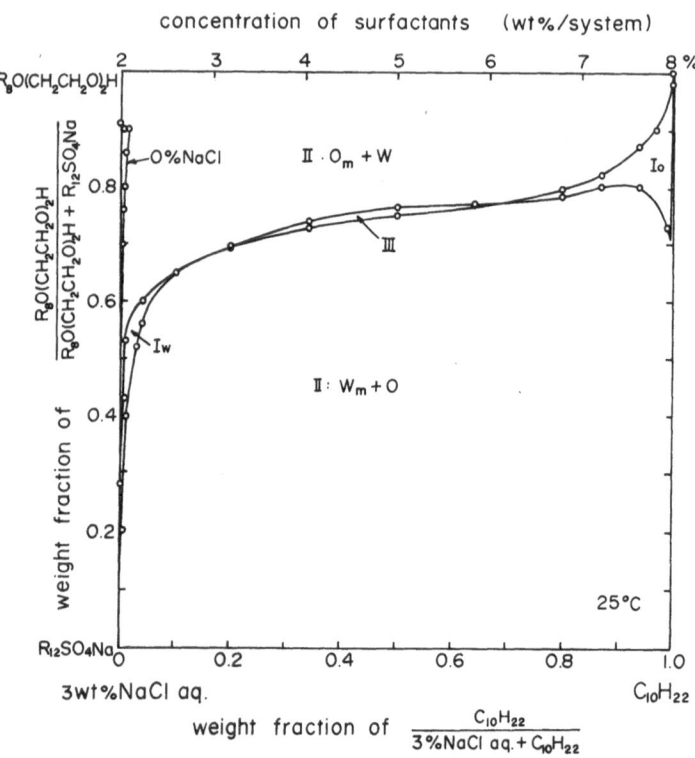

Fig. 5. Phase diagram of $R_{12}SO_4Na/R_8(OCH_2CH_2)_2H/3$ wt% NaCl aq./$C_{10}H_{22}$ at 25 °C. Abscissa and ordinate are the weight fractions of solvents and surfactants, respectively. Phase diagram without salt is also plotted

Fig. 6. Schematic illustration of the probable structure of surfactant phase. Black and white stripes represent the bound water and oil layers, respectively. There exist surfactant monolayers between bound water and oil layers

Black and white stripes represent the bound water and oil layers, respectively. Surfactant monolayers exist between bound water and oil layers.

Interfacial tension

A lower critical solution temperature (LCST) is found in the nonionic surfactant-water system and upper critical solution temperature (UCST) is observed in the nonionic surfactant-hydrocarbon system [19, 20]. Although these critical points vary with the types of oils, the LCST is lower than the UCST in three component system composed of water-nonionic surfactant-hydrocarbon [21, 22]. A three phase region composed of water, nonionic surfactant and oil phases exists between two critical solution end points [12, 13, 21, 22]. At critical end points the interfacial tensions of water-surfactant, γ_{W-D}, and oil surfactant, γ_{O-D} are zero regardless to the amount of oil and water, respectively [23]. Since surfactant solution consists of small molecules (solvent) and large aggregated micelles, the critical composition is close to pure solvent [24].

For example, the aggregation number of micelles is about 400 in aqueous solution of $C_{12}H_{25}O(CH_2CH_2O)_6H$ at 25 °C [25]. Since the volume ratio of micelle to water molecule is large (~ 10000), the critical composition of the solution estimated by equation (1) is about 1 wt% or 0.04 mole per cent of component 2 [26].

$$\frac{\phi_{1c}}{\phi_{2c}} = \left(\frac{NV_2}{V_1}\right)^{1/2} = \frac{X_{1c}V_1}{X_{2c}V_2}. \qquad (1)$$

Where ϕ_{ic} is the volume fraction of ith component at the critical composition, N the aggregation number, V_1 the molal volume of water and V_2 the molal volume of surfactant.

Interfacial tension γ_{int} between two phases is very small close to a critical solution and it is zero at the critical point [22].

$$\gamma_{int} = 0 \text{ (at critical point).} \qquad (2)$$

The Gibbs adsorption isotherm in dilute solution is

$$\left(\frac{\partial \gamma}{\partial \mu_2}\right)_T = \frac{1}{RT}\left(\frac{\partial \gamma}{\partial \ln a_2}\right)_T$$

$$= \frac{1}{RT}\left(\frac{\partial \gamma}{\partial \ln X_2}\right)_T \left(\frac{\partial \ln X_2}{\partial \ln a_2}\right)_T = -\Gamma_2 \qquad (3)$$

where Γ_2 is the surface excess of 2nd component, the

change of activity with concentration is zero at the critical temperature and composition.

$$\left(\frac{\partial \ln a_2}{\partial \ln X_2}\right)_T = 0 \text{ (at critical point).} \qquad (4)$$

Moreover, the activity of surfactant does not change significantly for concentrations in excess of cmc [1, 24]. Applying the thermodynamics of small system,

$$\left(\frac{\partial \ln a_2}{\partial \ln X_2}\right)_T = \frac{\overline{N}}{\overline{N^2}} \text{ (above cmc)} \qquad (5)$$

over a wide concentration range in which otherwise ($\partial \ln a_2 / \partial \ln X_2$) = 1, where N is the aggregation number [27].

From relations (1), (3), (4) and (5), we can conclude that the change of interfacial tension with concentration close to the critical point is very small.

a) Nonionic surfactant

The temperature dependence of the interfacial tensions between oil-water, γ_{O-W}, water-surfactant, γ_{W-D}, and oil-surfactant, γ_{O-D} close to the three

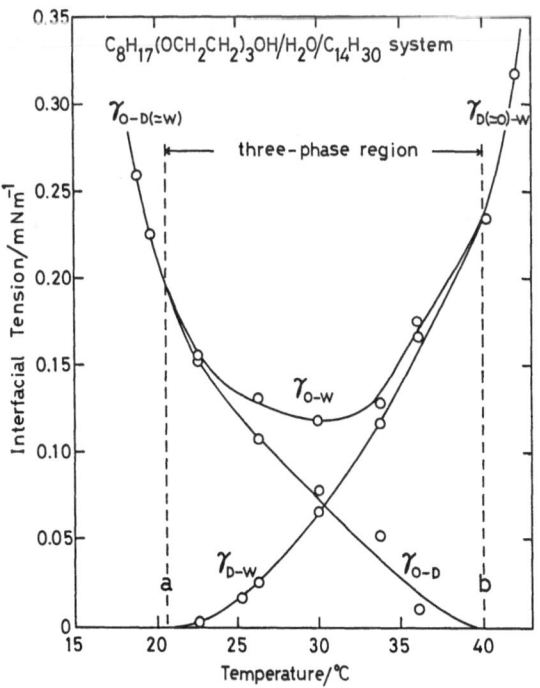

Fig. 7. The interfacial tension, γ_{O-W}, γ_{W-D} and γ_{O-D} as a function of temperature in $R_8(OCH_2CH_2)_3OH/H_2O/C_{14}H_{30}$ system, determined by the sessile drop method. The lower and upper critical solution end points are indicated as a and b, respectively (reference 28)

Table 2. The correlation among the types of surfactants, the types of hydrocarbons, upper and lower critical end points and oil-water interfacial tension minima. The minimum γ_{W-O} was obtained at the medium temperature

Surfactant	Hydro-carbons	LCST $W-D$	UCST $O-D$	UCST ΔT	min. γ_{W-O}
$R_4OCH_2CH_2OH$	C_7H_{16}	5.4	12.6	7.2	0.095
$R_4OCH_2CH_2OH$	C_8H_{18}	12.2	22.4	10.2	0.16
$R_8(OCH_2CH_2)_3OH$	$C_{10}H_{22}$	13.6	26.6	13	0.035
$R_8(OCH_2CH_2)_3OH$	$C_{12}H_{26}$	18.2	34.2	16	0.07
$R_8(OCH_2CH_2)_3OH$	$C_{14}H_{30}$	20.4	40.0	19.6	0.118
$R_8(OCH_2CH_2)_3OH$	$C_{16}H_{34}$	22.5	48.5	26	0.182
$R_{12}(OCH_2CH_2)_4OH$	$C_{12}H_{26}$	20.2	23.8	3.6	~0.001
$R_{12}(OCH_2CH_2)_4OH$	$C_{16}H_{34}$	27.5	34.8	7.3	0.009
$R_{12}(OCH_2CH_2)_4OH$	Squalane $(C_{30}H_{62})$	35.5	58.8	23.3	0.095
$R_{12}(OCH_2CH_2)_5OH$	$C_{14}H_{30}$	43.8	51.6	7.8	–

phase region of H_2O-$R_8(OCH_2CH_2)_3OH$-$C_{14}H_{30}$ system are shown in figure 7, [28].

Temperature *a* and *b* are the critical solution end temperatures of water-surfactant and hydrocarbon-surfactant in the presence of oil and water respectively. Since γ_{W-D} and γ_{O-D} are zero below and above the respective critical points, γ_{O-W} coincides with γ_{O-D} at LCST and γ_{O-W} with γ_{D-W} at UCST. In the three phase region, Antonoff's rule approximately holds, but γ_{O-W} is slightly smaller than $\gamma_{O-D} + \gamma_{W-D'}$ as shown in figure 7. Actually, flat lens of surfactant phase between oil and water phases is observed. Since the interfacial tension increases more than first order, a minimum is observed on the γ_{O-W} vs. temperature curve in the three phase region. If two critical end points approach each other the oil-water interfacial tension becomes smaller, because the sum of γ_{O-D} and γ_{W-D} become smaller. Correlation among the temper-

ature differences in two critical points, the interfacial tension minima and the types of nonionic surfactants is summarized in table 2 [28].

The narrower the gap of two critical end points, the lower the interfacial tension. The longer the hydrocarbon chain length of a surfactant, the lower the (oil-water) interfacial tension. The shorter the hydrocarbon chain length of oil, the lower the $\gamma_{O-W'}$. Correlation among the interfacial tension minima, the hydrocarbon chain length of alkane and the chain length of nonionic surfactants is shown in figure 8.

The effect of the alkyl chain length of surfactant to depress the γ_{O-W} is so remarkable.

b) Ionic surfactant

Temperature change does not affect much to the HLB of ionic surfactant. The HLB is changed by mixing an ionic surfactant with cosurfactant instead of varying the temperature. Figure 5 illustrated the continuous change of mixed surfactant solution from aqueous micellar solution to non-aqueous reversed micellar solution via surfactant phase with the composition of $R_8O(CH_2CH_2O)_2H$. The interfacial tension between oil and water in dilute solution of the same system is plotted as a function of the composition of surfactants vs. cosurfactant around the three phase region as shown in figure 9, [29].

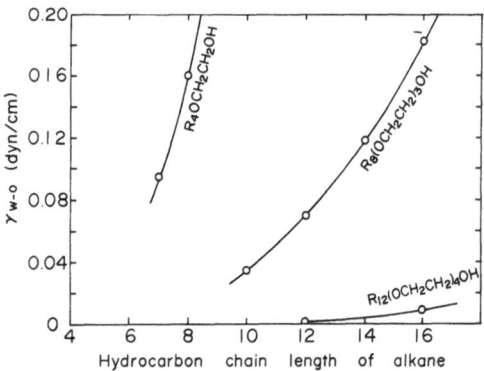

Fig. 8. Correlation among the interfacial tension minima, the types of nonionic surfactants and the hydrocarbon chain length of oil (alkane) (reference 28)

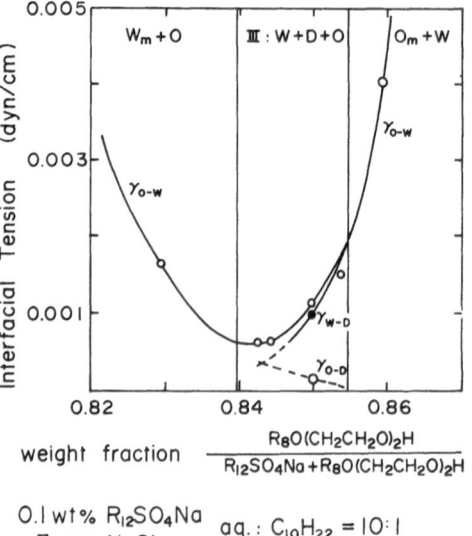

Fig. 9. The interfacial tension as a function of surfactant composition in $R_{12}SO_4Na/R_8(CH_2CH_2O)_2H/3$ wt% NaCl aq./$C_{10}H_{22}$ at 30 °C. Interfacial tension was determined by the spinning drop method

In a three component system composed of water-pure nonionic surfactant-hydrocarbon system, the upper (or lower) critical solution temperature means the upper (or lower) critical solution end point, and the interfacial tension between oil (or water) and surfactant, γ_{O-D} (or γ_{W-D}) is zero regardless of the amount of coexisting water (or oil) [28]. However, the upper or lower composition of cosurfactant at which surfactant phase disappears does not mean γ_{O-D} is zero in ionic surfactant solution containing salt, cosurfactant and oil.

The minimum interfacial tension was 0.64×10^{-3} dyn/cm. If longer chain surfactant is used, the interfacial tension would be much smaller.

Conclusion

In conclusion, we would like to emphasize that solution behavior of surfactants, such as, 1) micellar dispersion (pseudo-phase separation), surfactant phase separation, reversed micellar dispersion, 2) the change of solubility (solubilization) of solvents in pseudo-phase (micelle and reversed micelle) and true phase (surfactant phase) and 3) the change of interfacial tension between water-surfactant, surfactant-oil, and water-oil, etc., is inclusively understood as continuous phenomena with the change of the HLB of surfactant at the interface. The effect of temperature on the solution behavior of ionic and nonionic surfactant is completely different on one hand, but both surfactants behave similarly with the change of the HLB of respective surfactants on the other hand.

Acknowledgment

The author gratefully acknowledges the cooperation of Dr. H. Kunieda, Messrs. T. Arai and Y. Shibata who had obtained most of the data presented in this paper, and Prof. Stig Friberg for the revision of the manuscripts.

References

1. Shinoda, K., Nakagawa, T., Tamamushi, B., Isemura T., "Colloidal Surfactants" Academic Press Inc. N.Y. 1963, pp. 8–9.
2. Hato, M., Shinoda, K., J. Phys. Chem. **77**, 378 (1973).
3. Shinoda, K., Hirai, T., J. Phys. Chem. **81**, 1842 (1977).
4. Weil, J. K., Stirton, A. J., Nuñez-Ponzoa, M. V., J. Am. Oil Chemists' Soc. **43**, 603 (1966).
5. Murata, M., Tsumadori, M., Suzuki, A., Tsujii, K., Mino, J., XII Meeting of the Spanish Committee on Surface Active Agents, Barcelona (1981).
6. Shinoda, K., J. Phys. Chem. **85**, 3311 (1981).
7. Weil, J. K., Stirton A. J., Barr, E. A., J. Am. Oil Chemists' Soc. **43**, 157 (1966).
8. Shinoda, K., Minegishi, Y., Arai, H., J. Phys. Chem. **80**, 1987 (1976).
9. Tsujii, K., Saito, N., Takeuchi, T., J. Phys. Chem. **84**, 2287 (1980).
10. Shinoda, K., Kunieda H., J. Colloid Interface Sci. **42**, 381 (1973).
11. To be published.
12. Shinoda, K., Saito, H., J. Colloid Interface Sci. **26**, 70 (1968).
13. Saito, H., Shinoda, K., J. Colloid Interface Sci. **32**, 647 (1970).
14. Friberg, S. E., Lapczynska, I., Prog. Colloid Polym. Sci. **56**, 16 (1975).
15. Lindman, B., Kamenka, N., Kathopoulis, T., Brun, B., Nilsson P., J. Phys. Chem. **84**, 2485 (1980).
16. Healy, R. N., Reed, R. L., Stenmark, D. G., Soc. Pet. Eng. J., 147, (June 1976).
17. Healy, R. N., Reed, R. L., Soc. Pet. Eng. J., 129 (April 1977).
18. Bansal, V. K., Shah, D. O., Soc Pet. Eng. J., 167 (June 1978).
19. Shinoda, K., in: "Solvent Properties of Surfactant solutions," Marcel, Dekker, N. Y. 1967, pp. 29–33.
20. Shinoda, K., Arai, H., J. Colloid Sci. **20**, 93 (1965).
21. Kunieda H., Friberg, S. E., Bull. Chem. Soc. Jpn. **54**, 1010 (1981).
22. Kunieda, H., Shinoda, K., J. Dispersion Sci. and Tech. **3**, 233 (1982).
23. Atack, D., Rice, O. K., Discuss. Faraday Soc. **15**, 210 (1953).
24. Shinoda, K., "Principles of Solution and Solubility" Marcel Dekker Inc., N.Y. 1978, p. 160.
25. Balmbra, R. R., Clunie, J. S., Corkill J. M., Goodman, J. F., Trans. Faraday Soc. **60**, 979 (1946)
26. p. 144, equations (8.19) – (8.21) in Ref. 24.
27. Hall, D. G., Pethica, B. A., in: "Nonionic Surfactants" Edited by Schick, M., Marcel, Dekker, Inc., N. Y. 1967, pp. 516–557.
28. Kunieda H., Shinoda, K., Bull. Chem. Soc. Jpn. **55**, 1777 (1982).
29. To be published.

Received July 19, 1982;
accepted November 30, 1982

Author's address:

Kōzō Shinoda
Department of Applied Chemistry
Faculty of Engineering
Yokohama National University
Tokiwadai, Hodogayaku
Yokohama 240, Japan

Progress in Colloid & Polymer Science   Progr. Colloid & Polymer Sci. **68**, 8–13 (1983)

The adsorption of disodium and dipotassium salts of dodecyl phosphoric acid at the air-water interface*)

N. R. Pallas and B. A. Pethica

Clarkson College, Potsdam, New York, USA

Abstract: The surface tensions of aqueous solutions of the disodium and dipotassium salts of dodecyl phosphoric acid in the presence and absence of 0.1 M NaCl or 0.1 M KCl are reported at 10°, 25° and 35 °C. The adsorption isotherms and heats of adsorption are derived from the data and discussed.

Key words: disalts, dodecyl phosphate surface tension, adsorption

Introduction

For some time it has been known that monolayers of alkyl phosphates possess some unusual but useful properties. The use of these compounds as surfactants, in ore flotation, as antistatic agents, and in lubrication is supported by a small inconsistent body of literature. There is also considerable theoretical interest in the possibility of producing extremely high surface charge densities with monoalkyl phosphate salts.

Of the three synthetic routes available to the monoalkyl phosphates [1–5], only one assures the formation of the esters uncontaminated by di- or tri-alkyl compounds. In an early study it was found that monoalkyl phosphate salts may associate in the solid phase to form quarter salts [3]. It has been found that long chain phosphate monolayers interact with metal ions [6–9], and are said to show a pronounced tendency to associate through hydrogen bonding [9, 10]. More recently, studies of solutions of the doubly ionized monoalkyl phosphates appeared in the literature [11–15]. However, there are discrepancies between these studies, which this report should help to resolve.

Experimental

The preparation and characterization of the pure mono *n*-dodecyl dihydrogen phosphate used in this study were reported earlier, as were the preparations of sodium and potassium chlorides and hydroxides [3–11]. The Harkins-Brown method of drop-volumes was used for the surface tension determination [16]. Details of the apparatus are given elsewhere [24]. The main features are

accurate thermostating and control of the atmosphere, in this case N_2 saturated with water vapor after passing through beds of soda-lime and charcoal.

The glassware was cleaned by soaking in a dispersion of anhydrous chromate in sulfuric acid, followed by a thorough rinsing with twice-distilled water. The vessels were then soaked in a dilute solution of the appropriate hydroxide for one hour. The glassware was then rinsed again and finally steamed with three-times distilled water and allowed to dry on a clean metal rack in a plexiglas box through which was passed a stream of N_2 from the absorbent beds. The same box was used for preparing reagent solutions to reduce CO_2 absorption. Solutions of the disodium or dipotassium alkyl phosphates were made up by addition of the appropriate hydroxides to a 7% excess above disalt equivalence to reduce effects from any residual CO_2 at the high pH of these solutions. All concentrations are given as molarities at room temperature (20 °C).

Results

In measuring drop volumes of the alkyl phosphate solutions slow ageing effects were often observed, equivalent to a slow fall in the surface tensions. These ageing effects could be due to hydrolysis, impurities or residual CO_2. Hydrolysis can be effectively ruled out at the temperatures of these experiments on the basis of extrapolation of the known kinetics of phosphate ester hydrolysis at high temperatures [17]. Although impurities are always a problem in surface studies, all the available data suggest good purity [11]. Left exposed in laboratory air, the pH of these solutions falls rapidly, as expected. In the plexiglas box the pH dropped rather slowly and if the solutions were left stoppered inside the box no pH changes were observed – arguing against hydrolysis. We therefore conclude that there was always some absorption of CO_2 during our procedures even after up to one hour of flushing of the plexiglas box. A design in which the

---

*) Dedicated to the memory of Professor Dr. B. Tamamushi.

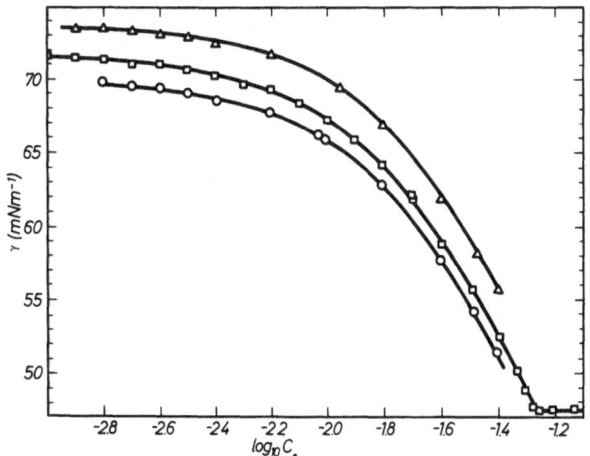

Fig. 1. Surface tension (γ mNm^{-1}) vs. logarithm molar concentration (log$_{10}$C) for Na$_2$ dodecyl phosphate without NaCl. $\triangle$ − 10 °C, $\square$ − 25 °C, $\bigcirc$ − 35 °C

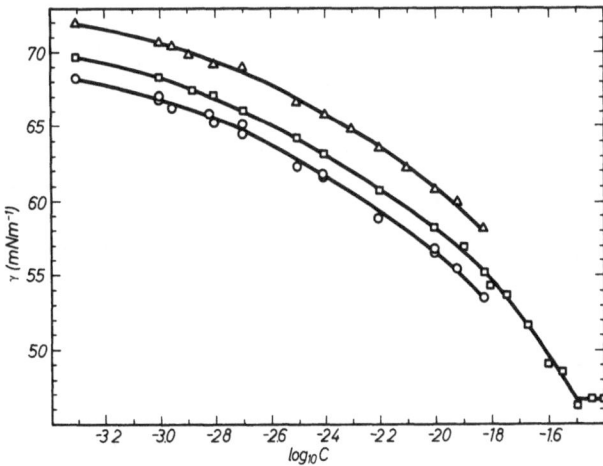

Fig. 3. Surface tension (γ mNm^{-1}) vs. logarithm molar concentration (log$_{10}$C) for Na$_2$ dodecyl phosphate with 0.1 M NaCl. $\triangle$ – 10 °C, $\square$ – 25 °C, $\bigcirc$ – 35 °C

making up of the solutions, filling the syringes and carrying out the measurements could all be done inside a permanently closed system would be desirable in future work in view of the sensitivity to CO$_2$. The pH of the collected expressed drops from the syringe inside the drop-volume apparatus was routinely measured, and usually showed negligible changes. If changes greater than 0.2 were observed, the results were rejected. However, trace CO$_2$ could be affecting the surface layers without measurable bulk pH shifts. In the worst cases, at concentrations just below the micelle points (cmc), surface tension ageing occurred up to two minutes, compared to the 30 seconds typically required for a drop-volume measurement as

recommended by Harkins and Brown [16]. In more dilute solutions the ageing effects were not observed. Above the micelle point (cmc), ageing effects disappeared, presumably due to solubilization of any traces of partially neutralized salts. We consider, in a worst-case analysis, that the surface tensions recorded here are good to ± 0.5 mNm^{-1} at high concentrations up to the cmc, and to ± 0.2 mNm^{-1} at low concentrations and above the cmc.

The first system studied was Na$_2$ dodecyl phosphate without added neutral electrolyte. The surface tension data are shown in figure 1, indicating a cmc of 0.056 M at 25 °C. The cmc's at the other temperatures were not investigated by the surface tension method as

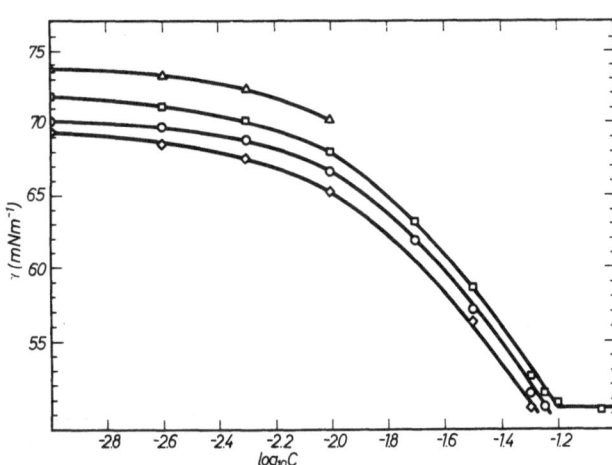

Fig. 2. Surface tension (γ mNm^{-1}) vs. logarithm molar concentration (log$_{10}$C) for K$_2$ dodecyl phosphate without KCl. $\triangle$ – 10 °C, $\square$ – 25 °C, $\bigcirc$ – 35 °C, $\Diamond$ – 40 °C

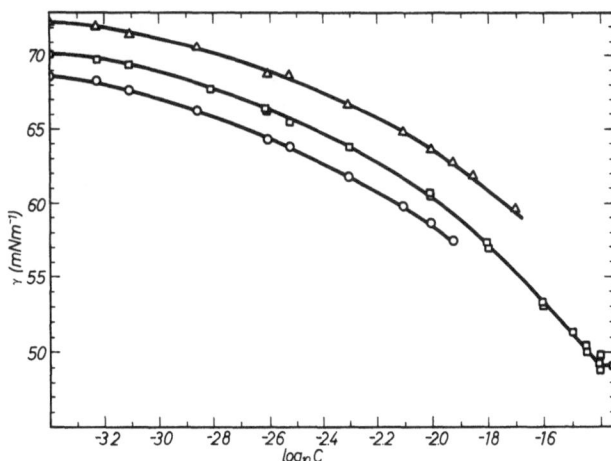

Fig. 4. Surface tension (γ mNm^{-1}) vs. logarithm molar concentration (log$_{10}$C) for K$_2$ dodecyl phosphate with 0.1 M KCl. $\triangle$ – 10 °C, $\square$ – 25 °C, $\bigcirc$ 35 °C

they had been measured from conductivities in another study from this laboratory [11]. The results for the K_2 salt are shown in figure 2, the cmc being 0.063 M at 25 °C. The results for the systems with added 0.1 M NaCl or KCl are shown in figures 3 and 4, giving micelle points of 0.040 M and 0.032 M at 25 °C. The values for the cmc's from surface tensions agree with those found previously by conductivity methods [11].

Discussion

The first data on the adsorption of disodium dodecyl phosphate were at 25 °C over a smaller range of concentrations than given here [6]. The surface tension values are somewhat lower than the present results, probably reflecting the slightly lower pH values in the absence of excess alkali in the earlier work. Nakagaki et al. [14, 15] report data on three disodium alkyl phosphates, including the dodecyl compound as prepared by reacting dodecanol with pyrophosphoric acid. The compound would appear to be impure, perhaps containing di- and triesters, as judged from the appearance of pronounced minima in the surface tension-concentration plots. The melting point recorded is 2 °C lower than that of the compound used in our study, and no equivalent weight was given. The micelle point given by Nakagaki et al. with a 2% excess of NaOH is 0.05 M as against 0.056 M for our compound with 7% of excess alkali. It is furthermore stated that differences were found between Wilhelmy plate and capillary rise measurements, perhaps reflecting a CO_2 effect for the Wilhelmy system.

The adsorption of the alkyl phosphates can be estimated from a modified form of the Gibbs Adsorption Equation, using a straightforward extension of arguments developed previously for ionized surfactants with or without added simple binary electrolytes [18, 25, 26]. For solutions of the disalts of dodecyl phosphate with the addition, as required, of alkali metal chlorides having the cation in common with the surfactant salt, the appropriate form of the Gibbs equation is

$$A = + kTf\frac{\text{d}\ln C_x}{d\gamma} \qquad (1)$$

where

$$f = 1 + \frac{4C_x}{2C_x + C_y} \qquad (2)$$

A is the area/molecule of the surfactant ion, C_x is the concentration of the alkyl phosphate salt, C_y is the sum of the concentrations of the added alkali halide and the excess alkali hydroxide, γ is the surface tension, k is Boltzmann's constant and T is the temperature. Equation (1) assumes that all activity coefficients are unity, and that at $C_y = 0$, no hydrogen ion exchange occurs at the ionized monolayer as C_x goes to zero. This latter assumption is entirely reasonable at the high pH of the solutions under study. The value of f can vary between 1 and 3. In the present experiments, in which a 7% excess of NaOH or KOH was present in all solutions, f takes the value of 2.93 in the absence of NaCl or KCl, and approaches 1.0 in experiments on the more dilute surfactant solutions with added halides. The surface pressure (Π) is defined as the lowering of the surface tension of the water or aqueous salt solution on adding the alkyl phosphate. The adsorption isotherms derived from the data using equations 1 and 2 are given in figures 5–8 in the form of Π as a function of A. The isotherms are only given for 25 °C in the figures since, as discussed below, the temperature coefficients for adsorption are small. The isotherms in figures 5, 6 and 7 were calculated from the tangents to both the linear (Π-C) and the logarithmic (Π-log C) plots using smoothed curves, with rather good agreement except for the Na_2 alkyl phosphate with 0.1 M NaCl (fig. 6). In fact, the trend of the results in this case strongly suggests that there is a linear Π-log C region over the Π range 8 to 12 mNm^{-1}. The deliberately smoothed curve through the data points which was used to give figure 7 is readily replaced by another within the experimental

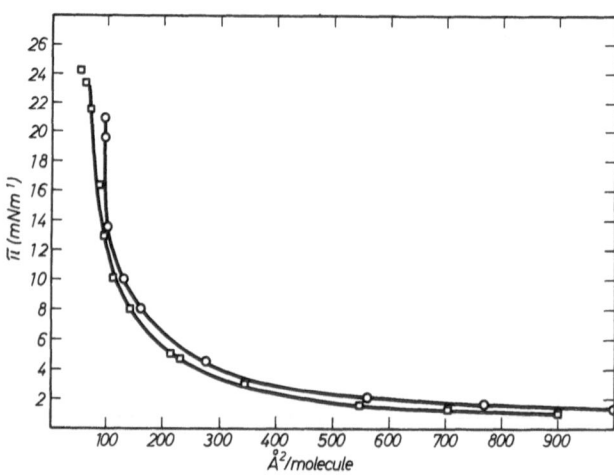

Fig. 5. Surface pressure (Π mNm^{-1}) vs. area (Å^2 molecule^{-1}) isotherms at 25 °C for Na_2 phosphate (□), and K_2 phosphate (○), without salt.

scatter to emphasize the "linear" region of the Π-log C results. If this emphasis is accepted, a degenerate phase transition results in the derived $\Pi-A$ isotherm, which is shown in figure 8. The calculations for the $\Pi-A$ isotherms shown in the figures neglect activity coefficients. The Güntelberg approximation to the Debye-Hückel theory [21, 22] has been used previously for long-chain phosphates [14]. It was checked against data on inorganic phosphates [23] and found to be a good approximation. When used in the calculations of the surface areas from the Gibbs equation, the Güntelberg approximation gave very small changes in the areas (~ 1 Å^2 molecule^{-1}) and is not helpful in interpretation of the apparent phase transition. Such a transition in adsorbed films of soluble surfactants at a liquid interface has not, to our knowledge, been demonstrated unambiguously in reliable and well-controlled experiments. In view of the aging effects in our own experiments, which reduce the precision of the isotherms despite the rather large number of data points, we do not claim to have established the phase transition in this case. The results serve as a useful commentary on graphical methods and may encourage other workers to examine the same system by alternative techniques, notably with radiotracers. What can be said from the $\Pi-A$ results is that the films of the Na$_2$ dodecyl phosphate are more condensed than those of the K_2 salt, and that addition of the halides condenses both monolayers. In the absence of added salt the monolayer densities at 22 mNm^{-1} correspond to 95 and 75 Å^2 molecule^{-1} for the K_2 and Na$_2$ salts. In the presence of 0.1 M KCl or NaCl, the molecular areas at 22 mNm^{-1} are 75 and 55 Å^2

respectively. These are large areas for a single-chain molecule. For comparison, Na dodecyl sulphate at 22 mNm^{-1} gives areas per molecule of 60 Å^2 without added salt [18] and 38 Å^2 in the presence of 0.1 M NaCl [25] at the nearby temperature of 20 °C. This suggests strongly that the double charge on the phosphate anion does give a substantial extra repulsion between the head groups. This repulsion is reduced on addition of neutral electrolyte, with the sodium ion more strongly bound than potassium to the anionic head group. This is the same trend as found for the ionized carboxyl group, and opposite to the trend for the sulphate group, as judged from the insoluble monolayer data of Goddard and his associates [13, 27, 28]. The fact that the molecular areas for the phosphate salts are so high indicates that counter-ion binding is not complete at the concentrations used in this study. A complicating factor is that if counter-ion binding is not large, the surface pH will be much lower than that in the bulk phase with the consequent possibility of a decrease in the degree of the acid-base ionization of the phosphate group. Surface potential studies are desirable to examine these ionic interactions. Of special interest will be surface potential experiments with insoluble spread films of very long-chain phosphate esters, where unusually high surface charge densities are likely to be available.

The surface tension data at different temperatures can be used to obtain the isosteric ($\Delta \bar{H}$) and integral (ΔH) heats of adsorption per molecule of alkyl salt. The Traube region for low C_x can be described by $\Pi = \alpha C_x$, where α is a constant for a given temperature and alkali halide concentration. In the Traube region f

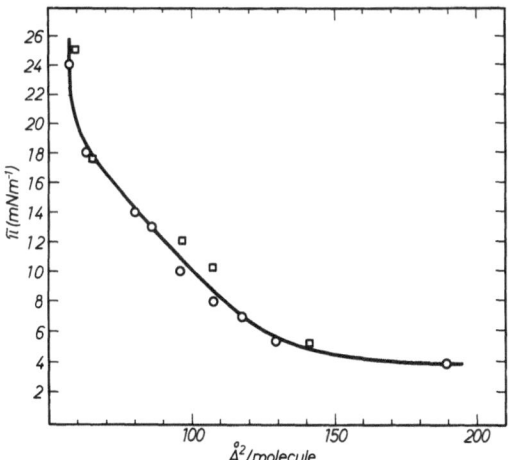

Fig. 6. Surface pressure (Π mNm^{-1}) vs. area (Å^2 molecule^{-1}) isotherms at 25 °C for Na$_2$ phosphate with 0.1 M NaCl, (○) from log plot and (□) from linear plot.

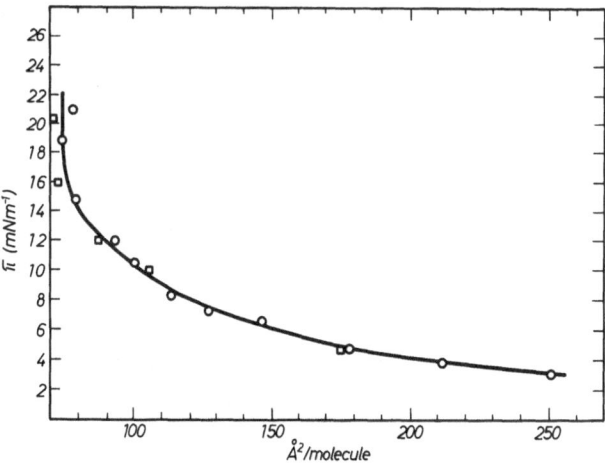

Fig. 7. Surface pressure (Π mNm^{-1}) vs. area (Å^2 molecule^{-1}) isotherm at 25 °C for K$_2$ dodecyl phosphate (○) from log plot and (□) from linear plot.

= 2.93 for $C_y = 0$, and when $C_y \neq 0$, f becomes 1.0 precisely. The temperature dependence of α gives the standard integral heats of adsorption (ΔH_o) corresponding to the ideal standard states of unit surface pressure and unit bulk concentration of the surfactant according to

$$\frac{d \ln \alpha}{dT} = \frac{\Delta H_o}{fkT^2} \qquad (3)$$

This equation is a simple extension of that derived previously [19, 20], and it is readily shown that the corresponding standard isosteric heat of adsorption ($\Delta \bar{H}_o$) is given by $\Delta \bar{H}_o = \Delta H_o + fkT$.

Away from the Traube region, the value of f for a given Π or for a given surface density ($\Gamma = 1/A$) will vary with temperature in the circumstances of our experiments whenever KCl or NaCl is added, but will remain constant in the absence of added salt. With no added salt, the heats are obtained, again by extending earlier arguments [25], to give

$$\left(\frac{d \ln C_x}{dT} \right)_\Gamma = - \frac{\Delta \bar{H}}{fkT^2} \qquad (4)$$

$$\left(\frac{d \ln C_x}{dT} \right)_\Pi = - \frac{\Delta H}{fkT^2} \qquad (5)$$

It will be seen later that the Π–C data show rather small dependence on temperature for all cases. Thus, in the presence of added KCl or NaCl, although f will strictly vary with T at constant Π or constant Γ, the variation will be quite small. Correspondingly, equations 4 and 5 can be applied to a reasonably approximation, with appropriate constant f values included. In general, however, the variation of f with T will need to be included in more complex forms of the Clausius-Clapeyron equation, a point overlooked in earlier work [26].

All the heats of adsorption calculated from the data are small. Taking first the results in the absence of KCl or NaCl, the most accurate values were obtained in the Traube region. For the Na$_2$ salt, ΔH_o is zero within experimental error over the temperature range 10 °C to 35 °C. For the K_2 salt, ΔH_o is approximately -10 KJ mole^{-1} averaged over the range 10° to 45 °C. Away from the Traube region, both heats of adsorption are difficult to estimate with any accuracy but are small and positive (< 4 KJ mole^{-1}) for both salts, with some evidence of a change in sign at high Π.

In the presence of 0.1 M salts, the ΔH_o are positive and not more than 4 KJ mole^{-1} for both Na$_2$ and K$_2$ salts. At higher concentrations and surface pressures,

the ΔH and $\Delta \bar{H}$ values for both Na$_2$ and K$_2$ salts do not vary significantly with temperature. Taken directly from the Π-log C plots, ΔH for the Na$_2$ salt is approximately 4 KJ mole^{-1} and $\Delta \bar{H}$ is -8 KJ mole^{-1}. For the K$_2$ salt ΔH is 8 KJ mole^{-1} and $\Delta \bar{H}$ is -12 KJ mole^{-1}. These differences in sign between ΔH and $\Delta \bar{H}$ are consistent with the changes in the Π–A curves with temperature that would be allowed by our data.

The heats of adsorption of the Na$_2$ dodecyl phosphate may be compared with those for the singly ionized sodium dodecyl sulphate and sodium dodecyl hydrogen phosphate reported earlier [6, 26]. Comparing the integral heats in the presence of 0.1 M NaCl shows that the effect of temperature is much less for the Na$_2$ compound than for either of the singly ionized surfactants. Interpolating the data for 25 °C, the median of the temperature range of the experiments reported here, ΔH for sodium dodecyl sulphate is of the opposite sign and twice the magnitude of the ΔH for Na$_2$ dodecyl phosphate. The ΔH for sodium dodecyl hydrogen phosphate is probably positive at 25 °C, and comparable in magnitude with the positive ΔH for the Na$_2$ dodecyl phosphate. Since all these heats are small, the adsorptions are clearly complex and involve a balance of positive and negative heat terms from loss of translational and rotational degrees of freedom, changes in chain conformation, solvation effects and ionic associations at the phosphate head group. No clear picture emerges from the results.

The ΔH for the K$_2$ dodecyl phosphate are positive and comparable with those for the Na$_2$ compound when 0.1 M added salt is present, or in the absence of added salt at higher surfactant concentrations. In the Traube region in the absence of salt, ΔH for K$_2$

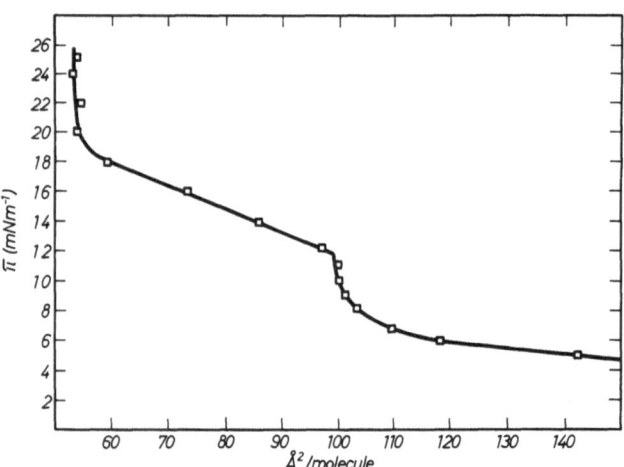

Fig. 8. Surface pressure (Π mNm^{-1}) vs. area (Å^2 molecule^{-1}) isotherm at 25 °C for Na$_2$ dodecyl phosphate from log plot emphasizing the "linear" region.

dodecyl phosphate is negative and comparatively large. At first sight, this would suggest that the potassium ion is more strongly bound than sodium to the phosphate head group in the monolayer. However, the Π-Å^2 isotherms clearly contradict such an interpretation, and the differences in ΔH for the K_2 and Na_2 is probably a reflection of the changes in solvation for the two cations in the adsorption process. Unambiguous evidence on the ion binding in these monolayers must await measurements of the surface potentials.

Acknowledgement

We wish to thank Brian Doran and Fred Englert for their assistance in the construction of the apparatus.

References

1. Stadtman, E., Lippman, F., J. Biol. Chem. 185, 549 (1950).
2. Uhlenbrock, J., Verkade, P.; Rec. Trav. Chim. Pays-Bas. 72, 395 (1950).
3. Brown, D., Malkin, T., Maliphant, G., J. Chem. Soc. 1584 (1955).
4. Peppard, C., Ferraro, J., Mason, G., J. Inorg. Nucl. Chem. 16, 246 (1960).
5. Nelson, A., Toy, A., Inorg. Chem. 2, 775 (1963).
6. Parreira, H., Pethica, B. A., 2nd Int. Cong. Surf. Act. Vol. 1, 44 (1957).
7. Parreira, H., J. Colloid Sci. 20, 742 (1965).
8. Kung, E., J. Colloid Interface Sci. 29, 105 (1969).
9. Muller, H., Friberg, A., Hellstrom, M., J. Colloid Interface Sci. 32, 132 (1970).
10. Gershfeld, N., Pak, C., J. Colloid Interface Sci. 23, 215 (1967).
11. Arakawa, J., Pethica, B. A., J. Colloid Interface Sci. 75, 441 (1980).
12. Tahara, T., Satake, I., Matuura, R., Bull. Chem. Soc. Japan 42, 1201 (1969).
13. Goddard, E. D., Kao, O., Kung, H. C., J. Colloid Interface Sci. 24, 297 (1967).
14. Nakagaki, M., Handa, T., Shimabayashi, S., J. Colloid Interface Sci. 43, 521 (1973).
15. Nakagaki, M., Handa, T., Bull. Chem. Soc. Japan 48, 630 (1975).
16. Harkins, W., Brown, D., J. Am. Chem. Soc. 41, 499 (1919).
17. Nakagaki, M., Handa, T., Yakagaku, Zasshi 92, 611 (1972).
18. Pethica, B. A., Trans. Faraday Soc. 50, 413 (1954).
19. Betts, J. J., Pethica, B. A., 2nd Int. Cong. Surf. Act. (London), 152 (1957).
20. Betts, J. J., Pethica, B. A., Trans. Faraday Soc. 56, 1515 (1960).
21. Güntelberg, E., Z. Phys. Chem. 123, 199 (1926).
22. Scatchard, G., Breckenridge, T., J. Phys. Chem. 58, 596 (1954).
23. Robinson, R., Stokes, R., "Electrolyte Solutions", Butterworths Scientific Pub., London, 1955.
24. Pallas, N. R., Pethica, B. A., Colloids and Surfaces, Colloids and Surface 6, 221 (1983).
25. Matijević, E., Pethica, B. A., Trans. Faraday Soc. 54, 1382 (1958).
26. Matijevic, E., Pethica, B. A., Trans. Faraday Sec. 54, 1390 (1958).
27. Goddard, E. D., Kung, H. C., J. Colloid Interface Sci. 37, 585 (1971).
28. Goddard, E. D., Kao, O., Kung, H. C., J. Colloid Interface Sci. 27, 616 (1968).

Received February 7, 1983;
accepted April 1, 1983

Author's address:

N. R. Pallas
Sohio 4440
Warrens Ville Center Road
Cleveland, Ohio 44128, U.S.A.

B. A. Pethica
Electro Biology Inc.
Fairfield, N. J. 07006, U.S.A.

Progress in Colloid & Polymer Science   Progr. Colloid & Polymer Sci. **68**, 14–19 (1983)

Stability of highly compressed spread monolayers of ^{125}I-labelled and cold BSA*)

F. MacRitchie and Lisbeth Ter-Minassian-Saraga**)

Physico-Chimie des Surfaces et des Membranes, Equipe de Recherche du CNRS associée à l'Université Paris V, UER Biomédicale, Paris, France.

Abstract: Compression-expansion cycles were performed on BSA spread monolayers at the air/aqueous interface using both unlabelled and radioiodinated protein. Irreversible losses in area were observed when films were held at surface pressures above 18 mNm^{-1} although the remaining films were qualitatively unchanged. Film was transferred quantitatively from the labelled monolayer to glass slides by a Blodgett-type technique and the measured specific activity used to evaluate the surface density of protein. Constancy of specific activity while films were maintained at high pressure confirmed that area losses were caused by desorption and that there was no selective desorption of tagged protein molecules. Specific activity measurements of aliquots of the substrate verified desorption and were in agreement with the amounts of protein calculated from monolayer losses. Rates of desorption increased with increasing surface pressure and were higher for the radiolabelled protein above a critical surface pressure. A decrease in activity of BSA monolayers, much greater than that corresponding to normal decay of ^{125}I, was observed when they were left for extended periods, apparently due to loss of iodine from the protein. The technique of transferring radiolabelled film to glass slides for analysis appears to have much potential for studying the stability of proteins at surfaces both alone and in association with other surface active substances.

Key words: protein monolayers, spread, stability

Introduction

Protein monolayers exhibit Π-A relationships which are characterized by a sigmoid curve in which the steep portion extrapolates to an area close to 1.0 m^2 mg^{-1} at zero surface pressure [1], independent of the molecular mass of the protein. This indicates that the configuration at the surface is similar for all proteins and corresponds to a highly unfolded state. As the surface pressure is increased, relaxation processes occur in the monolayer which have been interpreted in terms of displacement of segments from the surface [2]. An inflection point in the Π-A curve corresponding to a minimum in compressibility has often been described [3, 4]. For small molecules, a collapse pressure is usually found at or above the limit for monolayer stability when a monolayer $\rightarrow$ bulk phase transition occurs. For proteins, irreversible coagulation may occur [5] or desorption [6, 7].

Use of radiolabelled proteins allows the possibility to distinguish between coagulation and desorption since, in the former case, radioactivity should be conserved whereas desorption leads to decreasing total activity. However, one aspect requiring clarification in this work is whether the labelling process alters the properties, especially the surface properties of proteins.

Highly compressed films of proteins, either alone or in the presence of lipids, occur in stabilized foams and emulsions. Furthermore, their study is relevant to biological membrane structure and function. One obvious example is the stability of proteins inside compressing membranes of the lung alveolae [8].

Using ^{125}I-labelled BSA and a Blodgett-type technique to analyse the monolayer, we have compared the stability of cold and radiolabelled BSA in highly compressed films as well as their Π-A characteristics.

Materials and methods

Two samples of BSA were used, each crystallized and lyophilized, one from Sigma & Co and the other, labelled with ^{125}I from Sorin Biomedica, S.p.a. Italy. The number of BSA molecules

---

*) Decicated to the memory of Professor Bunichi Tamamushi. Que *l'importance* soit dans Ton regard, non dans la chose regardée.
André Gide.

**) To whom correspondence should be addressed.

carrying an I atom has been calculated from the initial specific activity of this sample (0.04 mC mg^{-1}) and knowing that the half-life of ^{125}I is 60 days. We obtained the result that one in ninety BSA molecules was iodinated on the average. The water was triply distilled, the first distillation being from KMnO$_4$. A phosphate buffer solution (pH 7.3 containing 0.15 M NaCl) was used in all experiments unless otherwise stated. All NaCl solutions were purified by foaming. A twenty-fold reduction in the surface area of the buffer solution after 30 min. standing gave no measureable surface pressure. Protein was spread from this same solution using an all-glass Agla microsyringe with the tip held at the surface. Under these conditions, quantitative spreading occurred as assessed by previously published Π-A isotherms [7]. The experiments were carried out at room temperature (20 $\pm$ 2 °C).

A Langmuir-type surface balance (MCN Lauda, Germany) was used. To measure stability of protein monolayers, a standard procedure was adopted. The protein was first spread on the maximum available area ($\sim$ 600 cm^2) and then compressed by moving the barrier to reduce the area at a constant rate of 0.8 cm^2 sec^{-1} until the required pressure was reached. The film was then held at this pressure for 20 min. after which the surface was rapidly expanded to the maximum area and again compressed. The cycles were then repeated as many times as required.

Films were transferred from the spread monolayer to small glass plates, cut from microscope slides, by a Blodgett-type technique. The clean dry slide was lowered slowly through the surface by an electrically driven holder. Constancy of surface pressure during this step showed that there was no transfer of film. The slide was then raised slowly with the film either maintained at constant area or constant surface pressure. During this stage, film was transferred to the slide. Changes in area of the surface film at constant pressure coincided closely with the immersed areas of slides (approx. 7 cm^2). The area on which film was deposited was clearly visible after removal of the slides and could be measured using Vernier calipers. Slides on which labelled protein was deposited were placed in vials and the radioactivity measured in a Packard Model 3002 scintillation counter. To allow quantitative determinations of the amounts of protein on the slides, small aliquots (5 μL) of the spreading solution were placed in several vials and their activities measured. These samples were used as calibration controls for the duration of the work, thus enabling the natural decay in activity of the ^{125}I with time to be allowed for. The amount of protein in solution was determined by assuming the Π-A relationships of labelled and unlabelled BSA to be identical and making use of the accurately known concentration of the unlabelled BSA solution. This is believed to be justified in view of the coincidence in properties of the films obtained with the two samples below 18 mNm^{-1}. Checks for leakage were carried out by compressing the surfaces outside the float and barrier and, in the case of spread labelled BSA, by taking slides in these regions and measuring the radioactivity. To detect the presence of desorbed BSA molecules in the substrate, the radioactivity of 5 ml samples of the substrate was measured.

Results

Compression isotherms

Figure 1 and figure 2 reproduce the recording of compression expansion (not shown) cycles up to pressures of 21 mNm^{-1} and 30 mNm^{-1} for the unlabelled BSA. Figure 3 and figure 4 reproduce similar recordings for the labelled protein at pressures

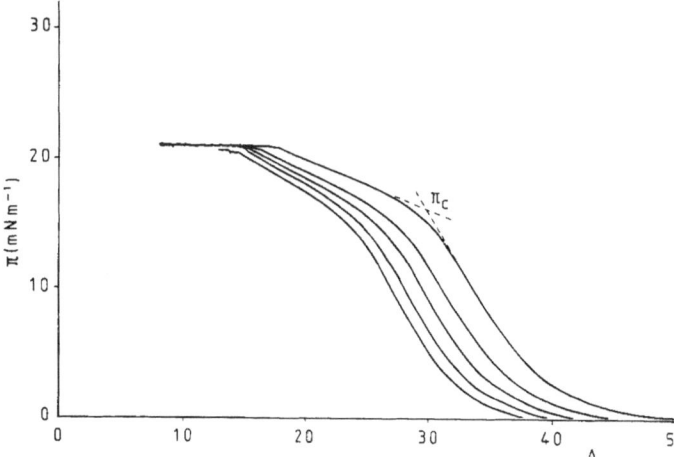

Fig. 1. Compression-expansion (not shown) cycles for cold BSA monolayer held for 20 min periods at 21 mNm^{-1}. A is in arbitrary units. Surface area may be calculated from the formula: area = 8.33 A + 10.7 cm^2

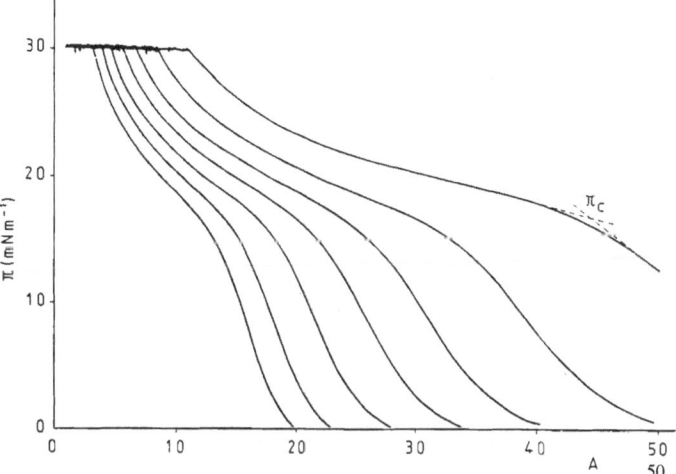

Fig. 2. Compression-expansion (not shown) cycles for cold BSA monolayer held for 20 min periods at 30 mNm^{-1}. A is in arbitrary units. Surface area may be calculated from the formula: area = 8.33 A $\pm$ 10.7 cm^2

of 21 and 27 mNm^{-1}. The critical pressure, Π_c, obtained by extrapolation of the two portions of the Π-A curve of different slopes, characterizes the critical Π and A values for the transition from a flat, extended conformation to one in which chains are buckled. The dynamic Π_c and A_c values are comparable for the labelled and unlabelled BSA, being 16 $\pm$ 1 mNm^{-1} and 0.74 $\pm$ 0.1 m^2 mg^{-1} respectively. Under equilibrium conditions (fig. 6), Π_c and A_c are slightly different: 17.5 mNm^{-1} and 0.6 m^2 mg^{-1} from the dynamic Π_c and A_c

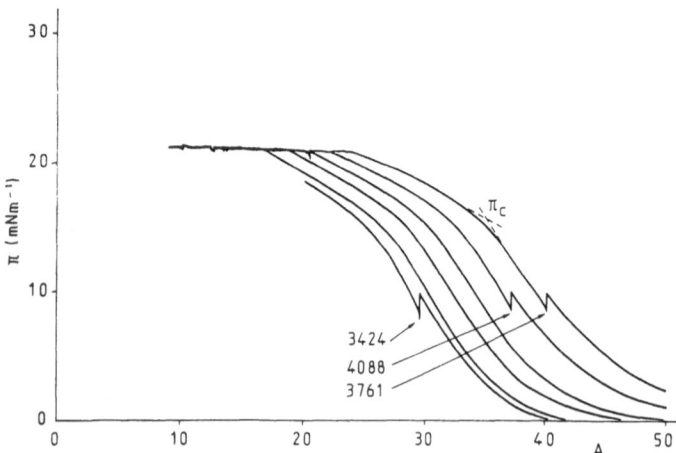

Fig. 3. Compression-expansion (not shown) cycles for ^{125}I-labelled BSA monolayer held for 20 min periods at 21 mNm^{-1}. A is in arbitrary units. Surface area may be calculated from the formula: area = 8.33 A + 10.7 cm^2. The numbers correspond to the measured counts min^{-1} cm^{-2} for the films transferred onto the glass slides.

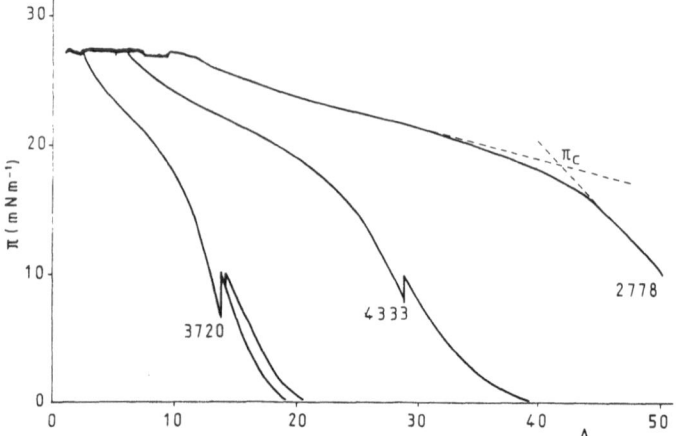

Fig. 4. Compression-expansion (not shown) cycles for ^{125}I-labelled BSA monolayer held for 20 min periods at 27 mNm^{-1}. A is in arbitrary units. Surface area may be calculated from the formula: area = 8.33 A + 10.7 cm^2. The numbers correspond to the measured counts min^{-1} cm^{-2} for the films transferred onto the glass slides

Film composition by radioactivity

In figure 3 and figure 4, the removal of film on slides can be seen as shifts in the Π-A curve. The numbers adjacent to the shifts represent the specific activity in counts min^{-1} cm^{-2}. The area corresponding to the Π-A curve shift agrees with the area of the film transferred on the slide within experimental error. Therefore the film transfer ratio is unity. There is a scatter of ± 8% between the specific activity of the three slides

Table 1. Irreversible losses of area at various surface pressures and corresponding radioactivity of films at Π = 10 mNm^{-1}

Desorption pressure Π_d (mNm^{-1})	irreversible decrease in film area (%)	specific activity (counts min^{-1} cm^{-2})
21 (1)	0	3761 (1)
	4.5	4088 (1)
	21.5	3424 (1)
24 (2)	0	3661 ± 200 (3)
	31.4	4050 ± 200 (3)
	43.2	3483 ± 120 (2)
	53.3	3403 ± 170 (2)
27 (2)	0	2770 ± 100 (2)
	41.5	4333 (1)
	68.6	3720 ± 400 (3)

The numbers in brackets indicate the number of slides taken.

which were taken at various times but no obvious decrease or increase. Other experiments (not shown) with films compressed at 24 and 34 mNm^{-1} and film transfer at 10 mNm^{-1} produced similar results. It can be seen in figure 3 and figure 4 and in table 1 that maintenance of the film at pressures above 18 mNm^{-1} reduced its area considerably and irreversibly without a systematic effect on the specific activity at Π = 10 mNm^{-1} except in one (see table 1, Π = 27 mNm^{-1}) out of nine experiments (not all shown).

From the irreversible loss in area ΔA at Π = 10 mNm^{-1}, measured for four successive experiments, and the known surface density (1.22 mg m^{-2}) at this pressure, we calculated a total loss of 0.077 mg. From the radioactivity contained in 1 ml of substrate (1,370 counts min^{-1} ml^{-1}) and the calibration factor equal to 3.3 × 10^{-8} mg count^{-1} min^{-1}, we obtain 4.5 × 10^{-5} mg ml^{-1} of substrate. The volume of substrate being 650 ml, the total amount of BSA desorbed into the substrate is 0,030 mg. The estimated and measured radioactivity in the substrate are thus of the same order of magnitude.

Figure 5 represents Π-log A plots of curves obtained for the unlabelled or labelled BSA. The parallel shifts of the various curves imply constancy of the film compressibility and desorption of compressed film molecules according to [8]. The compressibility of the labelled BSA film is similar to that of the unlabelled BSA at least up to 12 mNm^{-1} as shown by the corresponding Π-log A plots in figure 5. Typically, inside the range 8 mNm^{-1} < Π < 12 mNm^{-1}, the compressibility is equal to 0.02 mNm^{-1} ± 0.002.

Figure 6 represents the results of an experiment where film has been transferred to slides at various constant surface pressures. The data have been translated into BSA surface concentrations in mg m^{-2} on

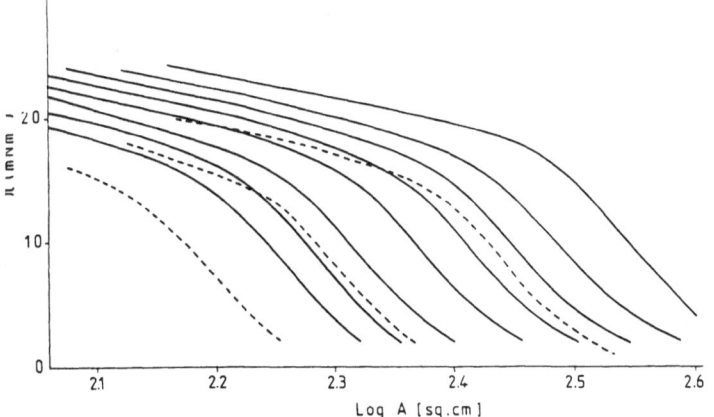

Fig. 5. Π-log A plots for compression isotherms of cold BSA compressed for 20 min periods at 24 mNm^{-1}. A [cm^2]. Full lines: cold BSA; dashed lines: ^{125}I – BSA

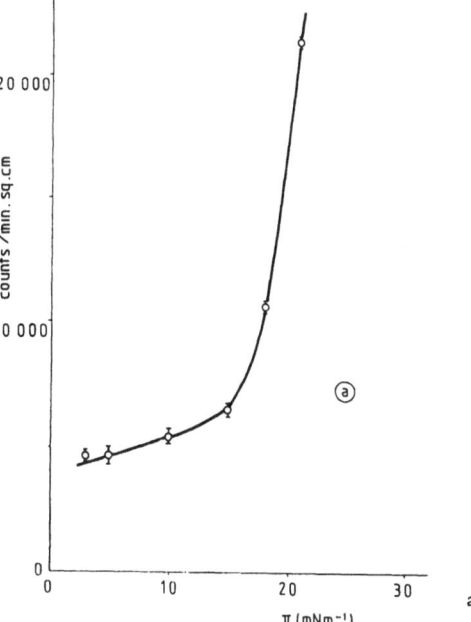

the slides using the calibration described in the methods section. Two independent experiments were carried out. The results are compared with the surface densities as a function of surface pressure for an equilibrated film of unlabelled BSA.

Kinetic experiments

All the films displayed irreversible loss in area when maintained. at constant pressure above 18 mNm^{-1}. Using a technique similar to that described in [6], the rate of BSA desorption was obtained from the irreversible loss in area observed in figures 1–4. Figure 7 shows some log A-time plots for BSA films at 10 mNm^{-1}. Except for the initial part, these plots are linear. This implies a constant rate of loss in area (shown to be desorption). The effect of film pressure on the rate constant for desorption $[k_d = -(d\ln A/dt)_\Pi]$ is shown in figure 8. The difference in behaviour of labelled and unlabelled BSA becomes measurable only at pressures above 18 mNm^{-1}, the rate of desorption of labelled BSA being greater in this region. Points for the rate of desorption of BSA with pure water as the substrate obtained from the present work and by calculation from data of Gonzalez and MacRitchie [6] are also plotted in figure 8.

Labelled BSA films were spread on the surface of pure substrate and a substrate containing 10^{-5} % unlabelled BSA. The films were maintained at 10 mNm^{-1} and transferred on to glass slides at various times. In each case, a decrease, much greater than that due to the natural decay of ^{125}I was observed although the rate of decrease appeared to slow down with time

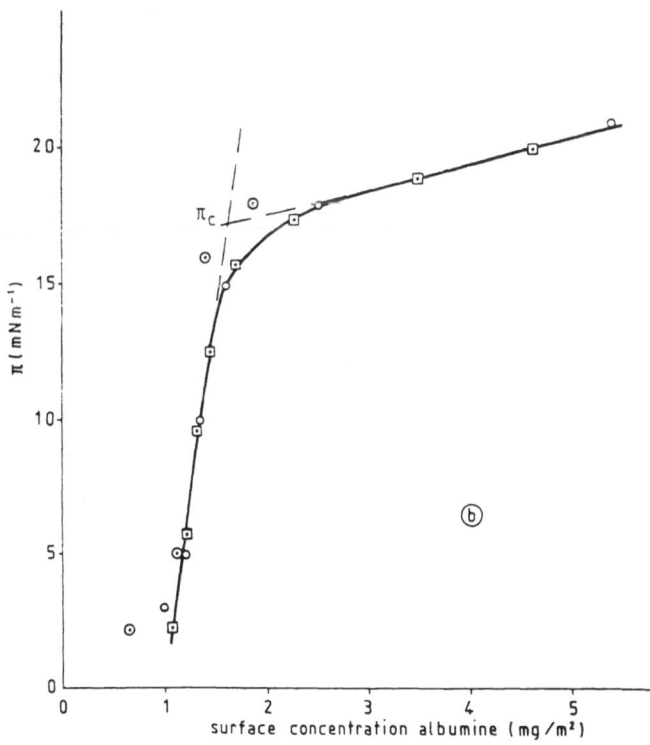

Fig. 6. a) Specific radioactivity (counts min^{-1} cm^{-2}) plot vs. Π. b) Π-surface concentration isotherm of BSA. Points measured directly $\boxdot$ are compared with those deduced $\bigcirc$, $\odot$ from radioactivity of transferred film at various values of Π (curve a of fig. 6)

in both cases. This behaviour is shown in figure 9. During the first 24 hours the radioactivity decreases by 40% to 50% of the initial value.

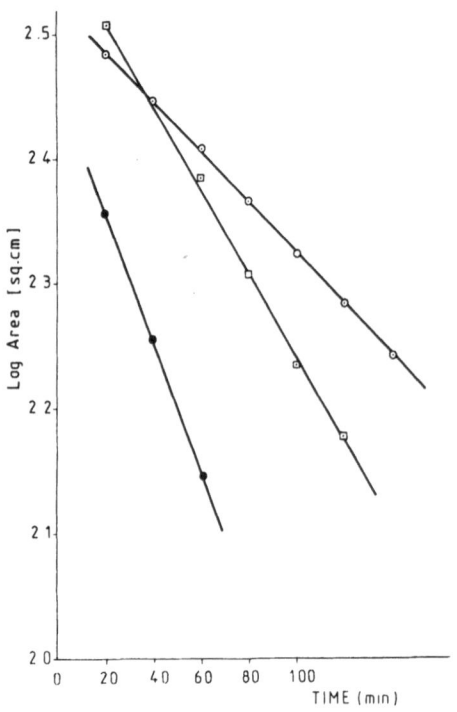

Fig. 7. Log area (at 10 mNm^{-1}) vs. time plots for cold and ^{125}I-labelled BSA monolayers held for 20 min periods at several pressures. ⊙, cold BSA, 24 mNm^{-1}; □, cold BSA, 27mNm^{-1}; ●, ^{125}I-labelled BSA, 24 mNm^{-1}

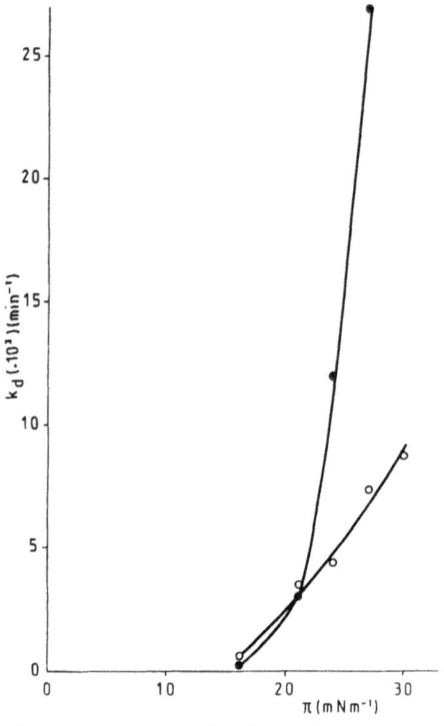

Fig. 8. Rate constants for desorption (k_d) as a function of surface pressure for cold and ^{125}I-labelled BSA monolayers. ●, ^{125}I-labelled BSA; ○, cold BSA

Discussion

BSA films show significant irreversible loss of area when held at constant surface pressures above 18 mNm^{-1}. At the same time, no significant change occurs in the compressibility or specific radioactivity (at a given pressure). This behaviour is consistent with a desorption of protein into the sub-phase. Confirmation of this was obtained by measuring the radioactivity of aliquots of the substrate after a known amount of labelled monolayer has been lost. The calculated concentration in the sub-phase was of the same order as that expected from the loss of monolayer, assuming homogeneous dispersion of the protein (see Methods and Results). The latter is thought to be favoured by the barrier movements during compression-expansion cycles. Where comparison with previous results for rates of desorption of BSA [6] could be made, good agreement was found (fig. 8).

The "constancy" of specific activity in the labelled BSA film after irreversible loss of the surface area implies that under our experimental conditions, the labelled molecules desorb "at the same rate" as the unlabelled ones. We have calculated (see Methods) that only about 1% of the molecules in the labelled

BSA are tagged with ^{125}I. However a difference in desorption behaviour of the labelled BSA at pressures above 18 mNm^{-1} has been observed (see Results). It may be due to a change in the sample brought about by the labelling procedure rather than as a result of BSA molecules labelling with ^{125}I.

The average surface properties of the protein films seem to be conserved at low surface pressures so that radio-iodinated protein may be used with confidence for surface chemical work inside this range of pressure. One factor which must be allowed for, however, is the relatively rapid decrease in activity of the protein (about 2%/hour) when it remains at the surface for long periods of time (fig. 9). From the results already discussed (i.e. no significant desorption below 18 mNm^{-1} and constancy of specific activity of labelled protein during desorption), the decrease in activity of the film cannot be ascribed to loss of protein. It therefore appears that the surface catalyses a reaction in which iodine is lost from the protein. We have no clear explanation to this phenomenon. Some workers [9] have reported that protein molecules although they do not readily desorb into a protein free substrate, are able to exchange with protein molecules

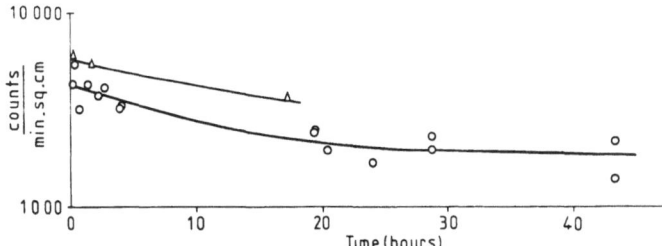

Fig. 9. Specific radioactivity (at 10 mNm^{-1}) of ^{125}I-labelled BSA monolayers (held at 10 mNm^{-1}) as a function of time. $\triangle$, control (pure substrate); $\bigcirc$, 10^{-5} % cold BSA

when these are present in the substrate. From our present results, it would appear that little desorption (and perhaps exchange) would occur when the film is at surface pressures below 18 mNm^{-1} (i.e. surface concentrations below $\simeq$ 1.5 mg m^{-2}).

Combined with the monolayer techniques described in references [6] and [8] the technique of transferring labelled protein from monolayers onto glass slides for subsequent analysis appears to have great potential for study of the stability of proteins at surfaces both alone and in admixture with other surface-active substances. For mixed lipid-macromolecular films, the potentiality of an analysis utilizing radioactive components has been stressed previously [10].

Acknowledgements

One of us, F. MacRitchie thanks "Naturalia & Biologia", the "Fondation pour la Recherche Médicale" and the French Government for financial assistance which allowed this collaborative research to be undertaken.

References

1. Bull, H. B., Advan. Protein Chem. **3**, 95 (1947).
2. MacRitchie, F., J. Colloid Interface Sci. **79**, 461 (1981).
3. Yamashita, J., Bull, H. B., J. Colloid Interface Sci. **24**, 310 (1967).
4. Thomas, C., Ter-Minassian-Saraga, L., Bioelectrochem. Bioenergetics **5**, 369 (1978).
5. MacRitchie, F., Owens, N. F., J. Colloid Interface Sci. **29**, 66 (1967).
6. Gonzalez, G., MacRitchie, F., J. Colloid Interface Sci. **32**, 55 (1970).
7. MacRitchie, F., J. Colloid Interface Sci. **61**, 223 (1977).
8. Ter-Minassian-Saraga, L., J. Colloid Interface Sci. **79**, 222 (1981).
9. Brash, J. L., Samak, Q. M., J. Colloid Interface Sci. **65**, 495 (1978).
10. Ter-Minassian-Saraga, L., J. Colloid Interface Sci. **70**, 245 (1979).

Received February 7, 1983

Authors' addresses:

Lisbeth Ter-Minassian-Saraga
Physico-Chimie des Surfaces et des Membranes
Equipe de Recherche du CNRS
associée à l'Université Paris V
UER Biomédicale
45, rue des Saints-Pères
75270 Paris cedex 06, France

F. MacRitchie
CSIRO, Wheat Research Unit,
c/o Bread Research Institute
Private Bag, P.O.
North Ryde, N.S.W., Australia

Progress in Colloid & Polymer Science

Progr. Colloid & Polymer Sci. **68**, 20–24 (1983)

Study on the size of reversed micelles of anionic and cationic surfactants*)

K. Kon-No, H. Asano, and A. Kitahara

Department of Industrial Chemistry, Science University of Tokyo, Tokyo, Japan

Abstract: The size of reversed micelles and its distribution formed by some anionic surfactants having different polar groups and by chiral and achiral cationic surfactants in apolar solvents were investigated by means of the vapor pressure osmometry. The study was also carried out for Aerosol OT purified carefully. The micellar size depended on the kind of polar groups of surfactants and the magnitude of the size was order of sulfonate > carboxylate > sulfate. However, solvent dependence of the micellar size was hardly found for any anionic surfactants. In case of cationic surfactants, the micellar size was independent of the chirality of surfactants and was larger for chiral surfactants than for achiral one. The distribution of micellar size was estimated from the analysis of the dependence of apparent aggregation number upon surfactant concentration by the curve-fitting method.

Key words: Reversed micelle, Reversed micellar formation, Reversed micellar size, Oil-soluble surfactants, Oil-soluble optically active surfactants.

Introduction

The formation of reversed micelles of surfactants has been studied from the different surfactants [1–3], solvents [4–6] and temperatures [2, 7] by various experimental techniques. The interaction between polar groups and the bulkiness of alkyl groups of surfactants have been known as the determining factors of the size of reversed micelles, respectively. However, the formation of reversed micelles by the surfactants having different polar groups has scarecely been studied. In addition of the bulkiness effect of alkyl groups of surfactants, the existence of asymmetric carbon in alkyl groups might be one of the control factors in the reversed micellar formation, because the chirality of the amino acid residue in the polygluta-mate determines the asymmetric structure of higher ordered aggregates in organic solvents [8]. On the other hand, different data have been reported on the micellar size of sodium 1,2-bis-(2-ethylhexyloxycar-bonyl)-*l*-ethane sulfonate, Aerosol OT, which has been used extensively for the study of the physico-chemical properties of reversed micelles, in spite of using the same kind of solvent and identical method. This may be due to the impurity containing in the surfactant because of different purification methods by workers.

In this work, the anionic surfactants having different polar groups and the chiral and achiral cationic surfactants having similar alkyl groups were synthesized and their reversed micellar size and its distribution in various apolar solvents were studied by means of the vapor pressure osmometry. The reversed micellar size of Aerosol OT purified carefully according to the method described previously [9] was also studied by means of the vopor pressure osmometry.

Experimental

Materials

Sodium 1,2-bis-(2-ethylhexyloxycarbonyl)-*l*-ethane sulfonate (AOT), sodium 1,3-bis-(2-ethylhexyloxy)-2-propyl sulfate (EPA), sodium 1,3-bis-(2-ethylhexyloxy)-2-propane sulfonate (EPO) and sodium 1,3-bis-(2-ethylhexyloxy)-2-propoxymethane carboxylate (EPC) were used as anionic surfactants. AOT was furnished from the Nakarai Yakuhin Co. and purified according to the previous method [9], i. e., AOT was treated twice with activated charcoal before and after the water extraction procedure in the Kunieda and Shinoda method [10]. Other surfactants were synthesized from 1,3-bis-(2-ethylhexyloxy)-2-propanol (154–155 °C/1.3 mmHg), which is prepared from 2-ethylhexylalcohol and epichlorohydrin according to the method by Kuwamura et al. [11]. EPA: The 1,3-bis-(2-ethylhexyloxy)-2-propyl sulfate was prepared by the sulfation of 1,3-bis-(2-ethylhexyloxy)-2-propanol with chlorosulfuric acid in dry ether at −10 to −5 °C and was neutralized with 10% sodium carbonate. EPO: The 1,3-bis-(2-ethylhexyloxy)-2-propyl bromide was prepared by bromination of 1,3-bis-(2-ethylhexyloxy)-2-propanol by the usual method and further thiolated with thiourea in

---

*) Dedicated to the memory of Professor Dr. B. Tamamushi.

95% ethanol. After the thiol was oxidated by 30% perhydroxide with sulfuric acid as a catalyst, it was neutralized with 10% sodium carbonate. EPC: The sodium 1,3-bis-(2-ethylhexyloxy)-2-propoxide was prepared by the treatment of 1,3-bis-(2-ethylhexyloxy)-2-propanol with metal sodium and further treated with chloroethyl acetate. The ester obtained was hydrolyzed with sodium hydroxide in water-ethanol mixture. The purification of these surfactants was similar with that of AOT [9].

The structure and purity of these surfactants were ascertained from IR, NMR, and elementary analyses. Anal. Calcd. for AOT: C, 54.02 %; H, 8.39 %; S, 7.21 %; Na, 5.17 %. Found: C, 53.92 %; H, 8.51 %; S, 7.08 %; Na, 5.06 %. Calcd. for EPA: C, 54.52%; H, 9.37 %; S, 7.66 %; Na, 5.71 %. Found: C, 55.77 %; H, 9.90 %; S, 7,08 %; Na, 5.60 %. Calcd. for EPC: C, 62.80 %; H, 10.52 %; Na, 5,80 %. Found: C, 62.13 %; H, 10.78 %; Na, 7,45%. Calcd. for EPO: C, 56.69 %; H, 9,76 %, S, 7.96 %; Na, 5.49 %. Found: C, 56.34 %; H, 9.89 %; Na, 5.05 %.

(+)- and (−)-α-phenylethyldodecyldimethylammonium bromides ((+)- and (−)-PDDAB) and dodecylbenzyldimethylammonium bromide (DBDAB) were used as cationic surfactants. (+)- and (−)-N,N′-dimethyl-α-phenylethylamiens were prepared from (+)- and (−)-α-phenylethylamiens, fromaldehyde and formic acid [12] and purified by distillation under reduced pressure. PDDAB was prepared by quaternizing of N,N′-dimethyldodecylamine with benzylbromide. These surfactants were recrystallized from acetone or ethyl acetate several times. The melting point was 100.5–101.5 °C for (+)-PDDAB, 101.0–101.5 °C for (−)-PDDAB, and 42–44 °C for DBDAB.

The structure and purity of these surfactants were ascertained from IR, NMR, and elementary analyses. Anal. Calcd. for (+)-PDDAB: C, 66.31%; H, 10.12 %; N, 3.52 %. Found: C, 66.61 %; H, 10.29 %; N, 3.51 %. Calcd. (−)-PDDAB: C, 66.31 %; H, 10.12 %; N, 3.52 %. Found: C, 66.93 %; H, 10.52 %; N, 3.51 %. Calcd. for DBDAB: C, 65.61 %; H, 9.96 %; N, 3.64 %. Found: C, 63.82 %; H, 10.37 %; N, 3.69 %.

The structure formulae of these surfactants are shown in figure 1.

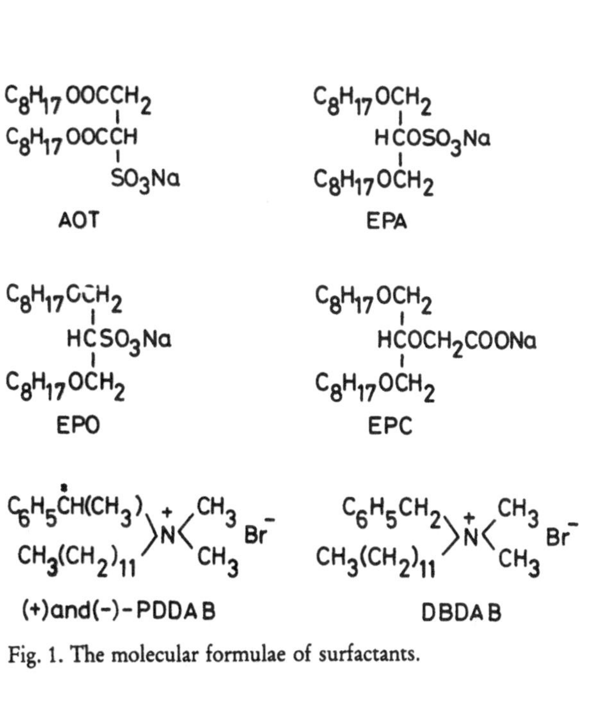

Fig. 1. The molecular formulae of surfactants.

Method

The measurement was carried out by use of the Hitachi-Perkin-Elmer Molecular Weight Measuring Apparatus 115. The calibration curve between ΔR and C was made with diphenyl at 30 °C, where ΔR and C are the resistance differences between the pure solvent and the solution and the molal concentration of the surfactants, respectively. The surfactant solutions were prepared in dry grobe. They were kept at the measuring temperature for 24 hr in a desiccator. The measurement was repeated until the ΔR reading gave a stationary value for each solution. The apparent molecular weight of surfactants was calculated using the molal constant obtained from the calibration curve. The molar ratio of water to surfactants in the surfactant solutions was below 0.09 throughout this study.

Results and discussion

Dependence of apparent aggregation number upon surfactant concentration

Representative examples of the dependence of apparent aggregation number upon the concentration of surfactants in anionic and cationic surfactant solutions are shown in figures 2 and 3, respectively. The apparent aggregation number of the anionic surfactants approaches a saturation value above about 5 mmole kg^{-1} (fig. 2), whereas that of cationic surfactants gradually increases with increase of the surfactant concentration (fig. 3). Similar concentration dependence of the apparent aggregation number for anionic surfactants was also observed in other solvents. Hence, the true aggregation number of anionic surfactants was taken as the apparent saturating aggregation number. On the other hand, the true aggregation number of cationic surfactants was calculated according to the previous method [13].

The change of apparent aggregation number with surfactant concentration was also utilized to estimate the distribution of micellar size. As seen in figure 2, the apparent aggregation number of all the anionic surfactants in any solvents was independent on the concentration of surfactants. This means the formation of reversed micelles having sharp size distribution. However, the size distribution of reversed micelles formed by cationic surfactants may be more broader from the dependence of apparent aggregation number on the concentration of surfactants (fig. 3). Then the change of apparent aggregation number with the concentration of cationic surfactants was simulated with the curve-fitting method proposed by Kertes et al. [14]. The plots of the fraction of each aggregate vs the concentration of (+)-PDDAB are presented in figure 4 as a representative result. Three different size of aggregates, i. e., trimer, pentamer and tridecamer could be estimated. Similar three aggre-

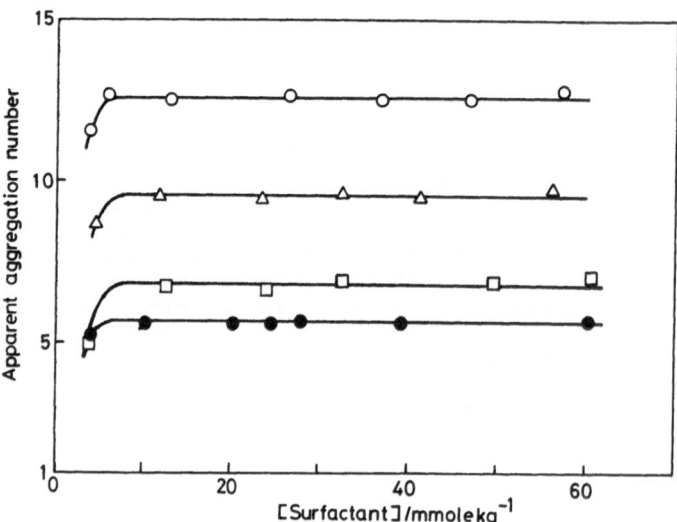

Fig. 2. The change of apparent aggregation number with concentration of anionic surfactants. ○: AOT/isooctane; △: EPO/carbon tetrachloride; □: EPC/carbon tetrachloride; ●: EPA/isooctane

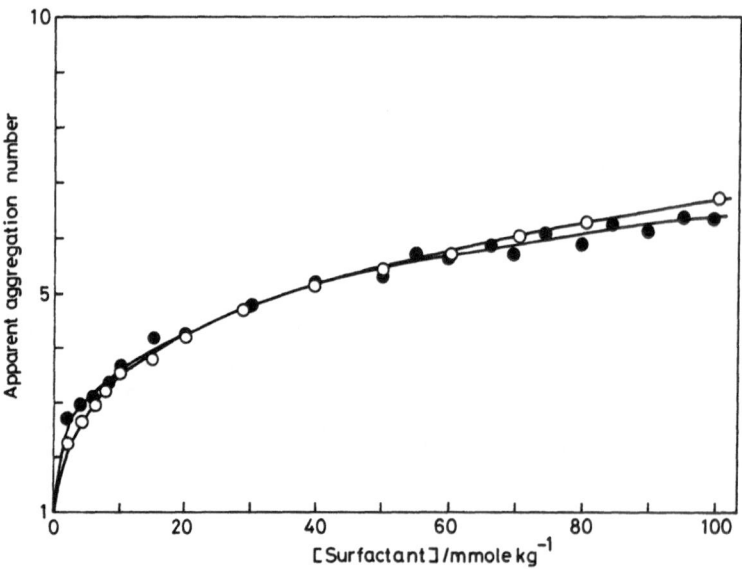

Fig. 3. The change of apparent aggregation number with concentration of cationic surfactants. ○: (+)-PDDAB/Benzene; ●: DBDAB/
Benzene

gates were also estimated in the benzene solution of (–)-PDDAB, but in DBDAB solution tetramer, nonamer and undecamer were estimated. Table 1 summerizes the overall stability constants of these aggregates, which mean the equilibrium constant between monomer and aggregates, in benzene solutions of cationic surfactants.

Effect of polar groups of surfactants

The values of aggregation number of EPA, EPC, EPO and AOT obtained in various solvents are presented in table 2. It is seen in table 2 that the magnitude of the aggregation number of surfactants having same alkyl groups is order of EPO > EPC >

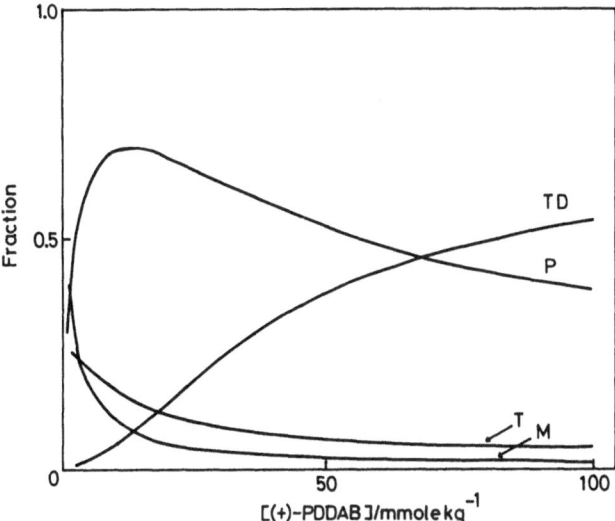

Fig. 4. The plots of fraction of each Aggregates vs. concentration of (+)-PDDAB in benzene. M, Monomer; P, Pentamer; T, Trimer; TD, Tridecamer

Table 1. The Stability constants of each aggregate of cationic surfactants

	$\log\beta$
(+)-and (−)-PDDAB	
Trimer	5.78
Pentamer	12.2
Tridecamer	34.9
DBDAB	
Tetramer	8.78
Nonamer	23.0
Undecamer	27.9

Table 2. The values of aggregation number of anionic surfactants in various apolar solvents

	C_6H_6	Cyclo-C_6H_{12}	C_7H_{16}	C_8H_{18}	CCl_4
EPO	10.4	11.6	12.2	11.7	9.7
EPC	7.5	8.6	8.3	8.7	6.9
EPA	5.8	6.2	6.0	5.7	5.8
AOT	10.9	13.2	13.0	12.8	9.7

Table 3. The values of aggregation number of cationic surfactants in benzene

	C_6H_6
(+)-PDDAB	8.7
(−)-PDDAB	8.9
DBDAB	6.3

EPA in any solvent. The micellar size of EPO having sulfonate group was comparable to that of AOT. Those facts show that the polar group of surfactant is one of the determinig factors of the size of reversed micelles. However, the order of the magnitude of micellar size by different polar groups could not be simply explained from the strength of the acid-base interaction between counterion and surfactant ions in micellar core proposed by Fowkes [15], because sodium ion is a hard Lewis acid, whereas SO_3^{2-} is a soft Lewis base and CH_3COO^- and SO_4^{2-} are hard Lewis bases. However, it is notable that the solvent dependence of the aggregation number of those surfactants was hardly observed.

Similar solvent independence was also observed for the aggregation number of AOT in table 2 and further the aggregation number in any solvent was lower than the values reported previously [1]. On the other hand, the agreement in the aggregation number of EPO with AOT show that the difference between the groups of ether and ester, which combine the alkyl group with polar group in the surfactant ions, affects hardly on the formation of reversed micelles.

Effect of chilarity of surfactants

It is known that AOT molecule has an asymmetric carbon. However, no data have been published on the aggregation number of reversed micelles of optically active surfactants. Then the aggregation number of chiral and achiral cationic surfactants having similar molecular structure was compared in order to examine the effect of asymmetric carbon in surfactant molecule on the reversed micellar formation.

As seen in table 3, the magnitude of aggregation number was independent of the chirality of surfactants, but it was larger for chiral surfactants than for achiral one. If the magnitude of reversed micellar size is controlled by the bulkiness of alkyl groups of surfactants [1], the aggregation number of (+) and (−)-PDDAB having methyl group attached to asymmetric carbon should be lower than that of DBDAB. However, the values presented in table 3 were inverse against to expection. This suggests that with reversed micellar formation of chiral surfactants the packing of alkyl group containing asymmetric carbon is more close than for achiral surfactant, leading to an increase in aggregation number. This suggestion might be substantiated from the fact that the amino acid residue in the polyglutamate and N-acylamino acids form higher ordered aggregates such as fibrous aggregates and cholesteric liquid crystals in organic solvents [8, 16]. Therefore it is considered from above results that the existence of asymmetric carbon in surfactant ion is one of the determining factors in the formation of reversed micelles similar to the polar groups.

References

1. Kon-No, K., Kitahara, A., J. Colloid Interface Sci. **35**, 636 (1971).
2. Kon-No, K., Jin-No, T., Kitahara, A., ibid. **49**, 383 (1974).
3. Eicke, H. F., Christen, H., Helv. Chim. Acta **61**, 2258 (1978).
4. Little, R. C., Singlettery, C. R., J. Phys. Chem. **68**, 3453 (1964).
5. Peri, J. B., J. Colloid Interface Sci. **29**, 6 (1969).
6. Eicke, H. F., Christen, H., ibid. **48**, 281 (1974).
7. Zulauf, M., Eicke, H. F., J. Phys. Chem. **83**, 480 (1979).
8. Tachibana, T., Kambara, H., Kolloid Z. Z. Polymere **219**, 40 (1967).
9. Kon-No., K., Katsuta, M., Nakamura, K., Mori, S., Kitahara, A., Nippon Kagaku Kaishi, 1980, 435.
10. Kunieda, H., Shinoda, K., J. Colloid Interface Sci. **70**, 577 (1979).
11. Kuwamura, T., Kogyo Kagaku Zashii **63**, 595 (1960), ibid. **64**, 1958 (1961).
12. U.S.P. 2, 776, 314; Chem. Abstr., 5112128 (1957).
13. Kon-No, K., Kitahara, A., Kogyo Kagaku Zashii **68**, 2058 (1965).
14. Kertes, A. S., Markovits, G., J. Phys. Chem. **72**, 4202 (1968)
15. Fowkes, F. M., Shinoda, K., Solvent Properties of Surfactant Solutions, p. 65. Marcel Dekker Inc., New York (1967).
16. Sakamoto, K., Yoshida, R., Hatano, M., Tachibana, T., J. Am. Chem. Soc.**100**, 6898 (1978).

Received February 10, 1983;
accepted April 21, 1983

Authors' address:

Kijiro Kon-No
Department of Industrial Chemistry
Science University of Tokyo
1–3, Kagurazaka, Shinjuku-ku
Tokyo 162, Japan

Progress in Colloid & Polymer Science   Progr. Colloid & Polymer Sci. **68**, 25–32 (1983)

On the influence of concentrated electrolytes on the association behaviour of a non-ionic surfactant*)

Th. van den Boomgaard, Sh. M. Zourab**), and J. Lyklema

Laboratory for Physical and Colloid Chemistry, Agricultural Univ. Wageningen, The Netherlands

Abstract: The micelle formation and viscosity are studied of aqueous solutions of Synperonic NPE-1800, a nonionic surfactant composed of nonylphenol, polypropylene and polyethylene blocks. The variables are the temperature (5–50 °C) and the nature and concentration of added electrolyte (NaCl up to 4M, Na_2SO_4 and $MgSO_4$ up to 0.4 M). The c.m.c. decreases with increasing temperature, upon addition of NaCl it first decreases or remains constant, depending on T but with further increased NaCl concentration it rises again. These two opposing trends are reflected in the changes of the enthalpy and entropy of micellization. For Na_2SO_4 or $MgSO_4$ no increase of c.m.c. at high concentration is observed. Viscometry supports the anomalous behaviour of concentrated NaCl solutions. The specificity is due to the Cl^--ion because 4M NaBr solutions exhibit no such exceptional features.

Key words: Nonionic surfactants; micelle formation, influence of temperature micelle formation, influence of concentrated electrolytes.

The effect of electrolytes on the formation and other properties of ionic micelles is as a rule more simple to describe in terms of basic physical principles than the same for nonionic systems. The reason for this difference is, that for ionic micelles the main effect of electrolytes is the compression of diffuse double layers, which is fairly well understood. In addition, there may be specific ionic effects, leading to lyotropic sequences in various phenomena, but theories are available for these, usually second order features. For instance, one could use Stern theory or site binding models.

For nonionics, double layers are all but absent and phenomenologically the explanation of the influence of electrolytes must be sought in terms of specific ion binding and the modification of the solvency by the salt, so called salting-out and/or salting-in. A more penetrating analysis requires information on the structure-forming and/or structure-breaking capacities of the ions involved or, for that matter, energetic and entropic contributions to the various interactions must be studied.

As a contribution to this interesting field we describe in this article some properties of Synperonic NPE-1800, a commercial nonionic surfactant on a nonylphenol-polyoxypropylene-polyoxyethylene basis. This substance has a relatively low molecular solubility in water, so that it is very surface active and has a low c.m.c. In addition, it is a powerful emulsifier for *O/W* emulsions. Of particular interest was our observation that paraffin oil-in-saline emulsions, stabilized by NPE-1800 were generally very stable, but that after some time substantial coalescence occurred in concentrated solutions of specific electrolytes, among which 4 M NaCl [1, 2]. This prompted us to study the effect of various electrolytes, up to high concentrations, on the c.m.c. and viscosity of NPE-1800 solutions. In the present investigation the temperature is another important variable, partly because of the practical relevance of emulsion stability at various temperatures, partly because it forms the basis of analyses in which enthalpic and entropic contributions are separated.

Although the main body of this work has been done with NPE-1800, a few additional data on three other Synperonics (NPE – A, B and C) with differing EO moiety length will also be reported for comparison purposes.

---

*) Dedicated to the memory of Prof. B. Tamamushi.
**) On leave of absence from the Univ. of Alexandria, Egypt.

Experimental

Materials

The general formula of the Synperonic NPE-1800 is

$$C_9H_{19}\text{—}\langle\bigcirc\rangle\text{—}\left[\text{O—CH}_2\text{—}\underset{\overset{|}{H}}{\overset{\overset{CH_3}{|}}{C}}\right]_m\left[\text{O—CH}_2\text{—CH}_2\right]_n\text{—OH} \quad (1)$$

The nonylgroup is homodisperse (i.e. it always contains nine carbon atoms) but branched in different ways. This followed from GC-MS analysis by Dr. M. A. Posthumus of the Dept. of Organic Chemistry of our University. There are at least 12 isomers. The branching was confirmed by n.m.r. Also from n.m.r. it could be inferred that the nonyl- and the PO-EO moieties are always in the para-position, irrespective of the extent of branching and polyether chain length.

In addition, it was found from n.m.r. that ethylene oxide, attacking at the para-position, has no preference for any isomer of the nonyl group. Considering the finding of Shachat and Greenwald [3] that the reaction mechanisms of ethylene oxide and propylene oxide with nonylphenol are identical, it may be concluded that NPE-1800 must be considered as a mixture with the polyether length distribution independent of the extent of branching. By virtue of the reaction mechanism (anionic polymerization) the distribution of the PO and EO parts has a Poisson character; this is also the case for the Synperonics NPE-A, B and C.

The molecular mass of NPE-1800 has previously been reported as 2750 [4] but we have evidence that the real value is lower. The value of M can be estimated by u.v. spectroscopy ($\lambda = 276.5$ nm), using NP molecules with shorter chains as the standards. We did so in water and methanol and found molar extinctions ε of 1.67×10^3 and 1.55×10^3 respectively. These values agreed very well with those reported by Gratzer and Beaven [5] ($\varepsilon = 1.67 \times 10^3$) and by Nadeau and Siggia [6] ($\varepsilon = 1.57 \times 10^3$). For M we found 2100 ± 50, independent of the solvent. N.m.r. proton counting yielded 2130 ± 260. The number ratio of PO/EO segments was found from n.m.r. to be $1:2$. On the basis of this information the actual values of m and n in (1) are 13 ± 2 and 26 ± 4 respectively, corresponding with $M = 2120$.

The Synperonics NPE-A, B and C were specially synthesized for us by ICI. In them m was the same as in NPE-1800, whereas n amounted to 50 ± 2, 85 ± 3 and 190 ± 5 respectively, corresponding with M-values of 3200, 4700 and 9300 respectively.

All water used was purified by filtration through a Millipore Milli RO 60 filtre combined with deionization with a Super Q system until the specific conductivity was below 10^{-4} Ωm^{-1}. Electrolytes were of analytical grade ex Merck and Baker and used without further purification.

Methods

Critical micelle formation concentrations at various temperatures T and electrolyte concentrations c_e were derived from surface tension (γ)-log c plots where c is the weight concentration of the surfactant, using a Dognon-Abribat-Prolabo Wilhelmy plate tensiometer. The liquid was thermostatted within 0.1 K. The Pt plate (width 1.956 cm) was cleaned prior to each measurement by rinsing with deionized water and glowing over an alcohol flame. For $c <$

c.m.c., surface tensions decreased slowly with time over several hours for reasons that need not be discussed here but it was observed that the obtained c.m.c.-values were insensitive to the measuring time. All data reported apply to a measuring time of 5 min. A similar procedure (3 min. waiting time) was adopted by Crook et al. [7].

Viscosities were measured with an Ubbelohde suspended level dilution viscometer. Solutions were filtered prior to the measurements.

Results and discussion

Micelle formation data

Figure 1 gives a typical collection of γ-log c plots. There were no minima in the curves around the c.m.c., pointing to the absence of solubilizable impurities. The breaking points are sharp, notwithstanding the heterodispersity of the surfactant. C.m.c.-values decrease with increasing T, a feature which is closely connected with the decreasing solubility of the surfactant in this direction. For NPE-1800 the Θ-temperature in the absence of electrolyte is about 90 °C.

C.m.c.-values at various T and c_e are tabulated in table 1. Accepting an accuracy of a few μM the following trends are observed. At low T, the first added electrolyte decreases the c.m.c. Phenomenologically this can be considered a salting-out feature. We have not enough and not sufficiently precise data to discriminate between the electrolytes but in first approximation they behave in a comparable way. With increasing concentration, the c.m.c. remains low for Na$_2$SO$_4$ and MgSO$_4$ but it increases again for

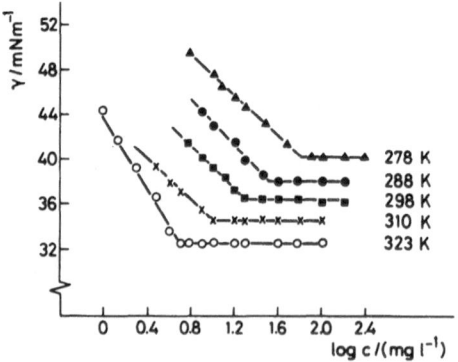

Fig. 1. Surface tension plots for aqueous solutions of NPE-1800 at various temperatures. No electrolyte added

Table 1. Critical micelle formation concentrations of NPE-1800. Values are expressed in $M \times 10^{-6}$

	Electrolyte/M	5 °C	25 °C	37 °C	50 °C
	0.0	29.0	9.5	4.8	2.4
NaCl	1.0	13.3	6.2	4.8	2.9
	1.5	19.0	9.5	8.1	5.7
	2.0	21.0	10.4	7.1	6.2
	2.5	20.0	13.8	11.9	10.5
	3.0	19.5	15.2		11.4
	3.5	25.2	19.0	13.3	
	4.0	31.9	19.0	12.4	
Na$_2$SO$_4$	0.1	22.4	7.6	6.7	4.3
	0.2	13.3	7.1	5.7	4.8
	0.3	13.3	9.0	8.1	4.8
	0.4	13.3	8.6	7.6	6.2
MgSO$_4$	0.1	19.9	5.7	4.8	4.3
	0.2	12.0	5.2	3.9	2.9
	0.3	11.4	6.1	4.8	3.3
	0.4	9.5	6.0	4.8	4.0

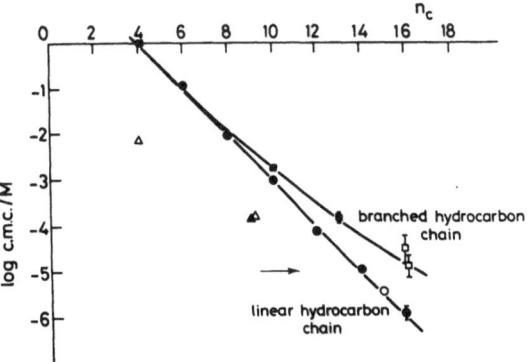

Fig. 2. Critical micelle concentrations at 25 °C of nonionic surfactants of the polyoxyethylene type as a function of the length of the hydrocarbon chain. Black symbols: values taken from the compilation by Mukerjee and Mysels [8].
Open symbols: more recent additions. o Meguro et al. [9], □ (top) Güveli et al. [14], □ (bottom) Barry and El Eini [11], ▲ and △ hydrocarbon chain plus a para-phenyl residue, △ Seng and Sell [13]. The arrow indicates NPE-1800

NaCl, ultimately exceeding the value at zero electrolyte ('salting-in'). At high T there is no significant trend observable with Na$_2$SO$_4$ or MgSO$_4$, but again NaCl leads to salting-in. Perhaps this salting-in is related to the reduction of the shelf stability of paraffin oil in water emulsions, stabilized by NPE-1800, in the presence of NaCl, reported above.

Comparison of c.m.c. with data for other nonionics

In NPE-1800 obviously the nonylphenol residue forms the hydrophobic core and the EO moieties may be expected to enrich the outer mantle. The PO segments assume an intermediate position. One way to get some insight in the structure of the micelle, with special reference to the effective size of the hydrophobic core, is to compare the c.m.c. with those for other nonionics of the hydrocarbon-EO type. To that end a plot has been made of the c.m.c. as a function of the length of the hydrocarbon tail n_c. Obviously, such a plot is also interesting *per se*.

Figure 2 gives the result. Most data are taken from the Mukerjee-Mysels compilation [8] (black symbols). Where various data were available, the preferred one was accepted. The open points refer to recent extensions.

A number of interesting features emerge. Three classes of compounds can be distinguished, depending on the nature of the hydrocarbon tail. If this tail is straight, the c.m.c. is insensitive to the length of the

EO-moiety and the relation between c.m.c. and n_c is almost exactly semi-logarithmical, i.e.

$$^{10}\log \text{c.m.c.} = a - bn_c \qquad (2)$$

with $a = 2.1$ and $b = 0.5$, similar to the data reported by Meguro et al. [9]. For the surfactants with a branched hydrocarbon chain, the trend is that the c.m.c. is higher. In (2), a is the same as for linear chains, but b decreases with n_c. The higher c.m.c. must be attributed either or both to a higher molecular solubility [10] and poorer packing of the molecules in the micelle. That packing does play a role may perhaps be inferred from the observation that for the straight chain nonionics the c.m.c. is independent of the EO length, whereas for the branched ones it depends systematically on it. Barry and El Eini, [11] Schick [12] and Seng and Sell [13] all report an increase of c.m.c. with increasing EO length, which is the expected trend. Remarkably enough, Güveli et al. found the reverse trend [14]. More systematic studies appear desirable.

Only few reliable data are available for alkyl phenol-containing surfactants (triangles in fig. 2). At least for the nonyl compound, the chain is branched so that comparison suggests that the benzene ring is equivalent to about four CH$_2$ groups, which is a bit higher than the value of three, reported by Tanford [15]. The arrow in figure 2 indicates the c.m.c. for NPE-1800. Extrapolation of the alkylphenol data points parallel to the branched chain curve suggests an equivalent total chain length $n_c \sim 13$, i.e. the 13 PO groups in the molecule are equivalent to only 4 CH_2 groups. This phenomenological comparison underlines the intermediate position of the PO group between EO and pure hydrocarbon. It is supported by the behaviour of PPO-PEO surfactants, so called Pluronics. PO chains become progressively less soluble with increasing length, so that PPO-PEO molecules of sufficiently long PO moiety exhibit surfactant character, although not with sharp c.m.c.'s [16, 17].

Most likely in the NPE-1800 micelle there is no distinct delimitation between a hydrophobic core and a hydrophilic mantle. Rather the transition is gradual and diffuse with the PPO part mixed partly with the NP, in part with the PEO and for the rest among itself. This is a general feature for all types of micelles. Modern techniques have shown the units to be in a highly dynamic state, continually exchanging monomers. Micelles are no static entities and only averaged values can be assigned to such geometrical features as shape, microstructure and boundary between core and mantle.

Finally we note that the sharpness of the c.m.c. (fig. 1), in contrast with the less sharp c.m.c. for Pluronics indicates that it is mainly the NP part that is responsible for the association into micelles.

Effects of electrolyte and temperature on c.m.c.

The decrease of the c.m.c. with c_e observed at low T and moderate c_e ($\omega \lesssim 1M$) (table 1) is a general feature for nonionics of the EO-type [18] and usually attributed to salting-out of the EO moiety. The same interpretation is commonly offered for the decrease of c.m.c. with T at low c_e. Adsorption studies of Pluronics on polystyrene latices corroborate these trends and interpretations: in the range under discussion the surface excess increases with T and c_e [19].

Very likely the increase of c.m.c. with c_{NaCl} at high concentrations, observed at any T is a reflection of a different phenomenon and may perhaps be attributed to salting-in of the hydrocarbon moiety (and also to some extent of the PO part). Generally, it may be expected that chaotropic ions reduce hydrophobic association in aqueous solutions. More details about these features and specific ionic effects will be reported elsewhere [1] but it may be repeated that the suggested explanation is in line with the decreased stability against *coalescence* (requiring removal of surfactants from the interface) of paraffin-in-conc. NaCl emulsions.

Thermodynamic considerations

In principle the c.m.c. is related to the Gibbs energy of micellization $\Delta_m G_i^\circ$ (i is the number of monomers per micelle) and from the temperature coefficient of the c.m.c. the corresponding enthalpy $\Delta_m H_i^\circ$ may be obtained, so that also $T\Delta_m S_i^\circ$ can be found. However, on closer inspection the situation is not entirely straightforward, the main problem being in $\Delta_m H_i^\circ$, which is an 'isosteric' enthalpy, i.e. a quantity applying to the fictive case of constant composition, size and shape as a function of T. In reality these properties depend on T; the way in which they are accounted for depends on the model adopted, but generally the functionalities can be written as [20–23]

$$RT \ln \text{c.m.c.} = \Delta_m G^\circ + \frac{1}{i} f(i, \sigma, y) \qquad (3)$$

$$\frac{d \ln \text{c.m.c.}}{dT} = -\frac{\Delta_m H^\circ}{RT^2} - \frac{1}{i^2} \frac{di}{dT} f'(i, \sigma, y) \qquad (4)$$

where $\Delta_m x$ stands for the change in x if one surfactant molecule is transferred from the bulk to the micelle, y is a measure of the various activity coefficients, in micellar systems and σ is the standard deviation of i. Alternatively, in stead of σ a quantity relating to the sharpness of the c.m.c. can be introduced (the two quantities are related because a more narrow distribution corresponds with a sharper c.m.c.). In (3) and (4) the c.m.c. must be expressed as mole fractions.

As we have no suitable information available on micellar size, size distribution and shape as a function of T, we can only evaluate the leading terms, i.e. the first terms on the RHS, of (3) and (4). Gibbs energies are collected in table 2. By their very nature, the data of this table exhibit the same trends as those of table 1. One trend is that in 4M NaCl $\Delta_m G^\circ$ is less negative than in the other electrolytes, otherwise the differences between the electrolyte are slight and hardly significant.

Table 2. Gibbs energy of micellization of NPE-1800 ($kJ \cdot mole^{-1}$)

	5 °C	25 °C	37 °C	50 °C
no electrolyte	−33.5	−38.5	−41.9	−45.6
NaCl 1M	−35.2	−39.6	−41.9	−45.1
NaCl 4M	−33.2	−36.8	−38.4	−41.4*)
Na₂SO₄ (average)	−35.0	−39.0	−40.9	−43.8
MgSO₄ (average)	−35.3	−39.8	−42.0	−44.5

*) 3M NaCl

Plots of ln c.m.c. vs. T^{-1} are reasonably linear (fig. 3), but the accuracy of our data does not allow us to discriminate between the various electrolytes. With respect to the slope the only significant difference is that between 'absence of electrolyte' and 'presence of electrolyte', the corresponding $\Delta_m H°$ values being +37.9 and +19.6 kJ mole^{-1} respectively. The positive sign corroborates that for NPE 1800 micelle formation is endothermic, i.e. it is entropically driven. Of the line for 4M NaCl the slope is the same as that for the other elctrolytes but the intercept is less negative, meaning that the (counteracting) enthalpy is the same but $\Delta_m S°$ is less. The driving force being the entropy increase due to hydrophobic bonding of the hydrocarbon- (and to a much lesser extent of the PO-) part of the surfactant, we conclude that 4M NaCl is more chaotropic than more dilute NaCl or the other electrolytes investigated and that this is the reason for the increase of c.m.c. observed at high c_{NaCl} (table 1). Entropical data are collected in table 3. Because of the uncertainties mentioned above, we can discriminate between three conditions only, viz. 'no salt', 'NaCl 4M' and 'all other electrolytes, NaCl 1M included'. The entropic differences between 4M NaCl and the other electrolytes are slight, but significant. We note also that $\Delta_m S°$ is independent of temperature. This means that the intrinsic probabilities of the various states of the system are insensitive to T.

It is interesting to compare the properties of salt-free and saline systems. For the second class, $\Delta_m H°$ is

Table 3. Entropies of micellization of NPE-1800 (J·mole^{-1}K^{-1})

	5 °C	25 °C	37 °C	50 °C
no electrolyte	+257	+256	+257	+259
NaCl 4M	+190	+189	+187	+189
other electrolytes	+197	+198	+197	+198

systematically less endothermal but $\Delta_m S°$ is also lower. A possible explanation may be sought in the binding of ions to the EO and/or PO segments and incorporation of electrolyte inside the micelle; this would reduce the EO-water and PO-water interaction in bulk and therefore reduce the enthalpy effect of transfer from solution to micelle. At the same time the hydrophobic bonding might be reduced. However before more information is available, this suggestion is not more than a speculation.

Finally, our thermodynamic data may be compared with those for (straight) alkyl chain polyoxyethylenes. According to figure 2, such a surfactant with $n_c = 14$ and a comparable number of EO groups has the same c.m.c. as NPE-1800 and hence the same $\Delta_m G°$ Meguro et al. [9] report for C$_{14}$ (EO)$_8$ $\Delta_m H° =$ +12.6 kJ. mole^{-1} and $\Delta_m S° = +172$ kJ. mole^{-1}K^{-1}. These data may be compared with $\Delta_m H° = +37.9$ kJ. mole^{-1} and $\Delta_m S° = +256$ J. mole^{-1}K^{-1} for NPE-1800. The main difference between the two surfactants is the presence of a PPO-part in NPE-1800, so that it may apparently be concluded that transport of ca 13 PO groups from aqueous solution to the micelle is enthalpically much more unfavourable and entropically more favourable than that of the equivalent ca. 4 CH$_2$-groups. In saline solutions, the difference with C$_{14}$(EO)$_8$ is much less: $\Delta_m H° = 19.6$ kJ. mole^{-1} and $\Delta_m S° = 190$–220 J. mole^{-1}K^{-1}. Perhaps this finding, in combination with that of the preceding paragraph may imply that it is the PO part that is most sensitive to the electrolyte.

Viscosities, effect of temperature

Reduced viscosities [η(solution) – η(solvent)]/η(solution)c with c in g/dl are plotted for NPE-1800 in figure 4. They depend linearly on c, with a slope decreasing with increasing T. At least formally the dependence can be represented by a Huggins-type equation

$$\frac{\eta_{sp}}{c} = [\eta] + [\eta]^2 K c \qquad (5)$$

where K is a measure of the interaction between the micelles and [η], the intrinsic viscosity is a measure of

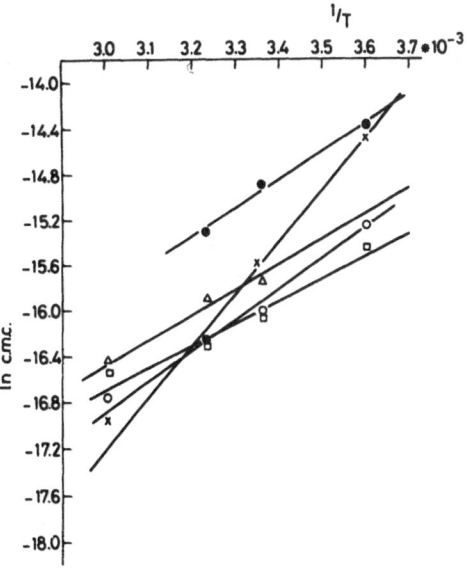

Fig. 3. Temperature dependence of the c.m.c. (in mol fractions) of NPE-1800 x no electrolyte ○ NaCl 1M ● NaCl 4M △Na$_2$SO$_4$ (average conc.) □ MgSO$_4$ (av. conc.)

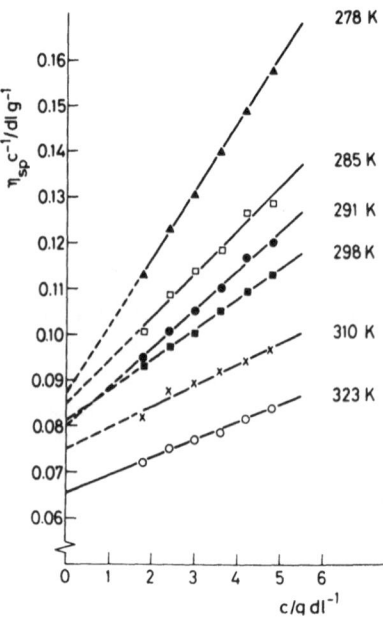

Fig. 4. Reduced viscosities of NPE-1800 solutions at various temperatures. No electrolyte added

the size of the micelle. The application of (5) must be considered with some reservation, because the average size of the micelle is not unique, but probably increases with c, hence $[\eta]$ is dependent on c [24]. Therefore, K is not a pure interaction energy parameter but contains also a size-factor. Moreover K is sensitive to the shape of the micelle because the hydrodynamic interaction between pairs of particles is sensitive to it [25]. Without independent information on micellar size and shape this difficulty can not further be unraveled. If the micelles are cylindrical, the rigidity of the particles may pose another problem [26] although Nagarajan, on the basis of model analyses of various types of surfactants concludes that rigidity prevails [27]. Irrespective of these complications, it appears that slope and intercept are suitable measures of the solvent quality: both decrease with T and this must be anticipated in view of the decreasing solvent quality. Table 4 summarizes the pertaining parameters.

Our data compare well with those reported in literature on similar compounds. For instance, El Eini et al. [28] found for $C_{16}(EO)_x$ intrinsic viscosities,

increasing from 6.3 to 12.1 ml.g^{-1} and K-values decreasing from 2.03 to 1.77 for x varying from 17–63. In literature many different values for K have been proposed. Bohdanecký and Kovár [26] list various values between 0.40 and 2.26 for rigid, non-interpenetrating spheres, with the probably best value being 0.69 [29]. The divergence between the different approaches is a consequence of the great difficulty in describing the hydrodynamics of interaction. Our K-values exceed the value of 0.69 and do so the more, the lower T, indicating hydrodynamic interaction decreasing with increasing T. This interaction is not likely determined by the shape of the particles because then one would rather expect the reverse order, because the surfactants become less soluble with increasing T, and therefore tend to make bigger, i.e. more elongated micelles at higher T. Hence, the observed trend must apparently be attributed to the decreasing solvation with increasing T in such a way that the reduction in effective size, as expressed through the coefficient $[\eta]^2$ is not enough to account for the whole effect.

The above interpretation is in line with experiences on Synperonics of higher EO-chain length. For instance, we found for NPE A and B the constant K to exceed the value for NPE-1800 by 10 and 38% respectively.

Finally, the reduction of $[\eta]$ with T must probably also be attributed to a decreasing hydration of the EO-mantle of the micelle, with its ensuing compaction. Also for pure solutions of polyethylene oxide in water, such a reduction has been observed [30].

Viscosity, effect of electrolytes

The electrolytes at the same concentration ranges as in table 1 reduce $[\eta]$ on the average by some 20%, with no clear ion specificity. This decrease is due to the decreased solvency. On K the electrolyte influence is more complicated in that K passes through a maximum, indicating that there are two counteracting effects on the hydrodynamic interaction (fig. 5). Information on the effect of these electrolytes on the size and shape of the micelles is needed to further analyse these two opposing trends. As stated before, ion incorporation into the micelle must also be considered.

Finally, we return to the deviating behaviour of 4M NaCl solutions, as opposed to more dilute (1M) NaCl and the other electrolytes at any concentration studied. We related this to ion incorporation into the micelle, probably into the PPO-part. By virtue of the higher polarizability of the anion, it must be expected

Table 4. Parameters of eq. (5) for NPE-1800 in aqueous solutions at various temperatures. No salt added

Temp.	5	12	18	25	37	50 °C
$[\eta]$/ml g^{-1}	8.6	8.4	8.2	8.1	7.5	6.5
K (—)	2.04	1.37	1.17	1.00	0.99	0.90

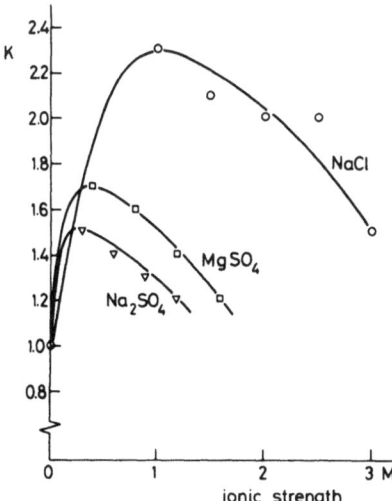

Fig. 5. Dependence of the Huggins constant K of NPE-1800 micelles on the nature and concentration of the electrolyte. $T = 25\,°C$

that this incorporation is primarily determined by the anion, so that the specific effect of conc. NaCl is a specificity of the Cl^- over the SO_4^{2-} ion.

Viscometric experiments by Mr. P. Peters in our department confirm this. Figure 6 shows reduced viscosities in 4M NaCl solution of NPE 1800, compared with those for NPE-A, B and C. There is a

dramatic difference between the 1800 and the other three, in that the former exhibits an extraordinarily high slope and irregular behaviour al low c. It looks as if extension of the EO part 'screens' the binding of Cl^--ions to the PPO moiety. In contrast to 4 M NaCl, a solution of 4M NaBr behaves quite 'normal'. The exact reason for this difference is difficult to establish; for one thing one would intuitively expect Br^--ions to incorporate more strongly than Cl^--ions because of their higher polarizability, but this is contrary to the facts. Whatever the reason may be, it is clear that viscometry is a sensitive tool to observe such differences.

Acknowledgements

The authors thank Dr. Th. F. Tadros for useful discussions, Dr. R. I. Hancock of ICI-PLC Ltd. (Great Britain) for providing us with the surfactants used and ICI-Millbank Ltd. (Great Britain) for financial support to TvdB. One of us (SMZ) wishes to thank DGIS of the Dutch Ministry of Foreign Affairs for financial support, enabling him to stay in the Netherlands and carry out part of this work.

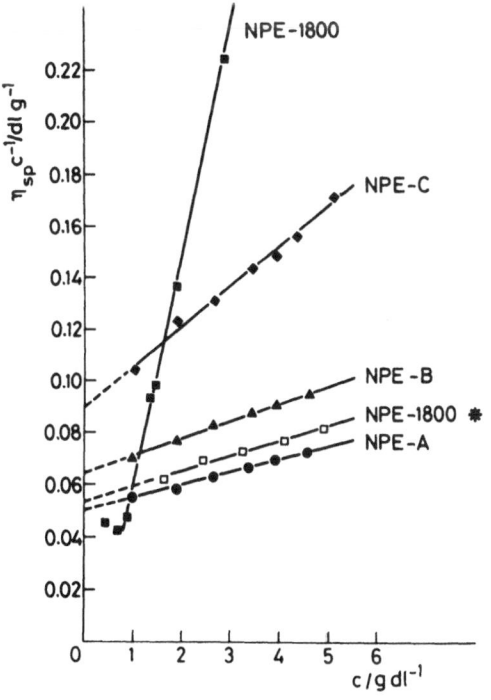

Fig. 6. Reduced viscosities in 4M NaCl solutions of various NPE surfactants. $T = 25\,°C$. For comparison, data for NPE-1800 in 4 M NaBr (*) are included

References

1. Boomgaard, A., van den, Thesis, Agricultural Univ. Wageningen Neth., in course of production.
2. Th. van den Boomgaard, Tadros Th. F., and Lyklema J., to be published.
3. Shachat, N. and Greenwald, H. L., ch. 2, in 'Nonionic Surfactants', Schick, M. J., Ed., Marcel Dekker, New York (1967).
4. Lee, G. W. J., Tadros, Th. F., Coll. & Surf. 5, 105 (1982).
5. Gratzer, W. B., Beaven, G. H., J. Phys. Chem. 73, 2270 (1969).
6. Nadeau, H. G., Siggia, S., ch. 26 in 'Nonionic Surfactants', Schick, M. J., Ed., Marcel Dekker, New York (1967).
7. Crook, E. H., Trebbi, G. F., Fordyce, D. B., J. Phys. Chem. 68, 3592 (1964).
8. Mukerjee, P., Mysels, K. J., 'Critical Micelle Concentrations of Aqueous Surfactant Systems', Natn. Bureau Standard U.S. Washington (1971).
9. Meguro, K., Takasawa, Y., Kawahashi, N., Tabata, Y., Ueno, M., J. Colloid Interfac. Sci. 83, 50 (1981).
10. McAuliffe, C., J. Phys. Chem. 70, 1267 (1966).
11. Barry, B. W., El Eini, D. I. D., J. Colloid Interface Sci. 54, 339 (1976).
12. Schick, M. J., J. Colloid Sci. 17, 801 (1962).
13. Seng, H. P., Sell, P. J., Tenside Detergents 14, 4 (1977).
14. Güveli, D. E., Davis, S. S., Kayes, J. B., J. Colloid Interface Sci. 86, 213 (1982).
15. Tanford, C., 'The Hydrophobic Effect. Formation of Biological Membranes and Micelles' John Wiley, New York (1973).
16. Schmolka, I. R., ch. 10 in 'Nonionic Surfactants' Schick, M. J., ed., Marcel Dekker, New York (1967).
17. Prasad, K. N., Luong, T. T., Florence, A. T., Paris, J., Vaution, C., Seiller, M., Puisieux, F., J. Colloid Interface Sci. 69, 225 (1979).
18. Becher, P., ch. 10 in 'Nonionic Surfactants', Schick, M. J., Ed., Marcel Dekker, New York (1967).

19. Tadros, Th. F., Vincent, B., J. Phys. Chem. **84**, 1575 (1980).
20. Phillips, J. N. Trans. Faraday Soc. **51**, 561 (1955).
21. Hall, D. G., Pethica, B. A., ch. 16 in 'Nonionic Surfactants', Schick, M. J., Ed., Marcel Dekker, New York (1967).
22. Holtzer, A., Holtzer, M. F., J. Phys. Chem. **78**, 1442 (1974).
23. Own analyses, not published.
24. Nagarajan, R., Shah, K. M., Hammond, S., Colloids & Surf. **4** 147 (1982).
25. Tanford, C., Nozaki, Y., Rohde, M. F., J. Phys. Chem. **81**, 1555 (1977).
26. See e.g. Bohdanecký, M., Kovár, J., 'Viscosity of Polymer Solutions', Jenkins, A. D., Ed., Elsevier Amsterdam (1982).
27. Nagarajan, R., J. Colloid Interfac. Sci. **90**, 477 (1982).
28. El Eini, D. I. D., Barry, B. W., Rhodes, C. T., J. Colloid Interfac. Sci. **54**, 348 (1976).
29. Peterson, J. M., Fixman, M., J. Chem. Phys. **39**, 2516 (1963).
30. Bailey F. E., Kolske, J. V., 'Poly(ethylene Oxide)', p. 48, Acad. Press New York (1976).

Received February 8, 1983

Authors' address:

Th. van den Boomgaard
Laboratory for Physical
and Colloid Chemistry
Agricultural University
De Dreijen 6
6703 BC Wageningen, The Netherlands

Progress in Colloid & Polymer Science   Progr. Colloid & Polymer Sci. **68**, 33–40 (1983)

Micellar effects of surfactants on the photoredox reaction with 3,3'-dialkyl thiacarbocyanines as sensitizers*)

T. Handa, H. Komatsu, and M. Nakagaki

Faculty of Pharmaceutical Sciences, Kyoto University, Sakyo-ku, Kyoto 606, Japan

Abstract: Photoelectron transfer from ethylene diamine tetraacetate to Methyl Viologen (MV^{2+}) is studied with 3,3'-dialkyl thiacarbocyanines as sensitizers. Reaction efficiency is remarkably enhanced when the sensitizer dye is solubilized in the micellar phase of heptaethylene glycol monododecyl ether. Experimentally, it is found that the dyes locate in the less polar and more viscous micellar environment than in aqueous phase. The enhancement of reaction efficiency is interpreted by the stabilization of excited state of the dyes in the micellar microenvironments.

The alkyl chain of the dye is found to have considerable effects on the reaction efficiency in the micellar phase. The effect of high microviscosity in the micelle is more intense on the dye with two longer alkyl chains, 3,3'-dioctadecyl thiacarbocyanine (C_{18}) than the dye with two shorter chains, 3,3'-diethyl thiacarbocyanine (C_2). Also, from the fluorescence quenching experiment, the higher accessibility of MV^{2+} to C_{18} than C_2 is found, which would reflect the difference in location and orientation of these dye chromophores in the micellar phase. Thus, the higher stability of excited state of C_{18} and the higher accessibility of MV^{2+} to C_{18} are regarded to be responsible for the higher reaction efficiency with C_{18} than C_2 in the micellar solution.

Key words: Photoredox reaction, micelle, thiacarbocyanine, micropolarity, microviscosity, fluorescence quenching.

Introduction

Surfactant micelles have been used as reaction media to understand the complex nature of biological membranes and also to develop the reaction systems in pharmaceutics and industry [1, 2]. As an explanation for the enhancement of the reaction rate in the micelles, the electrostatic and hydrophobic effects on the partitioning of reagents into micellar phase have been used [1, 3, 4]. In enzyme reactions, the involvements of hydrophobic interaction of enzyme with the micelle, steric effects of micelles and specific physical states of substrates in micelles have been suggested [5, 6, 7].

For photoinduced reactions, the micellar effects on the charge separation in excited complexes (exciplexes) and on the retardation in recombination of the generated ions have been elucidated [2b, 8, 9]. On the other hand, the stability of photoexcited state of sensitizer and, also, the accessibility of electron donor and acceptor to the sensitizer could be important factors in

determining the reaction effeciencies. The former would be intimately correlated with the microenvironment around the sensitizer, that is, the local effective polarity and viscosity in the close vicinity of the sensitizer.

In this study, the photoinduced electron transfer from ethylene diamine tetraacetate to Methyl Viologen sensitized by 3,3'-dialkyl thiacarbocyanine incorporated in nonionic micelle, heptaethylene glycol monododecyl ether in aqueous solution are measured. The results observed are discussed with the two factors: 1. the stabilization of excited state of sensitizing dye in micellar microenvironment. 2. The accessibility of electron acceptor, Methyl Viologen, from aqueous phase to sensitizing dye in micellar phase.

Experimental

Materials. Nonionic surfactant, heptaethylene glycol monododecyl ether (HED) was obtained from Nikko Chemicals Co., LTD. The surface tension and concentration curve for aqueous HED solution gave no minimum around the critical micelle concentration, and the value of the latter (8×10^{-5} M) agreed well with the reported value [10]. Sodium dodecyl sulfate (SDS) supplied

---

*) Dedicated to the memory of Professor Dr. B. Tamamushi.

from Nakarai Chemicals Co., LTD. was recrystalized twice from ethanol. 3,3'-diethyl thiacarbocyanine iodide (C_2) and 3,3'-dioctadecyl thiacarbocyanine bromide (C_{18}) were purchased from Japan Research Institute for Photosensitizing Dye Co., LTD. Methyl Viologen dichloride (MV^{2+}) was obtained from Sigma Chem. Co. Disodium ethylenediamine tetraacetate (EDTA) of analytical grade was obtained from Dojindo Laboratories.

Measurements. 0.5×10^{-6} M solution of C_2 or C_{18} was prepared by solubilizing with 20 mM HED in buffer solution of 20 mM EDTA-Tris and 2 mM MV^{2+} (pH = 7.0). An aliquot of the solution contained in a quartz cell ($1 \times 1 \times 4$ cm) with stop cock was degassed by nitrogen purging and by aspiration at 2700 Pa (20 mmHg) prior to irradiation. The cell was contained in a quartz jacket and kept at 25 °C by water circulation. These procedures gave rise to an excellent reproducibility in the photoinduced production of Methyl Viologen cation radical. The sensitizer dyes, C_2 and C_{18} have absorption bands around 560 nm in aqueous and micellar solutions. The irradiation source used was a 1 kW slide projector (tungsten lamp) equipped with the Toshiba Y-46 cut-off filter (passing light > 460 nm). The development of the reaction product, Methyl Viologen cation radical, MV^{+} were monitored by measuring absorption spectra on the Shimadzu UV-180 spectrophotometer. Concentration of MV^{+} was determined by using $\varepsilon_{395nm} = 3.8 \times 10^4$ M^{-1} cm^{-1} for MV^{+}[11]. MV^{+} has an absorption at 603 nm besides 395 nm. The former band resulted in faint reduction of the irradiation intensity for the sensitizer dyes (in this study the reduction was, at most, 8% of the initial intensity). To determine the wavelengths of absorption maxima of C_2 and C_{18} in various solvents and micellar solutions, the first derivative absorption spectra were measured with Hitachi-220 spectrophotometer. The fluorescence spectra of the same solutions (excited at 2.000×10^4 cm^{-1}) were also observed with Japan Spectroscopic FP-550 spectrofluorimeter. The temperature was maintained at 25 °C by water circulation through the cuvet holders of the spectrometer. Viscosities of methanol, ethanol, 1-propanol, 1-butanol and 1-octanol were measured by Ostwald viscometer at 25 °C with pure water ($\eta = 0.89$ c poise) as a reference. For 1-propanol and 1-butanol, viscosities were also measured at various temperatures.

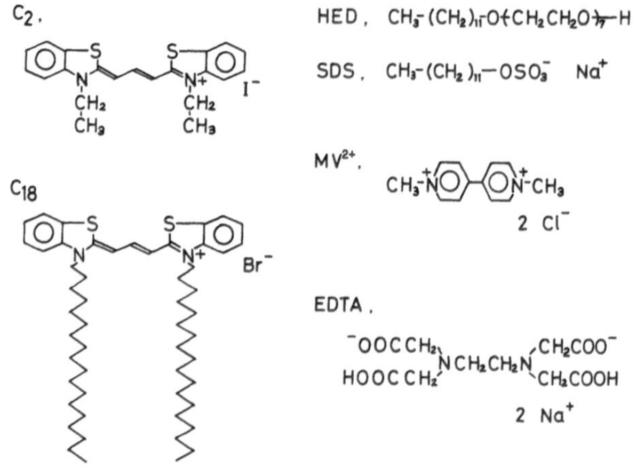

Results

Photoredox reaction with C_2 or C_{18}

The photoinduced electron transfer from EDTA to MV^{2+} was first observed with tris-(2,2'-bipyridine) ruthenium as sensitizer by M. Calvin et al. [12] as

$$EDTA + 2\,MV^{2+} + H_2O \xrightarrow{h\nu}$$

$$CH_2O + CO_2 + 2\,H^+ + \underset{HOOCCH_2}{\overset{HOOCCH_2}{\diagdown}} NCH_2CH_2N \underset{H}{\overset{CH_2COOH}{\diagup}} + 2\,MV^{+}$$

In our study, 3,3'-dialkyl thiacarbocyanines (C_2 and C_{18}) in nonionic micelle, HED, are adopted as sensitizers and, EDTA and MV^{2+} are dissolved in aqueous phases. The structures of C_2, C_{18}, MV^{2+}, HED and SDS are presented in figure 1.

In figure 2, the absorption spectra of reaction solution with C_{18}, which is practically insoluble in

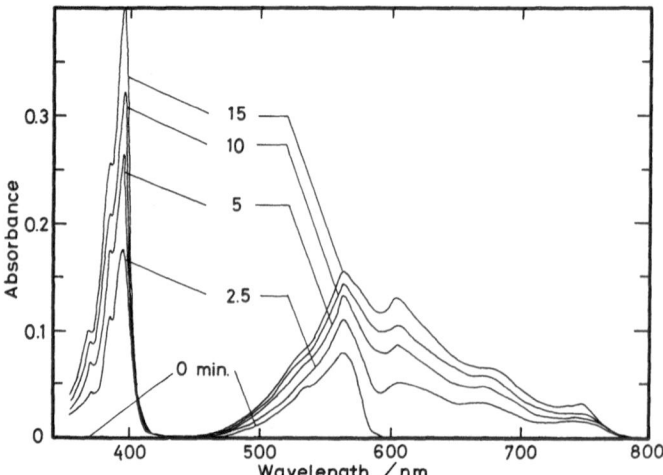

Fig. 1. Structures of sensitizers (C_2 and C_{18}), electron acceptor (MV^{2+}), electron donor (EDTA) and surfactants (HED and SDS)

Fig. 2. Spectral change of photoreaction solution with the irradiation time. 0.5×10^{-6} M of C_{18} is solubilized in the HED micellar solution (20 mM) at pH = 7.0. The absorption bands at 395 and 603 nm are contributed from Methyl Viologen cation radical, MV^{+}, produced by the reaction, and the band at 563 nm is attributed to C_{18} in the HED micellar phase

aqueous bulk phases and incorporated in HED micellar phase, are shown at various irradiation time. C_{18} has an absorption band at around 563 nm in HED micellar solution. The absorption intensity at 395 nm contributed from Methyl Viologen cation radical, MV^{+}, increases with time.

When C_2 is employed as the sensitizer, the problem arising from partitioning of the dye between micellar and aqueous bulk phases should be resolved. As will be shown in figure 3, C_2 exhibits a considerably large shift of the absorption maximum with the addition of HED in aqueous phase. The absorption maxima in the aqueous solution and 100 mM HED micellar solution are 554 and 562 nm, respectively. Further addition of HED in the solution did not give rise to appreciable change in the absorption spectra. The absorbances of C_2 at 554 and 562 nm in the HED micellar solutions, A_1 and A_2, respectively, are therefore,

$$A_1 = (\varepsilon_1^b n_b + \varepsilon_1^m n_m) / V \tag{1}$$

$$A_2 = (\varepsilon_2^b n_b + \varepsilon_2^m n_m) / V. \tag{2}$$

Here, n_b and n_m are moles of C_2 in the bulk and the micellar phases, ε_1^b and ε_2^b are the mole extinction coefficients in the aqueous phase at 554 and 562 nm, respectively, and ε_1^m and ε_2^m are those in the micellar phase. V is the volume of the solution containing the micelles, and the volume of micellar phase, v, is negligibly small as $v \ll V$. The total mole of C_2 in the solution, $n_0 = n_b + n_m$. The partition coefficient, K, for the partition equilibrium of C_2 between the bulk and the micellar phases is given as

$$K = (n_m / v) / (n_b / V). \tag{3}$$

From equations (1), (2) and (3), the ratio of absorbances, $R = (A_2/A_1)$, is represented [13] as follows:

$$R = R_m - (1/K) \varepsilon_1^b/\varepsilon_1^m [(R - R_b) (V/v)]. \tag{4}$$

Here, $R_m = (\varepsilon_2^m/\varepsilon_1^m)$ and $R_b = (\varepsilon_2^b/\varepsilon_1^b)$.

In figure 4, the experimental values of R for various amounts of HED in C_2 solutions are plotted against the values of $[(R - R_b) (V/v)]$, where the volume of micellar phase, v, is evaluated with the use of 1.0 g/cm^3 as the density of micellar phase in aqueous solution [14]. In figure 4, the mole ratio of C_2 to HED is, at most, 1/250 and the effect arising from the solubilized dye on the micellar volume is regarded to be negligible. The linear relation obtained in this figure, together with the experimental value of $(\varepsilon_1^b/\varepsilon_1^m) = 0.96$, gives the value of K as 9.6×10^2. The mole fraction of C_2 in the micellar phase (n_m/n_0), is calculated by equation (5),

$$(n_m/n_0) = \frac{K(v/V)}{1 + K(v/V)}. \tag{5}$$

The photoreaction is monitored at 20 mM HED (molecular weight, 494), i. e. $(v/V) = 0.01$. Therefore, 90% of C_2 is solubilized in the micellar phase.

In figure 5, the concentrations of photoinduced $MV^{\ddot{+}}$ by C_2 in the aqueous (curve 3) and the micellar (curve 2) phases, and by C_{18} in the micellar phase (curve 1) are shown as a function of the irradiation time. Since the curve 2 is higher than the curve 3, it may be concluded that the production of Methyl Viologen cation radical, $MV^{\ddot{+}}$, is enhanced more than twice, by the transfer of C_2 (90% of C_2 stated above) from the aqueous to the micellar phase. In the micellar phase, the thiacarbocyanine with two longer chains, octadecyl groups, C_{18}, exhibits about 10 times higher reaction efficiency in comparison with C_2.

Absorption Maxima of C_2 and C_{18} in Various Solvents. The solvent dependencies of absorption

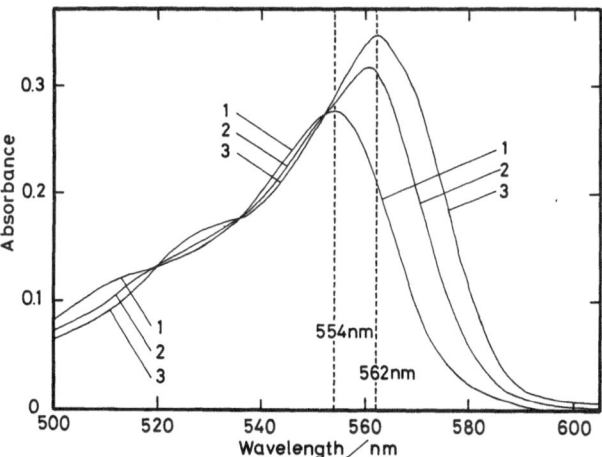

Fig. 3. Absorption spectra of C_2. 2×10^{-6} M of C_2. Curve 1: 0 M HED, curve 2: 5×10^{-3} M HED, curve 3: 1×10^{-1} M HED

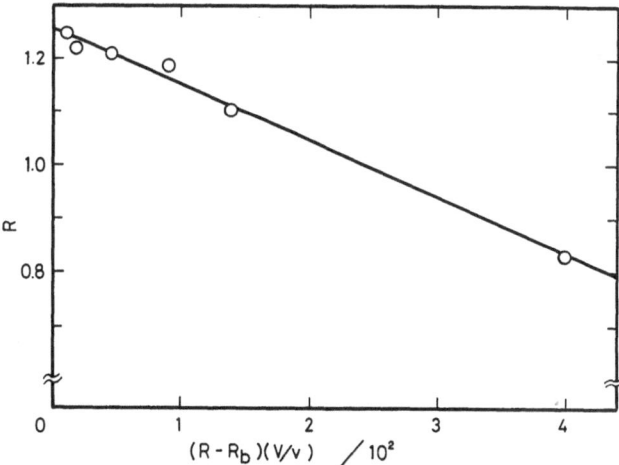

Fig. 4. Variation of R as a function of $(R - R_b) (V/v)$. From the linear relation obtained, the 90% of C_2 is judged to be solubilized in HED micelle, according to equations (4) and (5)

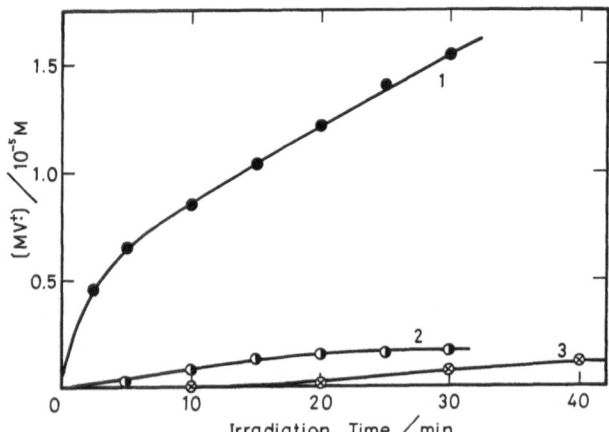

Fig. 5. Time course of Methyl Viologen cation radical, MV^+, production. 1 : C_{18} in 20 mM as sensitizer, 2 : C_2 in 20 mM HED as sensitizer, 3 : C_2 in aqueous phase as sensitizer

maxima of the dyes (solvatochromisms) of C_2 and C_{18} are measured. The dyes display considerably large shifts in the wave number at the absorption maximum, $\tilde{v}_{max}$, in various aliphatic alcohols and water as the solvents. In figure 6 and table 1, the values of the $\tilde{v}_{max}$ obtained by the observation of the solvatochromism are shown together with the values for HED and SDS micellar solution. The wave numbers of both C_{18} and C_2 increase linearly with the increase in the dielectric constant of alcohols. The values of the wave number of absorption maxima of C_2 in the micellar solutions (HED and SDS) are close to the value in 1-butanol. On the other hand, C_{18} in the micellar phases gives the values coinciding with those in 1-propanol.

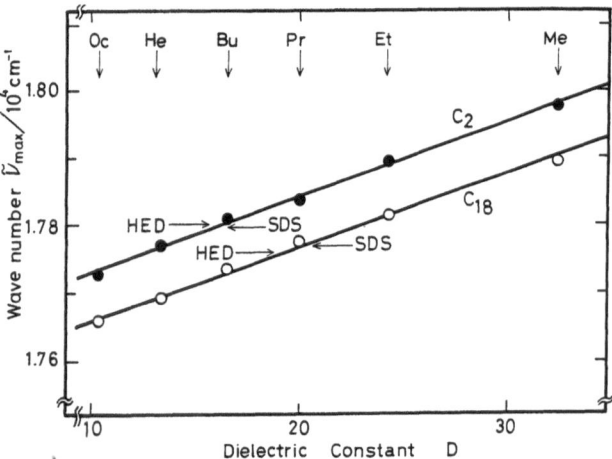

Fig. 6. Solvatochromisms for C_2 and C_{18} in aliphatic alcohols and micellar phases. Wave number of absorption maximum is plotted as a function of the dielectric constant of solvents. Me : methanol, Et : ethanol, Pr: 1-propanol, Bu: 1-butanol, He: 1-hexanol, Oc: 1-octanol

Table 1. Experimental values of wave number of absorption maxima ($\tilde{v}_{max}$) and relative fluorescence yields[a]), ϕ_f^R, for C_2 and C_{18} in various media at 25 °C

solvent or micelle	dielectric constant	viscosity (c poise)	$\tilde{v}_{max}$ (10^4 cm^{-1}) C_2	C_{18}	ϕ_f^R C_2	C_{18}
aqueous solution of 20 mM EDTA-Tris	78.5	0.89	1.806	—[c])	1	—[d])
methanol	32.6	0.56	1.798	1.789	1.54	2.18
ethanol	24.3	1.10	1.790	1.781	1.88	3.03
1-propanol	20.1	2.01	1.784	1.777	2.42	3.62
1-butanol	17.8	2.61	1.781	1.774	2.91	4.04
1-hexanol	13.3	4.37	1.777	1.769	3.92	5.35
1-octanol	10.3	7.7	1.773	1.766	4.87	6.66
1-decanol	8.1[b])	14.0[b])	1.764	1.762	6.12	7.69
HED (20 mM)			1.780	1.776	3.85	6.22
SDS (20 mM)			1.780	1.777	4.77	7.69

[a]) relative to the value for C_2 in the aqueous phase at 25 °C.
[b]) values at 20 °C. See reference 19.
[c]) insoluble.

Fluorescence Quantum Yields of C_2 and C_{18}. The fluorescence yields of C_2 and C_{18} in 20 mM EDTA-Tris aqueous solution, aliphatic alcohols and the micellar phases are evaluated as the fluorescence intensities integrated on the spectra and divided by the absorbance at the exciting wave number (2.000×10^4 cm^{-1}). The results (ϕ_f^R) are presented in table 1 as the values relative to that of C_2 in the aqueous phase at 25 °C. The following points are clarified : both C_2 and C_{18} exhibit the higher yields in the solvent of lower polarity, e. g. C_2 has the yield of about 5 times higher in 1-octanol than in water; C_{18}, which has two longer alkyl chains (octadecyl groups), has higher fluorescence yields than C_2 in all solvents studied; the yields in micellar phases are several times higher than in aqueous phase; in comparison with the values in the solvents which give close absorption maxima to those in the micellar phase (i. e. 1-butanol for C_2 and 1-propanol for C_{18}), the yields in micellar phases are remarkably higher.

Fluorescence quenching of thiacarbocyanine in micellar phase by MV^{2+}

The fluorescence quenchings of the sensitizer solubilized in the HED and HED-SDS (4 : 1) micellar phases by MV^{2+} are measured. In the aqueous phase and the HED micellar solution, the quenching is very slight, while in the mixed micellar solution, the intense quenchings are observed as shown in figure 7. Here,

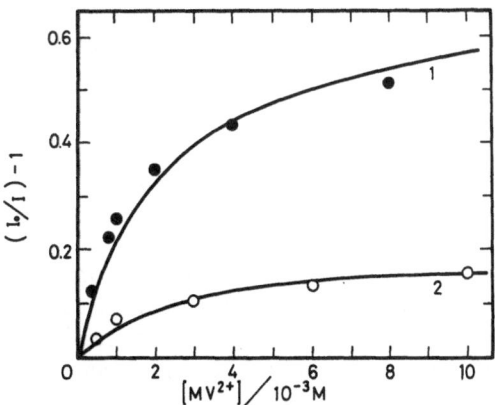

Fig. 7. Fluorescence quenchings of C_{18} (curve 1) and C_2 (curve 2) in micellar phase by MV^{2+}. The micellar phases are composed of 4:1 HED and SDS mixture. (I_0/I) is the ratio of fluorescence intensities without and with MV^{2+}. The solid lines in this figure are represented by equations (6a) and (6b)

(I_0/I) is the ratio of fluorescence intensities without and with the quencher, MV^{2+}, at the fluorescence maximum (580 nm), and is equal to the ratio of fluorescence yield. As seen in this figure, the quenchings exhibit the saturations which result in serious deviation from the simple Stern-Volmer equation [15]. The quenchings of C_2 and C_{18} in HED by MV^{2+} are represented by the following equations as shown by the solid lines in figure 7,

$$(I_o/I) - 1 = \frac{0.2 \times 4.5 \times 10^2 \,[MV^{2+}]}{1 + 4.5 \times 10_2\,[MV^{2+}]} \text{ for } C_2 \quad (6a)$$

$$(I_o/I) - 1 = \frac{0.7 \times 4.5 \times 10^2\,[MV^{2+}]}{1 + 4.5 \times 10^2\,[MV^{2+}]} \text{ for } C_{18} \,(6b)$$

indicating that C_{18} is 3.5 times more easily quenched by MV^{2+} than C_2 in the micellar solution.

Discussion

Photoredox reaction

The proposed mechanism of the reaction studied here is shown as follows,

$$^1C_n^+ \xrightarrow{\quad h\nu \quad} {}^1C_n^{+*} \quad (7)$$

$$^1C_n^{+*} \xrightarrow{\quad rl \quad} {}^1C_n^+ \quad (8)$$

$$^1C_n^{+*} \xrightarrow{\quad \text{fluorescence} \quad} {}^1C_n^+ \quad (9)$$

$$^1C_n^{+*} \xrightarrow{\quad isc \quad} {}^3C_n^{+*} \quad (10)$$

$$^3C_n^{+*} + MV^{2+} \xrightarrow{\quad\quad} (^3C_n - MV)^{3+*} \quad (11)$$

$$(^3C_n - MV)^{3+*} \xrightarrow{\quad\quad} C_n^{2+} + MV^+ \quad (12)$$

$$2\,C_n^{2+} + EDTA \xrightarrow{\quad\quad} 2\,^1C_n^+ + EDTA(ox) \quad (13)$$

Here, $^1C_n^+$ represents the sensitizer, C_2 or C_{18}, in the singlet ground state, $^1C_n^{+*}$ and $^3C_n^{+*}$ are the sensitizers in the singlet and triplet excited states. The radiationless relaxation (*rl*) and the intersystem crossing (*isc*) are presented by processes (8) and (10), respectively. $(^3C_n\text{-}MV)^{3+*}$ is exciplex formed of $^3C_n^{+*}$ and MV^{2+}, and C_n^{2+} is the sensitizer radical. EDTA(ox) shows the oxidated form of EDTA. According to the above mechanism, it is clear that the concentration of the excited state $^3C_n^{+*}$, generated from $^1C_n^{+*}$ by the process (10) plays an important role in the reaction efficiency. The yield of intersystem crossing (*isc*), ϕ_{isc}, is given as the result of the process (10) in competition with the processes (8) and (9):

$$\phi_{isc} = \frac{k_{isc}}{k_{rl} + k_f + k_{isc}}. \quad (14)$$

On the other hand, the fluorescence yield of the sensitizers, ϕ_f is:

$$\phi_f = \frac{k_f}{k_{rl} + k_f + k_{isc}} \quad (15)$$

and therefore

$$\phi_{isc} / \phi_f = k_{isc} / k_f. \quad (16)$$

Where, k_{rl}, k_f and k_{isc} are the rate constants for radiationless relaxation (process 8), radiative (fluorescence) relaxation (process 9) and intersystem crossing (process 10) of the excited state, $^1C_n^{+*}$. Because k_f and k_{isc} are considered not to be so much dependent on the environmental properties [16, 17], the increase in ϕ_f is accompanied by the increase in ϕ_{isc}, as is shown in equation (16). The increase would result from the decrease of k_{rl}, as is seen from equations (14) and (15). Therefore, the variation of the experimentally obtainable value, ϕ_f in the aqueous and micellar environments reflects that of ϕ_{isc}. In following sections, the effects of micellar microenvironment on ϕ_f, and therefore ϕ_{isc}, are investigated.

The process (11) shows the exciplex formation between the sensitizing dye in excited state, $^3C_n^{+*}$ and the electron acceptor, MV^{2+} and is examined by means of the quenching experiment of phosphorescence intensity of $^3C_n^{+*}$ by MV^{2+}. But this was very difficult because of very weak phosphorescence intensity of $^3C_n^{+*}$. Instead of this, the quenching of fluorescence $^1C_n^{+*}$ by MV^{2+} was examined. The quenching efficiency would reflect the local concentration of the quencher MV^{2+} around $^1C_n^{+*}$ and the strength of interaction between the excited sensitizer and the quencher. In the micellar phase, the strength of interaction depends on the locations and mutual orientations of the sensitizer and quencher, that is, the accessibility of the sensitizer and the acceptor. This problem and also the process (13) will be discussed in the separate section.

The process (12) represents charge separation between the sensitizer and the acceptor, and depends on the polarity in the close vicinity of the exciplex (^3C_n-$MV)^{3+*}$. As discussed later, thiacarbocyanines with two ethyl groups, C_2 and two octadecyl groups, C_{18} are found to be accomodated in the micellar environments of close polarities. Therefore, this process is considered not to be so sensitive to the alkyl chain length of sensitizing dye.

Micropolarity around sensitizing dye molecule

It has suggested that the rate of radiationless processes, k_{rl}, largely depends on the interaction of the excited dye molecule and the environment around it [16, 18], that is, the microenvironment in the close vicinity of the dye molecule. The microenvironment would actually mean the local effective micropolarity and the local effective microviscosity just around the dye molecule.

To estimate the polarity, the wave numbers at the absorption maxima of dyes solubilized in micellar phases are compared with their solvent dependencies. The reasons for employing of aliphatic alcohols and water to calibrate the micropolarities in micellar and membrane phases have already been discussed in detail [19, 20]. The solvatochromisms of 3,3'-dialkyl thiacarbocyanine have been predominantly ascribed to the interaction due to the dispersion force between the dye and solvent molecules [21]. As seen in figure 6 and table 1, the chromophores of C_2 and C_{18} accomodated in micellar phases experience remarkably less polar environment in comparison with aqueous phase, but seem to be in more polar environment than in liquid hydrocarbon. These results indicate that the chromophores are located in the surface region of the micelles. The chromophore with longer alkyl chains, C_{18}, seems to locate in a little more polar environment than C_2.

The correlations between the rate of radiationless relaxation of excited molecule and its environmental polarity have not been completely elucidated. The

relaxation processes through intramolecular charge transfer states of dyes and solvent-dye intermolecular complexes formed in the excited states have been suggested in the polar environments [22, 23]. The higher fluorescence yields, ϕ_f, of the dyes in the micellar phase are, therefore, partly due to the less polar environments around the dyes than in aqueous phase.

The microenvironmental effects would contain the polarity and viscosity factors. These are usually associated in a complex way and have not been investigated separately. Therefore, we will examine the problems in more detail.

Microviscosity in close vicinity of sensitizing dye molecules

In the radiationless relaxation process of the electronic excited state, a considerable part of the energy first goes into the intramolecular vibration and rotation modes and then transfers to the lattice modes of environmental molecules through the vibrational and rotational relaxation processes [16, 23, 24]. Carbocyanine dyes have been known to exhibit the intense twisting modes (the intrarotational modes) along the polymethine chain in the electronically excited state [25, 26, 27], because in the excited state, the conjugated π electronic structure is broken and this allows the twisting modes of the substituents at each end and destruction of the initial planar structure. The ability of such an intramolecular twisting, which may be strongly affected from the viscosity in the close vicinity of the dye molecule, is associated with the internal conversion process (radiationless relaxation) of the singlet excited state to the ground state and, therefore, with the fluorescence yield, ϕ_f.

The radiationless relaxation process of the excited state is divided as

$$k_{rl} = k_1 + k_2 \tag{17}$$

where k_1 is the rate of the usual internal conversion which is independent on the viscosity in the close vicinity of the dye but depends on the polarity around the dye, and k_2 is the rate of the relaxation via intrarotational (twisting) modes which is influenced by the microscopic viscosity around the dye molecule. By the use of the Stokes equation, k_2 is given [28] as

$$k_2 = A (T/\eta) \tag{18}$$

Here, A is a function of the effective volume associated with the intramolecular rotation, η is the viscosity of the medium and T is absolute temperature. From equation (15), (17) and (18), the following equation is derived [28b],

$$(1/\phi_f) = (1/\phi_f^o) + (A/k_f)\,(T/\eta)\,. \qquad (19)$$

ϕ_f^o is the fluorescence yield when the relaxation via intrarotational modes is absent as $\phi_f^o = k_f/(k_1 + k_{isc} + k_f)$. An equation similar to the equation (19) is also applicable to the relative fluorescence yield, ϕ_f^R, because, ϕ_f^R is proportional to the fluorescence yield, ϕ_f. It has already been known that the relation between the fluorescence yield of Auramine O and the solvent viscosities in glycerol at various temperatures and in dextrose-glycerol-water mixtures at 25 °C are correlated well by equation (19) [28]. The experimental results for oxadicarbocyanine in aliphatic alcohols at 25 °C also roughly fit the equation (19) [29].

The equation (19) is examined for C_2 and C_{18} in aliphatic alcohols and water at 25 °C. In these cases, however, no simple correlation as expected by equation (19), is obtained, and this is explained in terms of the effective polarity in these solvents. The equation (19) is, therefore, examined for the viscosity change of a single solvent by the change of temperature. In figure 8, the reciprocals of relative fluorescence yields (relative to the value of C_2 in aqueous phase at 25 °C, as before) of C_2 and C_{18} in 1-propanol and 1-butanol, and C_2 in aqueous phase are plotted as functions of (T/η). The experimental values of relative yields for C_2 and 1-propanol and 1-butanol are fitted on a straight line as expected from equation (19). The extrapolation to $(T/\eta) \to 0$ gives $\phi_f^{oR} = 13 \pm 2$. The results obtained with C_{18} also give a linear relation with $\phi_f^{oR} = 13 \pm 2$. In aqueous phase the linear relation for C_2 gives $\phi_f^{oR} = 1.9 \pm 0.3$. The lower value of ϕ_f^{oR} in the aqueous phase reflects the higher effective polarity in this phase in comparison with 1-butanol and 1-propanol. As described before, the effective micropolarities around C_2 and C_{18} in HED micelle are represented by those in 1-butanol and 1-propanol, therefore, these lower effective polarities provided by the micelle (in comparison with the value in aqueous phase) are considered to have important roles in the enhancements of fluorescence and intersystem crossing yields and, therefore, the reaction efficiency. If the fluorescence yields of C_2 and C_{18} in HED and SDS micellar phases at 25 °C are calibrated on the linear relation of ϕ_f^R vs. (T/η) for alcoholic solutions (see fig. 8), the effective microviscosity around the dye molecules is estimated to be 4–5 c poise in HED micelle, and 5–8 c poise in SDS micelle. For the microviscosities in micellar phase, values in a range of 8–30 c poise have

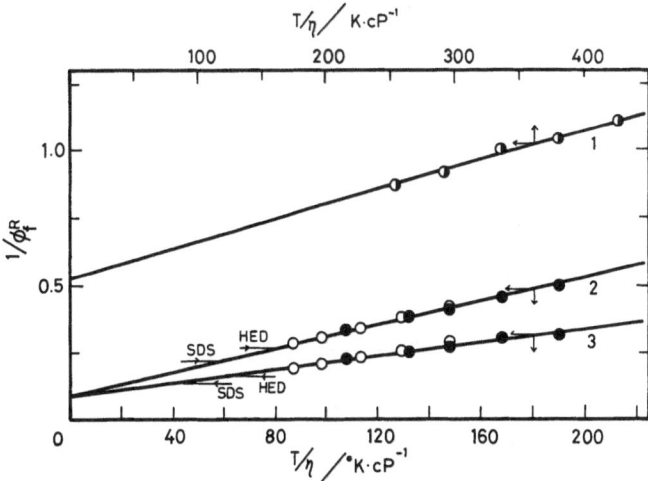

Fig. 8. Variation of reciprocal of relative fluorescence yield, ϕ_f^R, as a function of (T/η). See equation (19). 1: C_2 in aqueous phase (○), 2: C_2 in 1-butanol (○) and in 1-propanol (●), 3: C_{18} in 1-butanol (○) and in 1-propanol (●). The reciprocals of intercepts of these linear relations give the relative fluorescence yields in hypothetical glass states, 1: $\phi_f^{oR} = 1.9 \pm 0.3$; 2 and 3: $\phi_f^{oR} = 13 \pm 2$. The values of ϕ_f^R in micellar phases are calibrated with the lines 2 and 3 to give the effective microviscosities around C_2 and C_{18} molecules as 4–5 c poise in HED, and 5–8 c poise in SDS

been estimated by NMR and ESR relaxation times and fluorescence anisotropies [30, 31]. The more detail examinations on microviscosity are in progress in our laboratory.

On the basis of above discussions, it is concluded that 4–5 times higher viscosity in HED micelle than in aqueous phase also plays an important role in the enhancement of fluorescence yield. The values of A in equation (19), obtained from figure 8, are 2.3×10^{-3} c poise $\cdot$ K^{-1} for C_2 and 1.3×10^{-3} c poise. K^{-1} for C_{18}, respectively. This means that the effect of micellar viscosity on ϕ_f^R is larger for C_{18}, which has two octadecyl chains, than for C_2. This results the 60% higher fluorescence yield (table 1) and the higher reaction effeciency of C_{18} than C_2.

Accessibility of electron acceptor to sensitizer in micellar phase

As shown in the reaction process (11), the exciplex formation between the sensitizing dye in excited state $^3C_2^{+\,*}$ or $^3C_{18}^{+\,*}$, and the electron acceptor, MV^{2+}, approaching from aqueous phase to the micellar surface, is an important step in the photoreactions. The accessibility of MV^{2+} to the dye chromophores in the micellar surface may be examined by the estimation of the quenching efficiency of fluorescence of the dye by MV^{2+}. The quenching with the nonionic micelle, HED, is very weak and not clearly detected. In the mixed micellar (HED : SDS = 4 : 1) solution, however, the efficiency is largely amplified through the concentrating effects on MV^{2+} at the micellar surface due to the electrostatic interactions.

The dynamic quenching with micelle is represented by the following equation

$$(I_o/I) - 1 = k_q \theta_m = k_q \frac{K'\,[MV^{2+}]}{1 + K'\,[MV^{2+}]} \qquad (20)$$

where k_q and K' are the rate constant of quenching at the micellar surface and the partition coefficient of MV^{2+} to the micelle, respectively. θ_m is the degree of occupation of MV^{2+} at the micellar surface. $[MV^{2+}]$ is the concentration in aqueous phase. From the experimental results given by the equations (6a) and (6b), it is found that $K' = 4.5 \times 10^2$ M^{-1} for both C_2 and C_{18}, and $k_q = 0.2$ for C_2 and 0.7 for C_{18} in the mixed micelle. Thus C_{18} is 3.5 times more easily attacked by MV^{2+} at the micellar surface. A similar difference in quenching of C_2 and C_{18} is also observed in SDS micellar solutions. The difference in the micropolarities for C_2 (represented by 1-butanol) and C_{18} (represented by 1-propanol) would be understood in terms of the differences in the locations and the orientations of the dye chromophores in the micellar phase, which are responsible for the difference in the values of k_q, and, therefore, in the accessibilities of MV^{2+} to the sensitizers. The electron donor, EDTA, is also approaching from aqueous phase, and the difference in accessibilities to C_2 and C_{18} by EDTA is expected to be similar to that of MV^{2+}.

In conclusion, the higher photoreaction efficiency in HED micelle than in aqueous phase is explained by the larger stability in the excited state of the sensitizing dye in the less polar and more viscous micellar environment. In the micellar phase, the dye with two longer alkyl chains, C_{18}, exhibits the higher reaction efficiency than C_2. This fact is also interpreted by the two factors: the more intense effect from the micellar viscosity on C_{18}, and the higher accessibility of the acceptor, MV^{2+}, to C_{18} in the micellar phase, very probably because the location of the chromophore of C_{18} is nearer to the micellar surface as suggested by the higher micropolarity than C_2.

References

1. Mittal, K. L. ed., "Micellization, Solubilization, and Microemulsion" Vol. 2, Plenum Press, New York, 1977.
2. a) Fendler, J. H., Fendler, E. J., "Catalysis in Micellar and Macromolecular Systems" Academic Press, New York, 1975; b) Fendler, J. H., Acc. Chem. Res. 9, 153 (1976) and J. Phys. Chem. 84, usw. 1485 (1980).
3. a) Bunton, C. A., Ohmenzetter, K., Sepúlveda, L., J. Phys. Chem. 81, 2000 (1977); b) Bunton, C. A., Romsted, L. S., Smith, H. J., J. Org. Chem. 43, 4299 (1978); c) Quina, F. H., Chaimovich, H., J. Phys. Chem. 83, 1844 (1979).
4. a) Funasaki, N., J. Phys. Chem. 83, 237 and 1998 (1979); b) Nakagaki, M., Yokoyama, S., Yamamoto, I., J. Chem. Soc. Jpn. 1982, 1865.
5. Bhat, S. G., Brockman, H. L., Biochemistry 21, 1547 (1982).
6. De Araujo, P. S., Rosseneu, M. Y., Kremer, J. M. H., van Zoelen, E. J. J., de Hass, G. H., Biochemistry 18, 580 (1979).
7. Nakagaki, M., Yamamoto, I., Yakugaku Zasshi 101, 1099 (1981).
8. a) Grätzel, M., Ber. Bunsenges. Phys. Chem. 84, 981 (1980); b) Brugger, P. A., Infelta, P. P., Braum, A. M., Grätzel, M., J. Phys. Chem. 84, 2402 (1980); c) Kiwi, J., Grätzel, M., J. Am. Chem. Soc. 100, 6314 (1978).
9. Yamaguchi, Y., Miyashita, T., Matsuda, M., J. Phys. Chem. 85, 1369 (1981).
10. Lange, H., Proc. Inter. Cong. Surface Activity 3, 279 (1960).
11. Tunuli, M. S., Fendler, J. H., J. Amer. Chem. Soc. 103, 2507 (1981).
12. Ford, W. E., Otvos, J. F., Calvin, M., Nature 274, 507 (1978).
13. Mukerjee, P., Banerjee, K., J. Phys. Chem. 68, 3567 (1964).
14. Mukerjee, P., J. Phys. Chem. 66, 1733 (1962).
15. Stern, O., Volmer, M., Physical. Z. 20, 183 (1919).
16. Mataga, N., Kubota, T., "Molecular Interactions and Electronic Spectra" Marcel Dekker, New York, (1970).
17. a) Gouterman, M., J. Chem. Phys. 36, 2846 (1961); b) Robinson, G., Frosch, R., J. Chem. Phys. 38, 1187 (1964) and 37, 1962 (1962).
18. Radda, G. K., "Fluorescence Probes in Membrane Studies, Method in Membrane Biology" Vol. 4, Plenum Press, New York, 1975.
19. a) Mukerjee, P., Ramachandran, C., Pyter, R., J. Phys. Chem. 86, 3189 and 3198 (1982); b) Mukerjee, P., Cardinal, J. R., J. Phys. Chem. 82, 1620 and 1614 (1978).
20. Zachariasse, K. A., Van Phuc, N., Kozankiewicz, B., J. Phys. Chem. 85, 2676 (1981).
21. West, W., Geddes, A. L., J. Phys. Chem. 68, 837 (1964).
22. Kosower, E. M., Dodiuk, H., Tanizawa, K., Ottolenghi, M., Orbach, N., J. Am. Chem. Soc. 97, 2167 (1975).
23. McClure, W. O., Edelman, G. M., Biochemistry 5, 1908 (1966).
24. Stryer, L., J. Mol. Biol. 13, 482 (1965).
25. Dempster, D., Morrow, T., Rankin, R., Thompson, G. F., J. Chem. Soc. Faraday II 68, 1479 (1972).
26. Rulliere, C., Chem. Phys. Letters 43, 303 (1976).
27. Baker, R. H., Grätzel, M., Steiger, R., J. Am. Chem. Soc. 102, 847 (1980).
28. a) Oster, G., Nishijima, Y., J. Am. Chem. Soc. 78, 1581 (1956); b) Nishijima, Y., Koubunshi 13, 166 (1964).
29. Jaraudias, J., J. Photochem. 13, 35 (1980).
30. a) Menger, F. M., Jerkunica, J. M., J. Am. Chem. Soc. 100, 688 (1978); b) Menger, F. M., Acc. Chem. Res. 12, 111 (1979).
31. Fendler, J. H., J. Phys. Chem. 84, 1485 (1980).

Received February 8, 1983;
accepted April 12, 1983

Authors' address:

M. Nakagaki
Faculty of Pharmacentical Sciences
Kyoto University
Sakyo-ku, Kyoto 606, Japan

Microemulsions*)

S. E. Friberg

Chemistry Department, University of Missouri-Rolla, Rolla, MO, USA

Abstract: Microemulsions are transparent emulsions that form spontaneously when the ingredients are brought into contact. The droplet size varies for different compositions in the range 25–1500A.

The microemulsions may be stabilized by a combination of an ionic surfactant and medium chain length alcohol such as pentanol or with a nonionic surfactant.

In this article, the relation between micellar solutions and microemulsions in such systems is discussed and the basis for the stability of premicellar aggregates in W/O microemulsions is briefly evaluated.

Key words: Microemulsions, Liquid crystals, Surfactant association structures, Inverse and normal micelles, Solubilization.

Introduction

Water and hydrocarbon with a carefully balanced mix of surfactants may spontaneously form emulsions with such a small droplet size that the individual droplets cannot be observed in a microscope and the emulsion appears transparent to the eye. Hence they are colloidal dispersions and Dr. Tamamushi, in whose honor this article is written, always had an inspiring interest and insight in colloidal association structures [1–3]. These transparent emulsions have been industrially used for half a century [4]; their application in liquid fuels and in polishes [5] is well known.

Their scientific evaluation began in the forties [6], and the Schulman school developed a thesis about the importance of a temporary ultra-low or even negative interfacial tension as an explanation for the spontaneous formation of the microdroplets [7–13]. This postulate may at first be dismissed as an oversimplification of a rather complex phenomenon [14–18], but it should be emphasized that an extreme lowering of the interfacial tension [14] certainly is the key factor in the thermodynamics of microemulsion systems. Viewed in this light, the intuitive approach of the Schulman school was obviously correct in essence.

Apart from the theoretical evaluation of their stability, the properties of microemulsions have been the focus of extensive research efforts during the last decade. This research has been prompted by the economic potential of enhanced oil recovery [19, 20], and since an ultralow water/oil interfacial tension is a necessity for this application, numerous treatments of this phenomenon and its cause have been published, among others, by the Texas school of Schechter and Wade [21–23].

The research on the properties of microemulsions in general has employed a series of investigative methods such as light and neutron scattering [24–28], dielectric properties [29, 30], electron microscopy [27, 31, 32], positron annihilation technique [33, 34], phase equilibria and others [35–62]. Some problems concerning structure remain to be solved and the following treatment of the relations between micellar solutions and microemulsions has been limited to a reasonable selection of well established facts.

Microemulsions from micellar solutions

The traditional microemulsions owe the stabilization of the small water or oil droplets to a judicial combination of an ionic surfactant and a medium chain length alcohol as cosurfactant, giving a microemulsion with four components: water, oil, surfactant and cosurfactant. A superficial count of these components would lead to an initial conclusion of the microemulsions as being simple micellar systems; the micellar concept is well established among the association colloids.

---

*) Plenary lecture at the 56th National Colloid Symposium of the American Chemical Society, Blacksburg, 1982. Dedicated to the memory of Professor Dr. B. Tamamushi.

Aqueous micellar solutions

This conclusion is not correct as is shown by the results of the extensive investigations into soap systems by Per Ekwall [63]. Figure 1A and B demonstrate the limited solubilization that is obtained using normal micelles. Neither the hydrocarbon nor the alcohol will be solubilized to more than a few percent by the surfactant micelles in the aqueous phase. These micellar solutions do not merit the name microemulsions. The name of microemulsion should be reserved for systems that contain higher amounts of dispersed phase. A further condition would be that the droplets should possess space, which may be characterized as interior. In a micelle, practically all the molecules are engaged in the surface layer. In order to find high solubilization and with it the microemulsions, it is necessary to examine the role of the cosurfactant. Figure 1 obviates the necessity of its presence for microemulsion formation.

The solubility of the different components in the cosurfactant is a useful study in order to understand the generic relations between microemulsions and micellar solutions. The results from Ekwall's [63] numerous investigations serve as an excellent basis for such an exercise.

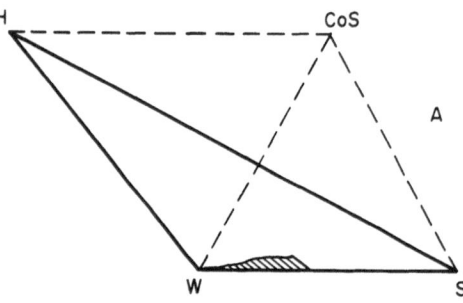

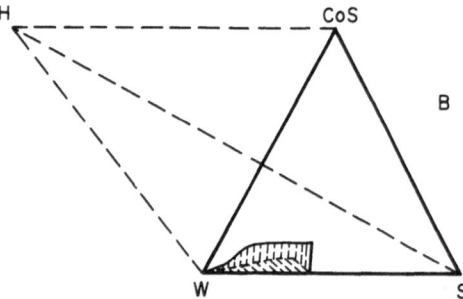

Fig. 1. Normal micelles display limited solubilization of hydrocarbon (*A*) and cosurfactant (*B*). These micellar solutions do not deserve the name microemulsions. W = water, S = surfactant, CoS = cosurfactant and H = hydrocarbon

W/O microemulsions

The solubility of the hydrocarbon and the cosurfactant are mutually complete, but the solubility of water and surfactant is highly limited as is found in figure 2A. As a matter of fact, both these solubilities are less than 5% by weight (*P, Q*, fig. 2A). High solubilities are observed first when water and surfactant are combined. The combinations show two features of interest of which the first one may appear surprising at first. The solubility of the ionic surfactant is strongly enhanced by the presence of water. An increase of the water content from zero to ten percent gives a corresponding augment of the surfactant solubility from five to fifty percent (*T*, fig. 2B). In a similar manner, a careful adjustment of the cosurfactant/surfactant ratio will bring forward a spectacularly high solubility of water; in the figure a maximum content of water is given as 66% by weight (*R*, fig. 2B).

The water is maximally solubilized into an isotropic solution with the following composition. Water 66%, cosurfactant 22% and surfactant 12% (by weight). This solution is not water continuous in spite of its high water content, and the solubility of hydrocarbon in it appears to be a promising approach to obtain a *W/O* microemulsion.

Figure 3 shows the result of this exercise. The solubility area at 50% hydrocarbon retains identical shape to the one at 0% hydrocarbon within reasonable approximations. This means that a composition with 50% hydrocarbon, 33% water, 11% cosurfactant and 6% ionic surfactant has been obtained in the form of a stable isotropic solutions. This is *W/O* microemulsion.

Figure 4 shows the general development of the *W/O* area with increasing hydrocarbon content [40]. The area remains proportionally identical for hydrocarbon content up to 50%; compositions in excess of this

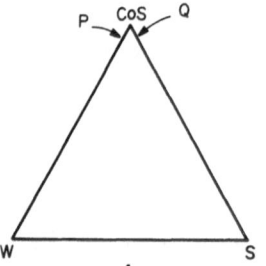 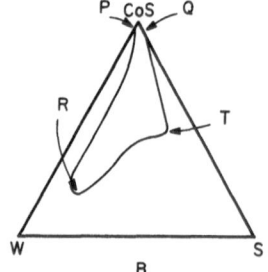

Fig. 2. The individual solubility of water (*P* in *A*) and surfactant (*Q* in *A*) in the cosurfactant (pentanol) are both small. First with their combinations are large solubilities of the surfactant (*T* in *B*) and water (*R* in *B*) found

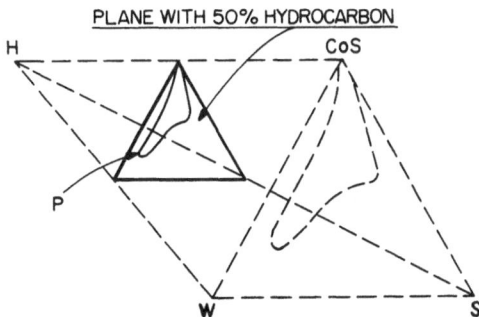

Fig. 3. The *W/O* microemulsions are formed as a direct continuation of the cosurfactant solution of water and surfactant as shown in figure 2

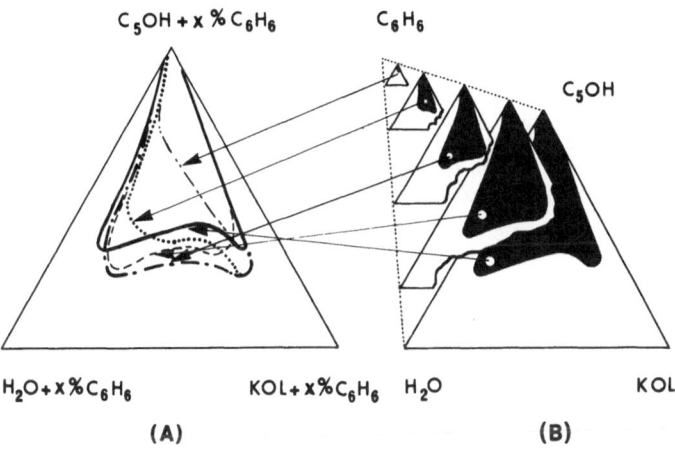

Fig. 4. The form of the cosurfactant solubility area remains constant for hydrocarbon contents less than 50%. For higher fractions of the hydrocarbon the maximum water solubility is reduced [40]. KOL = potassium sulfate

value lead to strongly reduced water content. This reduction in water content accompanies the dissociation of the cosurfactant dimers to monomers as shown by infrared spectroscopy [27]; the consequence is easily comprehended of the cosurfactant being removed from the microemulsion droplet interface due to the consecutive equilibria involved.

With this knowledge, the preparation of *W/O* microemulsions become a straight forward and simple procedure. The solubility area with no hydrocarbon present is determined by titration and hydrocarbon is added to desired concentration taking into consideration the adjustments that may be necessary as exemplified by the diagrams in figure 4.

The size of the aggregates vary in a non-trivial manner in the *W/O* microemulsion area. Determination of the dielectric constant [64], light scattering and electron microscopy observations [27] as well as

studies of interdroplet transfer [65] agree with the following interpretation: At low water content, approximately below a water/surfactant molecular ratio of 10, the aggregates are extremely small, probably involving only one surfactant molecule with a few associated water and alcohol molecules. With increase of the water content above this level, a gradual association takes place to larger aggregates; no sudden onset of association similar to the one in water to micelles of a certain size is found.

The structure of the aggregates have from tradition been perceived as well defined with the sudden change from internal to external properties that characterizes an interface. Recent NMR determinations of the diffusion coefficients of the components by Lindman and collaborators [37] have resulted in a modification of this view. The results rather lend themselves to an interpretation in the form of locally higher concentrations of the polar components. The transition to the parts of the system that contain only the hydrophobic substances does not take place over a distance of the minute thickness of a traditional surface, but extends over a considerable range forming an extremely diffuse interface.

O/W microemulsions

The extension of *O/W* microemulsion regions from the aqueous micellar solutions shows similarities to the solubility regions for *W/O* microemulsions. For these the attempts to dissolve either water of surfactant in cosurfactant/hydrocarbon solutions gave only insignificant solubilities; the large solubilization of water was found first when the surfactant and the water were combined (Cfr. figs. 2A and B). In the same manner, it has been found [63] that hydrocarbon and cosurfactant displayed only limited solubilization when added separately to the water/surfactant solution (fig. 1).

However, a combination of the cosurfactant and the hydrocarbon gave the desired extension of solubility regions to justify the name microemulsions. Figure 5 [41] clearly shows the pronounced growth of the solubility when both the components are added in comparison with the addition of only one of them. It is also found that a certain surfactant/water ratio produces a maximum hydrocarbon solubilization; exactly the same trend as found for the surfactant/cosurfactant ratio for *W/O* systems.

In contrast with the conditions for the *W/O* microemulsions, the solubility regions for the *O/W* one are sensitive to order of addition of components and to the nature of the hydrocarbon. This and the

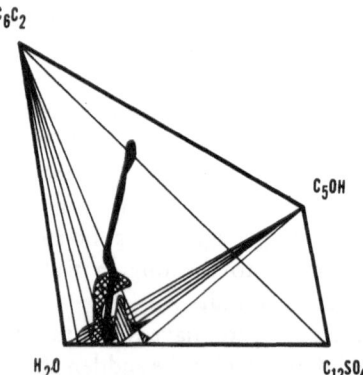

Fig. 5. The *O/W* microemulsion area [41] strongly depends on the surfactant content. $C_{12}SO_4$ = sodium octenyl sulfate, C_5OH = pentanol and C_6C_2 = *p*-xylene

fact that they may not be thermodynamically stable lead to rather complicated conditions [66] with too much detailed information to be suitable for a review article of this kind. Instead, the total picture and the conditions when the *W/O* and *O/W* regions are united will be discussed.

Overlapping regions

Typical *O/W* and *W/O* microemulsions areas are shown in figure 6 and a section in order to find optimum conditions for the *O/W* systems may lead to a use of such high surfactant concentrations that the plane will also include part of the *W/O* system. The three solubility areas in that plane, figure 7, will include the *O/W* microemulsion (I), the *W/O* microemulsion (II) and the hydrocarbon/cosurfactant solution with the small surfactant/water aggregates (III).

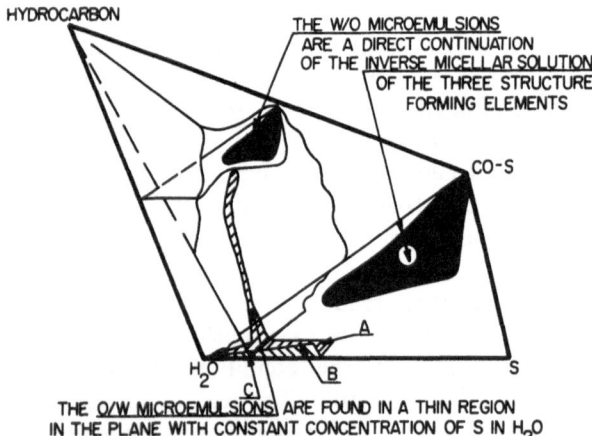

Fig. 6. The general shape of *W/O* and *O/W* microemulsion regions

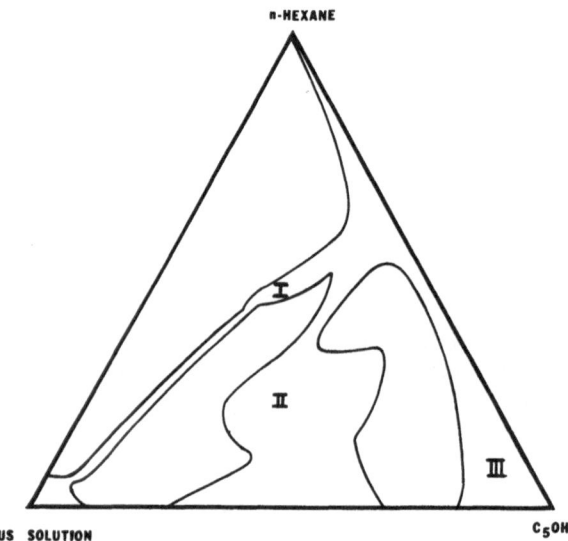

Fig. 7. For sufficiently high surfactant concentrations, (compare fig. 6) a triangular diagram of an aqueous surfactant solution, a cosurfactant and a hydrocarbon will include three regions: (I) the *Q/W* microemulsion, (II) the *W/O* microemulsion and (III) the hydrocarbon/cosurfactant *solution*

For some systems, the *W/O* and *O/W* regions will coalesce and a single solubility region will reach from the hydrophobic parts to the aqueous corner and the structure in the transition zone between *O/W* and *W/O* systems is of interest. A bicontinuous disordered structure has been suggested [67] as well as a more ordered bicontinuous structure with zero radius of curvature at each point [68]. Recent investigations by Clausse and collaborators [69, 70] have followed this transition zone from the properties of the *W/O* and *O/W* areas. An interesting observation was that these characteristics also were found after increased chain length of the cosurfactant caused a separation of the two distinct areas from each other. The solubility regions close to the limit of water (or oil) solubility gave results indicating bicontinuous behavior. Structure modifications in this direction have earlier been proposed by Sjoblom [44].

Stability theory

The theory for microemulsions with well defined droplets has been covered in a long series of papers by Ruckenstein [14, 16, 35] and the structure and stability of the aggregates in the transition zone has been discussed by the Minnesota group [68, 71, 72]. One of the remaining problems to be covered is the surprising effect of water on the solubility of surfactants in the region of low water content in the *W/O* microemulsion areas. The right hand limit of the solubility region in figure 2B shows the solubility of the surfactant in the cosurfactant to be increased from 5% to 40% by the addition of 10% water.

Lightscattering and other methods [37] have shown the aggregates in this part of the system not to be of colloidal size; presumably, they are surfactant monomers with a few associated water and alcohol molecules. The reason for the stability of such an aggregate versus the components is an intriguing problem.

A preliminary evaluation has been made of the relative importance of the different thermodynamic factors that comprise the total free energy for such an aggregate [73]. The thermodynamic cycle was estimated in the following manner. The components, soap and water, were brought to the gaseous state from the solid (soap) and the liquid state (water) by using known values for the free energy change for the water and by an estimation using a crystalline alkane plus the value of an electrolyte for the soap. The enthalpy of formation of the soap/water complex was estimated from CNDO/2 calculations [74]. The entropy for the distribution of these complexes in space in a conformation similar to the one in pentanol at reasonable concentrations was calculated using the methods described by Ruckenstein [14] and Reiss [15].

The results (fig. 8) showed the free energy of the complex to be unstable relative to the components for the number of water molecules per soap less than 5 and in excess of 12. The agreement with experimental values is excellent [37], but fortuitous considering the rather crude approximations used. On the other hand,

the importance of the results does not lie with the number of water molecules giving stability to the aggregate; the important feature is *the shape of the curve* for the free energy versus the number of water molecules.

The curve minimum automatically leads to a minimum in the number of water molecules to be neccessary for the stability of the aggregate and it also leads to a conclusion of too many water molecules also leading to instability.

The kind of structures obtained at the high limit will be discussed in the next section; the remaining part of this section will discuss the relative importance of the different thermodynamic factors comprising the total free energy of the aggregate.

It turns out that of the different enthalpic and entropic terms, the important one is the enthalpy of water binding to the polar part of the aggregate. Table 1 shows the enthalpy of the first two water molecules to be considerably greater than the one from the subsequent four ones due to the combined interactions of the hydrogen bond to the carboxylic group, in addition to the interactions from the water/metal ion liquid formation. Water molecules in excess of six are limited to hydrogen bond interactions with the associated water molecules; an interaction of lower energy than the one in liquid water. The high enthalpy values of the first six water molecules and low values for the following ones is the main factor to give shape of curve in figure 8 resulting in stability for a limited range only of water molecules per surfactant.

Consequences of instability

These stability conditions leave an interesting question to be answered. What kind of structures will be formed when the low and the high water concentrations are exceeded that lead to instability of the monomeric aggregate?

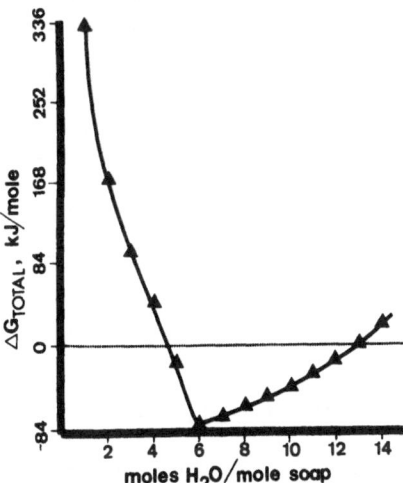

Fig. 8. The free energy for the formation of a surfactant/water aggregate displays a minimum as function of the number of water molecules per surfactant [73]

Table 1. Hydration enthalpy for water molecules bound to a sodium carboxylate (73).

Water molecule number	Hydration enthalpy kcal/mole
1	-52.0
2	-49.8
3	-26.2
4	-25.0
5	-25.5
6	-24.8
7	-8.4
8	-8.4
9	-8.3

Exceeding the low water limit leads to separation of crystalline soap or alcohol/soap molecular compounds. There has been comparatively few studies on these equilibria due to poor reversability, but the molecular structures have been extensively investigated [63].

The conditions when the high limit is exceeded are better known. Experimental evidence [40] shows three structures to be obtained depending on the cosurfactant/surfactant ratios. The highest value of this ratio leads to the separation of an isotropic aqueous liquid phase, while the lowest ratio gives an optically anisotropic liquid crystal with lammelar structure. Intermediate ratios do not give a phase separation; instead a step-wise association to inverse micelles or *W/O*-microemulsion droplets takes place.

These three developments may be understood in a qulitative manner from the Ninham approach [75, 76]. It emphasizes the importance of goemetrical restraints for the formation of different structures in addition to the classical thermodynamic conditions.

The key element in the considerations of geometrical restraints is the ratio $v/a_o l_c$ in which v is the volume of the hydrocarbon chain, a_o is the cross-sectional area of the polar group and l_c is approximately 90% of the fully extended length of the hydrocarbon chain. A value exceeding one implies inverse structures, while a value between one and one half means the formation of a lamellar liquid crystal.

Experimental values [63] show the separated lamellar liquid crystal to be within the 0.5–1 range and the distinction between the lamellar liquid crystal and the *W/O* microemulsion droplet structure is obviously a consequence of the packing conditions in the amphiphilic part of the structure. The alcohol/surfactant ratio will influence the packing conditions in the following manner. A relative increase of the alcohol amount will increase the $v/a_o l_c$ ratio for two reasons. The more hydrophobic alcohol does not only possess a small polar head giving a reduction of the average a_o value, but may also be envisaged as penetrating the outer hydrophobic part of the amphiphilic layer *in realiter* causing a disproportionate increase of the v value.

A quantitative evaluation of this combined effect is difficult since information about the average radial location of the alcohol versus the interface is not available, and since the ratio $v/a_o l_c$ strongly depends on this location.

The remaining problem, an explanation for the conditions distinguishing between the separation of an isotropic aqueous solution and the formation of *W/O* microemulsion droplets within the cosurfactant/hydrocarbon phase is not as directly amenable to be analized using geometrical restraints. At present, a qualitative explanation using chemical equilibria appears the only possibility. An attempt to a quantitative description of these leads to rather extensive investigations of fairly trivial nature. The interdependent free/hydrogen bonded water/alcohol molecular equilibria and the subsequent interaction with the water/surfactant association structure leads to equations involving a great number of equilibrium constants; a verification of the hypothesis would obviously require a series of independent investigations into water/alcohol interactions in dilute systems.

Summary

The microemulsion regions have been discussed in relation to micellar solutions and liquid crystalline structures.

The formation of *W/O* microemulsions was perceived as an intermediate state between separation of an aqueous solution and a lamellar liquid crystalline phase for water content in excess of the one tolerated by a small surfactant/water aggregate in a cosurfactant solution.

References

1. Tamamushi, B. and Matsumoto, M. in: "Liquid Crystals and Ordered Fluids" (J. F. Johnson and R. S. Porter, Eds.) vol. 2, p. 711, Plenum Press, New York (1974).
2. Tamamushi, B., Biorheology **18**, 667 (1981).
3. Tamamushi, B. and Watanabe, N., Colloid and Polymer Sci. **258**, 174 (1980).
4. Kokatner, V. R., U. S. Patent 2,111,100 (1935).
5. Prince, L., Ed., "Microemulsions: Theory and Practice," Academic Press, New York (1967).
6. Hoar, T. P. and Schulman, J. H., Nature **152**, 102 (1943).
7. Schulman, J. H. and Riley, D. P., J. Colloid Sci. **3**, 383 (1948).
8. Bowcott, J. L. and Schulman, J. H., Z. Electrochem. **11**, 117 (1955).
9. Sears, D. I. and Schulman, J. H., J. Phys. Chem. **68**, 3529 (1964).
10. Shah, D. O. and Hamlin, R. M., Science **171**, 483 (1971).
11. Rosano, H. L., Schiff, N. and Schulman, J. H., J. Phys. Chem. **92**, 366 (1961).
12. Prince, L. M., J. Colloid Interface Sci. **23**, 165 (1967).
13. Rosano, H. L., J. Soc. Cosmetic Chem. **25**, 609 (1974).
14. Ruckenstein, E. and Chi, J. C., J. Chem. Soc. Trans. Faraday Soc. II **71**, 1690 (1975).
15. Reiss, H., J. Colloid Interface Sci. **53**, 61 (1975).
16. Ruckenstein, E., J. Dispersion Sci. Technol. **2**, 1 (1981).
17. Bellocq, A. M., Bourbon, D. and Lemanceau, B., Ibid. **2**, 27 (1981).
18. Biais, J., Bothorel, P., Clin, B. and Lalanne, P., Ibid. **2**, 67 (1981).
19. Shah, D. O. and Schechter, R. S., Eds. "Improved Oil Recovery by Surfactant and Polymer Flooding," Academic Press, New York (1967).
20. Shah, D. O., Ed. "Surface Phenomena in Enhanced Oil Recovery," Plenum Press, New York (1981).

21. Cayias, J. L., Schechter, R. S. and Wade, W. H., Soc. Petr. Eng. J. **16**, 351 (1976).
22. Doe, P. H., Wade, W. H. and Schechter, R. S., J. Colloid Interface Sci. **59**, 525 (1977).
23. Doe, P. H., El-Amary, M., Wade, W. H. and Schechter, R. S., J. Am. Oil Chem. Soc. **55**, 505 (1978).
24. Gulari, E., Bedwell, B. and Alkhnafaji, S., J. Colloid Interface Sci. **77**, 202 (1980).
25. Bellocq, A. M. and Fourche, G., Ibid. **78**, 275 (1980).
26. Finsy, R., Devriese, A. and Lekkerkerker, H., J. Chem. Soc. Faraday Trans. II **76**, 767 (1980).
27. Sjoblom, E. and Friberg, S. E., J. Colloid Interface Sci. **67**, 16 (1978).
28. Ober, R. and Taupin, C., J. Phys. Chem. **84**, 2418 (1980).
29. Boned, C., Clausse, M., Lagourette, B., McClean, V. E. R. and Sheppard R. I., Ibid. **84**, 152C (1980).
30. Bansal, V. K., Chinnaswamy, K., Ramachandran, C. and Shah, D., J. Colloid Interface Sci. **72**, 524 (1979).
31. Talmon, Y., Davis, H. T., Scriven, L. E. and Thomas, E. L., Rev. Sci. Instrum. **50**, 698 (1979).
32. Biais, J., Mercier, M., Lalanne, P., Clin, B., Bellocq, A. M. and Lemanceau, B., C. R. Hebd. Seances Acad. Sci., Ser. C **285**, 213 (1977).
33. Boussaha, A., and Ache, H. J., Colloid Interface Sci. **78**, 257 (1980).
34. Boussaha, A., Djermouni, B., Fucugauchi, L. A. and Ache, H. I., J. Am. Chem. Soc. **102**, 4654 (1980).
35. Ruckenstein, E. and Krishnan, R., J. Colloid Interface Sci. **76**, 201 (1980).
36. Tondre, C. and Zana, R., J. Dispersion Sci. Technol. **1**, 179 (1980).
37. Lindman, B., Kamenka, N., Kathopoulis, T., Brun, B. and Nilsson, P. G., J. Phys. Chem. **84**, 2485 (1980).
38. Vrij, A., Nieuwenhuis, E. A., Fijnaut, H. M. and Agterof, W. G. M., Faraday Discuss. Chem. Soc. **65** 101, (1978).
39. Overbeek, J. Th. G., Ibid. **65**, 7 (1978).
40. Friberg, S. E. and Buraszewska, I., Prog. Colloid Polym. Sci. **63**, 1 (1978).
41. Rance, D. G. and Friberg, S. E., J. Colloid Interface Sci. **60**, 207 (1977).
42. Biais, J., Bothorel, P., Clin, B. and Lalanne, P., Ibid. **80**, 136 (1981).
43. Eicke, H.-F. and Markovic, J., Ibid. **79**, 151 (1981).
44. Sjoblom, J., Colloid Polym. Sci. **258**, 1164 (1980).
45. Podzimek, M. and Friberg, S. E., J. Dispersion Sci. Technol. **1**, 341 (1980).
46. Eicke, H.-F., Pure Appl. Chem. **52**, 1349 (1980).
47. Bansal, V. K., Shah, D. O. and O'Connell, J. P., J. Colloid Interface Sci. **75**, 462 (1980).
48. Borys, N., Holt, S. and Barden, R., Ibid. **71**, 526 (1979).
49. Eicke, H.-F., Ibid. **68**, 440 (1979).
50. Talmon, Y. and Prager, S., Nature **267**, 333 (1977).
51. Scriven, L. E., Proc. Int. Sympl. Solubilization, Microemulsions **2**, 877 (1977).
52. Lindman, B., Kamenka, N., Brun, B., Nilsson, P. G., Microemulsions, (Proc. Conf. Phys. Chem. Microemulsions); Robb, I. D., Ed.; Plenum, New York, N. Y.; 115 (1982).
53. Clausse, M., Heil, J., Peyrelasse, J., Boned, C., J. Colloid Interface Sci. **87**, 584 (1982).
54. Kubik, R. and Eicke, H.-F., Helvetica Chimica Acta **65**, 170 (1982).
55. Muller, N., J. Phys. Chem. **86**, 2047 (1982).
56. Benton, W. J., Natoli, J., Qutubuddin, S., Mukherjee, S., Miller, C. A., Fort, T., Jr., SPEJ, Soc. Pet. Eng. J. **22**, 53 (1982).
57. Lianos, P., J. Phys. Chem. **86**, 1935 (1982).
58. Eicke, H.-F., Microemulsions, (Proc. Conf. Phys. Chem. Microemulsions); Robb, I. D., Ed., Plenum, New York, N. Y.; 17 (1982).
59. Linse, P., Gunnarsson, G., Jonsson, B., J. Phys. Chem. **86**, 413 (1982).
60. Kratohvil, J. P. and Aminabhavl, T. M., Ibid. **86**, 1254 (1982).
61. Miller, R., Colloid & Polym. Sci. **259**, 1124 (1981)
62. Cameron, D. G., Umemura, J., Wong, P. T. T., Mantsch, H. H., Colloids Surf. **4**, 131 (1982).
63. Ekwall, P. in: "Advances in Liquid Crystals" (Brown, G. H., Ed.), Academic Press, Vol. 1, p. 1 (1975).
64. Clausse, M., Sheppard, R. J., Boned, C. and Essex, C. G. in: "Colloid and Interface Science," Vol. II, p. 233 Academic Press, N.Y. (1976).(65. Eicke, H.-F., Ibid. Vol. II, p. 233 Academic Press, N. Y. (1976).
65. Eicke, H.-F., Ibid. Vol. II, p. 319 (1976).
66. Podzimek, M. and Friberg, S. E., J. Dispersion Sci. Technol. **1**, 341 (1980).
67. Friberg, S., Lapczynska, I. and Gillberg, G., J. Colloid Interface Sci. **56**, 19 (1976).
68. Scriven, L. E., Nature **263**, 123 (1976).
69. Clausse, M., Peyrelasse, J., Heil, J., Boned, C. and Lagourette, B., Ibid. **293**, 636 (1981).
70. Lagourette, B., Peyrelasse, J., Boned, C. and Clausse, M., Ibid. **281**, 60 (1979).
71. Talman, Y. and Prager, S., Ibid. **267**, 233 (1977).
72. Idem, J. Chem. Phys. **69**, 2984 (1978).
73. Bendiksen, B., Ph. D. Thesis, University of Missouri-Rolla (1981).
74. Pople, J. A., Santry, D. P. and Segal, C. A., J. Chem. Phys. **48**, 129 (1965).
75. Israelachvili, J. N., Mitchell, D. J. and Ninham, B. W., J. Chem. Soc. Faraday Trans. II **72**, 1525 (1976).
76. Ninham, B. W. and Mitchell, D. J., Ibid. **76**, 201 (1980).

Received January 17, 1983;
accepted February 4, 1983

Author's address:

Stig E. Friberg
Chemistry Department
University of Missouri-Rolla
Rolla, MO 65401 USA

Progress in Colloid & Polymer Science   Progr. Colloid & Polymer Sci. **68**, 48–52 (1983)

Phase equilibria in the system soybean lecithin/water*)

B. Bergenståhl**) and K. Fontell

Department of Food Technology, University of Lund, Lund, Sweden

Abstract: The binary phase diagram of a system of a commercially available purified soybean lecithin and water was determined by X-ray diffraction and polarization microscopy studies. A lamellar liquid crystalline phase exists at room temperature in a region between 35 and 7% (w) of water and this is transformed into an L2-phase at 240 °C. At low water contents there occur at room temperature several crystalline and liquid crystalline structures, which transform above 55 °C into a cubic liquid crystalline structure. This phase exists up to about 130 °C and may contain at most about 15% (w) of water. At still higher temperatures there is a transition into a liquid crystalline hexagonal structure which in turn is transformed into the L2-phase above 240 °C.

Key words: Plant lecithin, phase equilibria, liquid crystalline structure.

Introduction

Large quantities of lecithin of plant origin are used in the food industry. The study by Small and coworkers [1] on the system of egg lecithin and water is often used as reference in industrial discussions about the probable phase behaviour of aqueous systems containing plant lecithins. However, depending on the parent plant oil there are large variations in the fatty acid pattern of the lecithin and there may furthermore be a wide variation in this pattern for different batches of lecithin from nominally the same source. Recent work by Rydhag [2, 3] has shown the phase behaviour at room temperature of aqueous soybean lecithin systems at high water content. We have therefore considered it worthwhile to determine the phase diagram of a system containing a well characterized soybean lecithin and water.

Materials and methods

Pure soybean phosphatidylcholine with the trade name Epikuron 200 was obtained from Lucas Meyer (Hamburg, West-Germany). This quality is well-characterized as no other phospholipids were found by thinlayer chromatography. The water content of the compound was estimated to be ~ 2.5% (w) based upon weight loss after freeze-drying for a week. (All concentrations given in the following will include a correction for this amount of water). The

content of unsaturated fatty acid chains is high, the main component being the C–18 acid with two double bonds [3] (table 1).

The chain melting occurs much under 0 °C for the major part of the concentration region and minor variations in the fatty acid composition have therefore a limited influence on the phase behaviour. Samples were prepared by weighing appropriate amounts of phospholipid and water into ampoules which were flame-sealed. Mixing of the samples was carried out by intermittent centrifugation of the ampoules back and fort until it was deemed that equilibrium had been obtained.

In the X-ray diffraction studies was used a Guinier-camera after Luzzati [4] in which camera the temperature of the specimens could be controlled between room temperature and about 125 °C. The textures of the samples were studied in a polarizing microscope equipped with a Koffler heating stage. Non-sealed specimens were mostly used and in the temperature studies heating rates between 1 and 3 °C per minute were employed. Above the boiling point of water the specimens were kept in sealed glass capillaries and the temperature readings given are more approximate.

Results and discussion

The binary phase diagram for the system soybean phosphatidylcholine/water is presented in figure 1. Phosphatidylcholines have a very low solubility in water (~ 10^{-10} M) [5]. An extended homogeneous lamellar liquid crystalline phase whose water content

*) Dedicated to the memory of Professor Dr. B. Tamamushi.

**) Present address: Institute for Surface Chemistry Box 5607, S-114 86 Stockholm, Sweden.

Table 1. The fatty acid pattern of the soybean phosphatidylcholine "Epikuron 200", % (w) (after ref. 2).

C_8	C_{16}	$C_{16:1}$	C_{18}	$C_{18:1}$	$C_{18:2}$	$C_{18:3}$
0.8	12.2	0.4	2.7	10.7	67.2	6.0

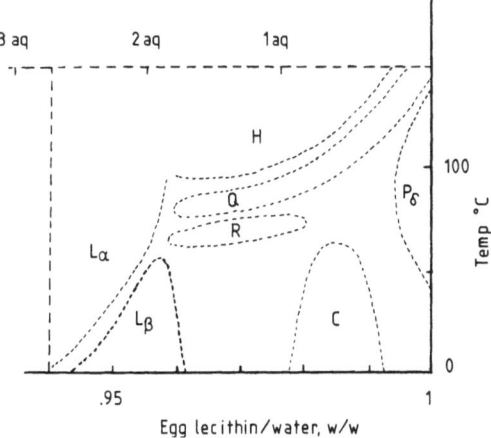

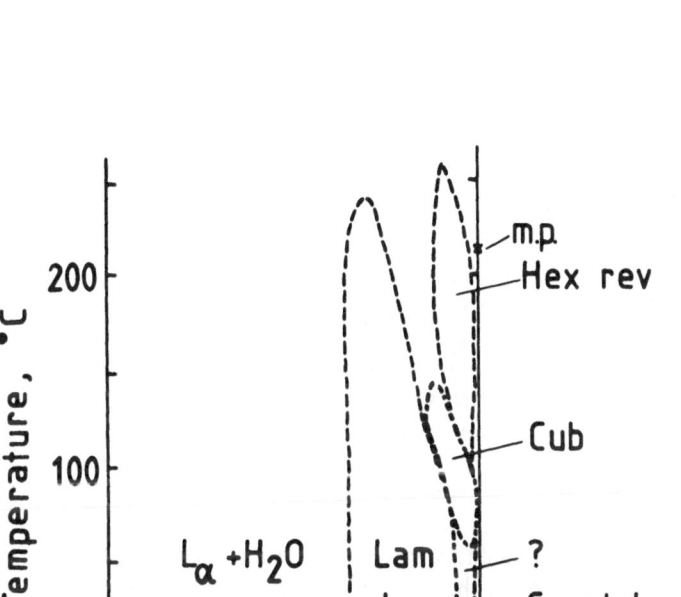

Fig. 5. The "dry" region of the phase diagram for the system egg lecithin/water, redrawn after Luzzati et al. [4, 8]. Notations: L_α, L_β and P_δ; Lamellar phases, the lamellae are infinite planar sheets in L_α and L_β and ribbons in P_δ, the conformation of the paraffin chains is liquid in L_α, partly ordered in the other phases. H; Reversed hexagonal liquid crystalline phase. Q; Cubic liquid crystalline phase. R; Rhombohedral liquid crystalline phase. C; Crystalline phase

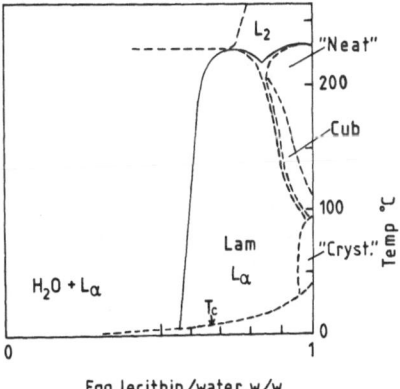

Fig. 1. The phase diagram for the binary system of soybean lecithin/water. Notations: Lam.; Liquid crystalline phase with lamellar structure. Cub.; Liquid crystalline viscous isotropic phase with cubic structure. Hex. rev.; Liquid crystalline phase with reversed hexagonal structure. m. p.; Transition temperature to isotropic solution for non-aqueous samples

Fig. 7. The phase diagram of the system egg lecithin/water, redrawn after Small [1]. Notations: L_α, lamellar liquid crystalline phase; Cub., "viscous isotropic" cubic liquid crystalline phase; "Neat", the phase in the temperature region above the cubic phase, the notation is incorrect, should be Hex. rev. (compare with figs. 1 and 5, see also p. 1032 in ref. 8); "Cryst"., region with ill-defined crystalline and/or liquid crystalline structures; L_2, isotropic "oil" solution; T_c, crystalline/liquid crystalline transition

lies between 7 and 35% (w) exists at room temperature. This phase is stable (in sealed ampoules) up to 240 °C. At water contents below 7% (w) there exist other phase structures (see below). The liquid crystalline samples with 7 to 35% (w) of water were slightly turbid to the eye and birefringent, the consistency was semi-fluid. The lamellar liquid crystalline nature was ascertained by the mosaic texture of the specimens in

the polarizing microscope [6, 7] and by the nature of the X-ray diffractograms [4, 8]. The diffraction patterns contained in addition to a diffuse 4.5 Å reflection in the wide-angle region several "crystalline" reflections in the low-angle region, which reflections indicated the existence of a one-dimensional repeat with values depending on concentration between 40 and 55 Å.

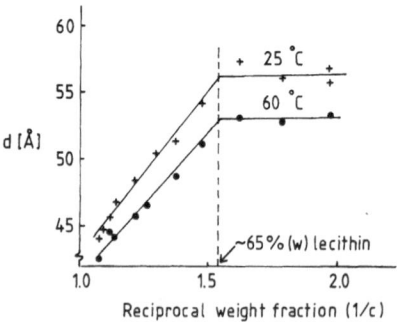

Fig. 2. The fundamental repeat, *d,* versus the reciprocal weight fraction of lecithin (25 and 60 °C, respectively). The phase boundary towards water is shown by the break in the curve

Plots of the fundamental repeat *versus* the reciprocal weight fraction of the lecithin showed a rectilinear behaviour and furthermore they confirmed the ocular observation that the boundary of the phase towards higher contents of water was located at about 65% (w) of lecithin (fig. 2). The values of the thickness of the lecithin bilayers (including the choline groups) ranged between 36 and 41 Å, and the area per polar head group at the interfaces between polar and non-polar regions obtained values of 70 to 63 Å². (The mean molecular weight of the lecithin was estimated to be 773 and it was assumed that the partial specific volumes of water and lecithin were unity, and additivity of volumes on mixing was implied. One should properly use the correct values for the partial specific volumes, experimental or estimated ones, in the calculation of the structure parameters. However, the numerical values will not be greatly affected [1, 9, 10]).

Samples with a water content below 7% (w) were very stiff, they were transparent and anisotropic. The textures in the polarizing microscope differed from that exhibited by lamellar samples but gave no clear guidance for a structure proposal (fig. 3a). At very low water contents one could discern the presence of crystalline entities.

The X-ray diffractograms of "non-aqueous" samples showed in addition to sharp reflections in the low-angle region several sharp reflections in the wide-angle region. This points to the existence of (hydrated?) crystals. When such a sample was heated on the microscope stage several morphological transformations were observed before the sample melted at 212 °C. It is to be noted that during such a temperature run no indication of the occurrence of a cubic or hexagonal structure could be observed.

Samples containing between 2 and 7% (w) of water gave at room temperature X-ray diffractograms with a diffuse 4.5 Å reflection and several spotty "crystalline" reflections (spacing values of 40, 44 and 48 Å) (fig. 4). It has been suggested that the structure in this concentration region is the liquid crystalline reversed hexagonal one [3] but neither the microscopical appearance nor the spacing relations support such a conclusion. Another previous suggestion that in this concentration region a cubic structure occurs at room temperature has not been substantiated [11]. For the egg lecithin/water system Luzzati et al. have demonstrated in this concentration region at room temperature and above the existence of several crystalline and liquid crystalline structures (fig. 5) [4, 8].

 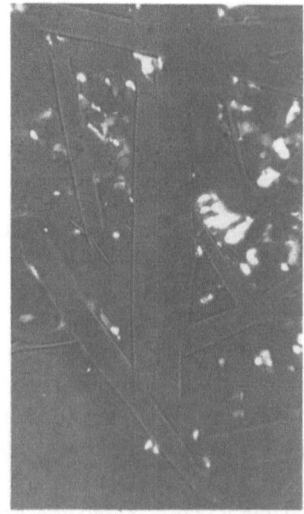 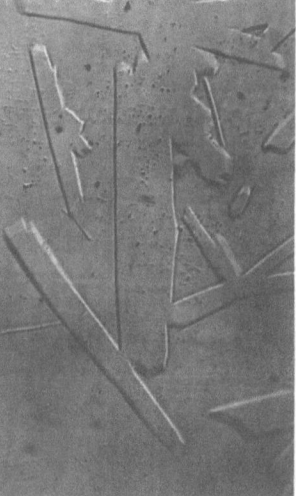

Fig. 3. The microscopic appearance of a sample containing 94% lecithin and 6% water (w) at different temperatures (75 X). a) 20 °C, polarized light; b) 83 °C, polarized light; c) 83 °C, non-polarized light

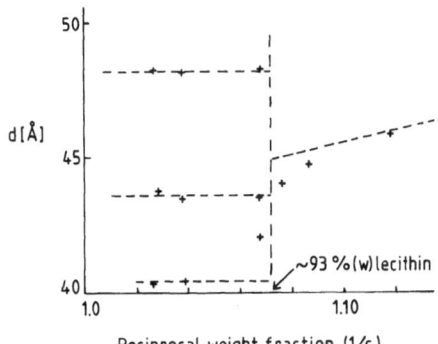

Fig. 4. A plot of the spacings *versus* the reciprocal weight fraction of lecithin showing the upper boundary of the lamellar phase located at a water content of about 7% (*w*) (25 °C)

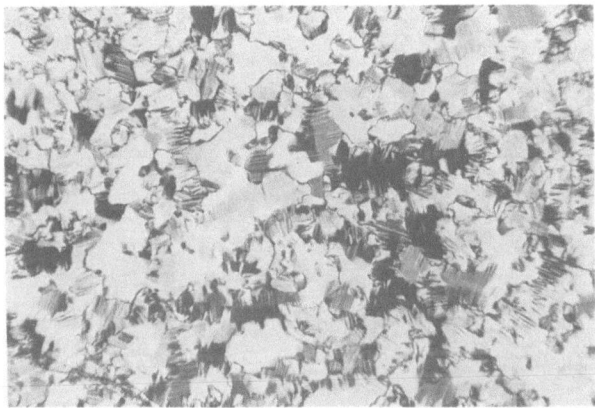

Fig. 6. The fan-like texture of the reversed hexagonal phase (145 °C, composition 97% lecithin, 3 % water (*w*). Polarized light, (75 *X*)

When samples with a water content between 2 and 7% (*w*) were heated they became above 55–60 °C isotropic and extremely stiff. A peculiar observation was that in the polarizing microscope one could discern ribbon-like entities, which in non-polarized light became more visible (fig. 3b and c). The X-ray diffractograms in the low-angle region were so spotty that it was fruitless to attempt to determine any spacing values. A possible interpretation of the findings is that the structure is cubic. This conclusions was supported by ^{2}H and ^{31}P NMR spectroscopy studies. Below 55 °C one obtained NMR spectra showing a quadrupole splitting (indicating the presence of an anisotropic liquid crystalline structure elements) while above that temperature only a singlet was obtained (indicating the presence of an isotropic phase).

Small claims that the cubic structure in the corresponding temperature/concentration region in the egg lecithin system is face-centered [1]. However, he does not present any X-ray diffraction data. On the other hand, Luzzati et al. have interpreted their X-ray findings from the same system as evidence for a body-centered structure of the same type as that they have found for the cubic phases of some strontium soaps [14]. Accordingly to this interpretation the cubic structure will consist of short rod-like aggregates joined three and three into two independent intertwining networks.

On further rizing the temperature there occurred depending on the concentration in the soybean lecithin system a new phase transition between 110 and 145 °C. The substance became again birefrigent and its microscopical appearance was the fan-like texture which is typical of a hexagonal structure (fig. 6) [6, 7]. No X-ray diffraction studies were performed at that temperature in this concentration region. The NMR spectra showed a quadrupole splitting whose value was about half of that obtained for samples below 55 °C.

The location of the samples in the phase diagram indicated that the hexagonal structure above 110 °C is the reversed variant where the structure consists of long rod-like aggregates with a polar core of water and choline groups in a non-polar hydrocarbon environment. The last fragments of the hexagonal structure disappeared at 235 °C and the samples obtained the appearance of an oil isotropic liquid.

The same phase transitions were noticeable by differential scanning calorimetry. The formation of a cubic phase was revealed by an endothermic peak at about 39 °C and the cubic-hexagonal transition by a second peak at 114 °C. These transitions involve enthalpy changes in the magnitude of 80 and 400 J/mol lecithin, respectively.

There are thus large similarities in the phase behaviour of aqueous systems of egg lecithin and pure soybean lecithin (figs. 1 and 7). On the "dry" side there exists at room temperature liquid crystalline and crystalline structures. In contrast to the reports by Small and by Luzzati we could not find any indications of cubic or hexagonal structure when non-aqueous samples were heated up to the melting point. On the other hand these structures were observed in samples containing 1 to 4 moles of water per mole of lecithin. (Small considers the samples in the temperature region above that for the cubic phase to have a lamellar structure [1]).

On the water-rich side the phase diagram is dominated by the lamellar liquid crystalline phase which for samples containing more than about 35% (*w*) of water is in equilibrium with almost pure water.

In the consideration of the probable phase behaviour of aqueous systems of technical lecithins of plant origin one has to keep in mind that they usually, besides phosphocholines, contain other phospholipids (see for instance table 2 in ref. 3) that greatly influence the phase behaviour [15]. It has been shown that already small amounts of a charged component increase the capability of the lamellar phase to incorporate water [12, 13]. This property and the presence of an "oil" contaminant may also influence the emulsion behaviour.

Acknowledgements

Drs. Bodil Ericsson and Ali Khan are thanked for the measurement by DSC and NMR, respectively. The work has been supported by the Swedish Natural Science Council.

References

1. Small, D. M., J. Lipid Research **8**, 551 (1967).
2. Rydhag, L., Fette, Seife, Anstrichmittel **81**, 168 (1979).
3. Rydhag, L. and Wilton, I., J. Am. Oil Chemists Soc. **56**, 83 (1981).
4. Luzzati, V., in: Biological Membranes (Ed. D. Chapman), Chap. 4, Academic Press, New York, 1968.
5. Tanford, C., in: Hydrophobic Effect, J. Wiley, New York, 1973.
6. Rosevear, F. B., J. Am. Oil Chemists Soc. **31**, 628 (1954).
7. Rosevear, F. B., J. Soc. Cosmetic Chemists **19**, 581 (1968).
8. Luzzati, V., Gulik-Krzywicki, T. and Tardieu, A., Nature **218**, 1031 (1968).
9. Lundberg, B. and Sjöblom, L., Acta Acad. Aboensis, Ser. B, Math. Phys. **33**, 1 (1973).
10. Reiss-Husson, F., J. Mol Biol. **25**, 363 (1967).
11. Lindman, B., Ahlnäs, T., Almgren, M., Fontell, K., Jönsson, B., Kahn, A., Nilsson, P.-G., Olofsson, G., Söderman, O. and Wennerström, H., Finn. Chem. Lett., 1982, p. 74.
12. Gulik-Krzywicki, T., Tardieu, A. and Luzzati, V., Mol. Cryst. Liq. Cryst. **9**, 285 (1969).
13. Rydhag, L. and Gabran, T., Chem. Phys. Lipids **30**, 309 (1982).
14. Luzzati, V., Tardieu, A., Gulik-Krzywicki, T., Rivas, E. and Reiss-Husson, F., Nature **220**, 485 (1968).
15. Cullis, P. R., and de Kruiff, B., B.B.A. **513**, 31 (1972).

Received February 3, 1983

Authors' address:

K. Fontell
Department of Food Technology
University of Lund
Box 740
S-220 07 Lund, Sweden

Progress in Colloid & Polymer Science Progr. Colloid & Polymer Sci. **68**, 53–58 (1983)

Simulation of binary liquid mixtures

Part I: Association and coagulation*)

H. Christen, H.-F. Eicke, and Pei-jie Xia**)

Computer Center of the University and Institute of Physical Chemistry, University of Basel, CH-4056 Basel, Switzerland

Abstract: With the help of a computer program association and coagulation processes of binary liquid mixtures are simulated according to the Monte Carlo procedure. The program is based essentially on a two-dimensional lattice model where the potential energy of the particles are approximated by constant pair interaction energies. The kinetic energy is simulated by a Brownian motion. This Brownian motion is simulated by a site exchange process which is controlled by the environment of the particle. Extensive calculations have been performed for different mole fractions of the binary mixture and various interaction energies. As a result the time-dependence of the number of clusters formed (= interconnected areas of particles of the same kind) are discussed.

Key words: computer-simulation (PASCAL); Monte-Carlo-model; Brownian motion; binary solution; coagulation.

Introduction

A characteristic property of lyophobic colloids is their tendency to coagulate (or flocculate). In particular, surfactant solutions in water/oil mixtures forming *W/O*- or *O/W*-microemulsions may cream or sediment by changing temperature or adding suitable additives. In order to circumvent accidental experimental details which may occasionally obscure the basic physical processes, a computer simulation of association and coagulation phenomena based on the Monte Carlo Model [1–3] is believed to represent an interesting approach. In many respects such a simulation presents an "ideal" experiment since the experimental conditions may be modified deliberately. We are well aware that a more satisfactory simulation of binary liquid mixtures must be done with the molecular dynamics procedure [1], particularly if the time dependence of the respective molecular processes is of interest. It is well-known, however, that the calculative effort regarding this technique is quite considerable, even for high-speed computers. It turns out that the principal phenomena of association and coagulation processes can be reasonably well described by an empirical Monte Carlo procedure. Even a time dependence of the processes can – more indirectly – be simulated.

The Model

The simulation is based upon a model with the following properties:

i) Lattice model (cells of identical size): this implies that mixtures of about equally sized molecules of both components are to be considered.

ii) Two-dimensional model: the molecules are restricted to a plane lattice. Each particle has four neighbors (i. e. coordination number 4). Statements regarding polydispersity and shape factors of aggregates (clusters) may be considerably falsified since the topology of possible clusters is remarkably restricted compared with the 3-dimensional case (cf. network of gels).

iii) Pair interaction energies: only four nearest neighbors of each molecule contribute to the interaction energy. This corresponds to introducing a "box"-potential.

iv) Simulated Brownian motion: the kinetic energy of the particles are simulated by a mechanism of sites exchange, i. e. each molecule changes sites with a neighboring molecule according to certain selection rules.

---

*) To the memory of Professor Bun-ichi Tamamushi.

**) Present address: Institute of Photographic Chemistry, Academia Sinica, Peking, China.

These four properties characterize the model qualitatively. A quantitative description is feasible with the following three parameters:

x = the mole fraction of the binary mixture
$\omega = \omega_{11} + \omega_{22} - 2\omega_{12}$ the balance of the molecular interaction energies of neighboring molecular pairs $1-1$, $1-2$, $2-2$, and
T = the absolute temperature of the system.

Moreover, the size of the lattice, i. e. the number of lattice sites, may be varied, as well as the occupation of the boundary (lyophilic or lyophobic border).

The simulation procedure allows one to obtain statements regarding the time-dependence of the particle distribution across the lattice. If groups of neighboring particles of the same kind are considered as clusters, then it is possible to get an insight into the formation and decomposition of clusters. Hence sizes and numbers of all clusters can be controlled dynamically. Thus the time-dependence of particle numbers, number and weight averages of aggregates, distribution function (i. e. polydispersity) of the clusters and the (cross-)linkage (= form factor) as a function of the mole fraction (x) and molecular energy (ω) are accessible.

Programming of the Simulation

The simulation programs were written in PASCAL [4]. This language is particularly well adapted to the present problem. Problem-oriented data types may easily be introduced in PASCAL.

i) the lattice can be generated as a matrix with binary elements (e. g. "black" and „white" for solute and solvent). This matrix corresponds to a frame-buffer which is common in digital image processing with binary pixels. A dynamic representation with video terminals is thus easily possible.

ii) PASCAL permits the programming of recursive algorithms which facilitate the data processing. With recursive FILL-algorithms [5] size, shape and border line of the clusters can be determined.

iii) PASCAL-programs are, to a large extent, independent of the particular computer used, and, therefore, easily transferable. The simulations were carried out on DEC PDP 11/44 and PDP 11/60 computers using Swedish PASCAL [6].

The main simulation program MIXPIC is composed of three parts:

i) Generation of the lattice and occupation of the lattice sites with components 1 and 2 ("white" and "black").

ii) Simulation of the Brownian motion with site-exchange.

iii) Storing of the lattice after a certain time, i. e. after a definite number of exchange steps, into a "picture-file".

Steps ii) and iii) were repeated several times: hence the picture-file contains a large number of "snapshots" of the binary mixture. Secondary programs are necessary for subsequently analyzing these picture-files (HISTOGRAM, PRINTPIC, PLOTPIC).

In order to fill the components 1 and 2 into the lattice, a random number generator produces a number r within the interval, $\{O, \ldots, 1\}$ with a constant density distribution, i. e. equal probability for equally sized incremental intervals. With these random numbers each lattice site is occupied at a fixed mole fraction according to the following rule: if $r < x$ the lattice site is occupied by a "black" particle, otherwise by a "white" one, where "black" denotes solute and "white" solvent molecules. This procedure produces a distribution of components which corresponds to an ideal mixture since the occupation probability is independent of the environment. Figure 1 shows such an ideal mixture for $x = 0.5$ within a $(100)^2$-lattice.

Simulation of the Brownian motion

Since the model lattice was filled without any correlation an ideal mixture was obtained which

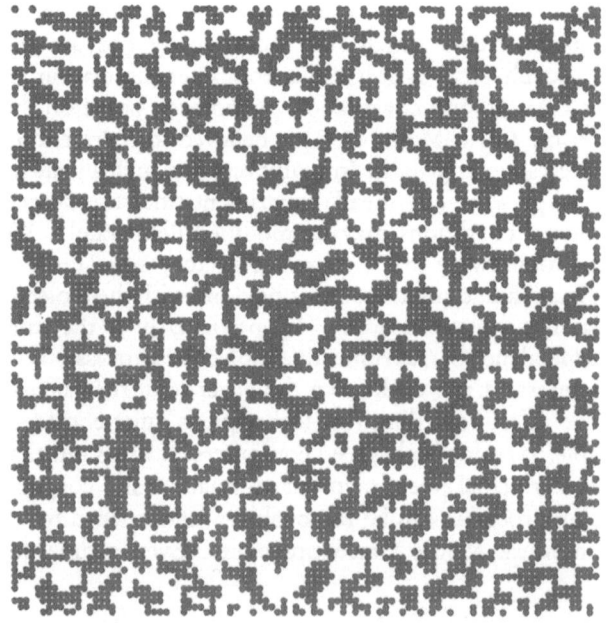

Fig. 1. Simulated ideal binary mixture of mole fraction $x = 0.5$ within a $(100)^2$-lattice

represents a state of a "real" mixture (i. e. with interactions) "far from equilibrium". In order to release the system to equilibrium a suitably simulated Brownian motion has to be introduced. This is done in the following way. At each lattice site a permanent exchange process occurs, i. e. an exchange with one out of four possible neighboring molecules is carried out. Which of the four exchange possibilities is realized depends on the proper environment-dependent exchange probabilities:

$$p(\uparrow), p(\downarrow), p(\rightarrow), p(\leftarrow) \text{ with } \sum_{i=1}^{4} p_i = 1. \quad (1)$$

This exchange process is constantly repeated for each lattice site with equal probability. One random number pair selects the momentary lattice site (where the exchange process is to occur), another one determines the exchange *direction* (see relation (1)).

In order to calculate the exchange probabilities p_i for each of the four site exchange processes, the proper differences of the interaction energies $\Delta\omega_i$ ($i = 1, \ldots, 4$) of the pair-interactions before and after an (imagined) exchange step must be determined. From these $\Delta\omega_i$ one obtains the probabilities p_i i. e.

$$p_i = \frac{e^{\frac{\Delta\omega_i}{kT}}}{\sum_{i=1}^{4} e^{\frac{\Delta\omega_i}{kT}}}. \quad (2)$$

For purpose of illustration $\Delta\omega$ of a single exchange process is determined. Without loss of generality let the selected lattice site (A) be occupied by a "black" particle and the exchanging neighboring particle at (B) be "white". The exchange process can now be split into four steps (see fig. 2):

α) Remove "black" from lattice site (A) to ∞ (= reservoir)

β) Fill "white" from ∞, into (A)

γ) Remove "white" from (B) to ∞

δ) Fill "black" from ∞ into (B)

Hence, the energy differences associated with these four exchange processes are

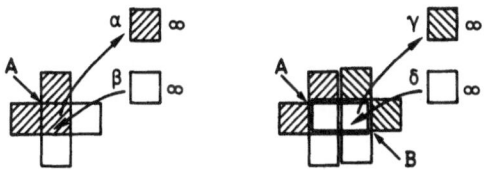

Fig. 2. Illustration of the site exchange process

$$\Delta\omega_\alpha = n_b(A)\omega_{bb} + n_w(A)\omega_{bw}$$
$$\Delta\omega_\beta = -n_b(A)\omega_{bw} - n_w(A)\omega_{ww}$$
$$\Delta\omega_\gamma = \{n_w(B)+1\}\omega_{ww} + \{n_b(B)-1\}\omega_{bw}$$
$$\Delta\omega_\delta = -\{n_w(B)+1\}\omega_{bw} - \{n_b(B)-1\}\omega_{bb} \quad (3)$$

where the following notations were used:

n_b (A) = number of "black"-neighbors of (A)
n_w (A) = number of "white"-neighbors of (A)
n_b (B) = number of "black"-neighbors of (B)
n_w (B) = number of "white"-neighbors of (B)

and ω_{bb}, ω_{ww}, ω_{bw} are the pair interaction energies for black/black, white/white, and black/white-pairs. The difference between $\Delta\omega_\alpha$ and $\Delta\omega_\beta$ on the one hand and $\Delta\omega_\gamma$ and $\Delta\omega_\delta$ on the other is due to the fact that the exchange process at (A) (steps α and β) modifies the environment with respect to (B) (i. e. in the present example one "black" particle less and one "white" more).

Finally, the balance of the energies corresponding to the exchange processes (= sum of the four steps) is given by

$$\Delta\omega_i = (n_b(A) - n_b(B) + 1)(\omega_{bb} + \omega_{ww} - 2\omega_{bw}), (i = 1, 2, 3, 4) \quad (4)$$

where the boundary conditions

$$n_w(A) = 4 - n_b(A) \text{ and } n_w(B) = 4 - n_b(B),$$

referring to the coordination number 4, hold.

The first bracket of equation (4) considers the difference with respect to the environment of (A) and (B). Equation (4) implies that $A \neq B$, otherwise $\Delta\omega_i$ has to be zero a priori. The second bracket corresponds to a molecular interaction energy balance, $\omega_i = \omega_{11} + \omega_{22} - 2\omega_{12}$, which is the same as used in many models of binary systems (e. g. strict regular solution, quasi-chemical theory, etc.).

If the four possible $\Delta\omega_i$ (considering the coordination number 4) are determined, the corresponding probabilities p_i according to equation (2) can be calculated. These p_i are used to subdivide the interval $\{0, \ldots, 1\}$ of the random number generator into four segments. Hence, one random number clearly decides which of the four exchange directions is to be selected.

An essential question concerns the magnitude of the molecular interaction energy balance in equation (4) in order to simulate "real" systems of binary liquid mixtures. Principally, the equation $\omega = \omega_{11} + \omega_{22} -$

$2\omega_{12}$ would appear suitable to obtain a first guess, where ω_{11} and ω_{22} are derived from cohesion energies of the pure components per molecule. ω_{12}, however, is not available in this way. Another possibility is offered by the theory of regular mixtures [7–9]. In particular the approximation of Heitler and Herzfeld [10] appears useful, where the enthalpy of mixing of the two components is determined by

$$\overline{\Delta H} = \frac{H_{before} - H_{after}}{N_1 + N_2}$$
$$= x_1 \cdot x_2 \cdot \frac{z}{2} \, (\omega_{11} + \omega_{22} - 2\omega_{12}) \qquad (5)$$

where z is the coordination number, x_1, x_2 the mole fractions, N_1, N_2 the number of molecules 1 and 2, and the pair interaction energies ω_{11}, ω_{22}, and ω_{12}. Equation (5) is derived assuming that, for example, each molecule of component 1 within its immediate environment (solvation shell) contains a mixture of molecules of types 1 and 2. The composition of this mixutre is determined by the mole fraction (local mole fraction = over-all mole fraction). This is valid, of course, only if $\omega = \omega_{11} + \omega_{22} - 2\omega_{12} = 0$, i. e. for ideal or athermic mixtures. In general, however, there will always be differences between local and overall mole fractions, either by association or solvation interactions [11]. Both phenomena produce an additional order which had to be considered by an entropy of mixing. Since the enthalpic terms already contain the above mentioned approximations, the Heitler-Herzfeld approximation employs an ideal mixing entropy only (the errors in the enthalpic and entropic terms partially cancel each other). Thus, the free energy of mixing with respect to one molecule, and in units of kT, is

$$g(x) = \frac{\overline{\Delta G}}{kT} = a \cdot x \, (1 - x)$$
$$+ x \cdot \ln x + (1 - x) \ln (1 - x) \qquad (6)$$

where $a = z\omega/2kT$. Starting from equation (6) the spinodal curve and the critical point may be calculated, which is simple in the present case. From the condition that the second derivative of $g(x)$ has to disappear, one finds

$$a = \frac{1}{2x \, (1 - x)} . \qquad (7)$$

The critical consolute point is found at $x = \frac{1}{2}$ and a at this point is equal to 2. If, for each mole fraction

(x) a particular a is determined, one has, vice versa, for a given (a) a particular mole fraction, where phase separation should occur. Because of the relation $a = z\omega/2kT$ and the condition $z = 4$, corresponding values for ω/kT can be determined, so that a table with suitabele pairs (x, ω) for the simulation can be derived:

Table 1

x	ω/kT					
0.1	0	0.1	0.5	1.	2.	← border line of phase separation
0.2	0	0.1	0.5	1.	2.	
0.3	0	0.1	0.5	1.	2.	
0.4	0	0.1	0.5	1.	2.	
0.5	0	0.1	0.5	1.	2.	

These values of ω/kT should exist – in the frame of the regular mixture theory – still below the border line of the phase separation.

The proper simulation calculations

The time-dependence of the coagulation was derived for five values of the interaction parameter ω/kT which correspond to a particular value of the mole fraction x (table 1). Starting point is the "ideal" system (without interaction) at time $t = 0$. The coagulation process is started by "switching on" the particle interaction. In constant time-intervals, i. e. after a constant number of iterations of the Brownian motion-simulation within the computer program, the number of clusters is determined according to their sizes.

10^4 ($100 \cdot 100$ area) lattice-sites were chosen to minimize rim effects (less than 5%). The number of monomers is then given by $N_1 = x \cdot 10^4$. Figure 3 displays the time dependence of coagulation processes for $x = 0.3$ and different interaction energies ω/kT: each line of figure 3 shows single "snapshots" at increasing time intervals from left to right at constant interaction energy. On the very left side is $t = 0$. The interaction energy increases from top to bottom indicated by the increased clustering.

For further evaluation only the total number of all particles, i. e. the sum of all clusters (including monomers), is determined. Counting the number of clusters at $t = 0$, i. e. z_o, for the ideal system ($\omega/kT = 0$) as a function of the mole fraction, the top curve in figure 4 is obtained. At small values of x (< 0.2) increasing mole fractions parallel growing particle numbers. For $x > 0.3$, however, the particle number decreases due to a coalescence of clusters. Because of $\omega/kT = 0$ a clustering in the top curve of figure 4 can only be due

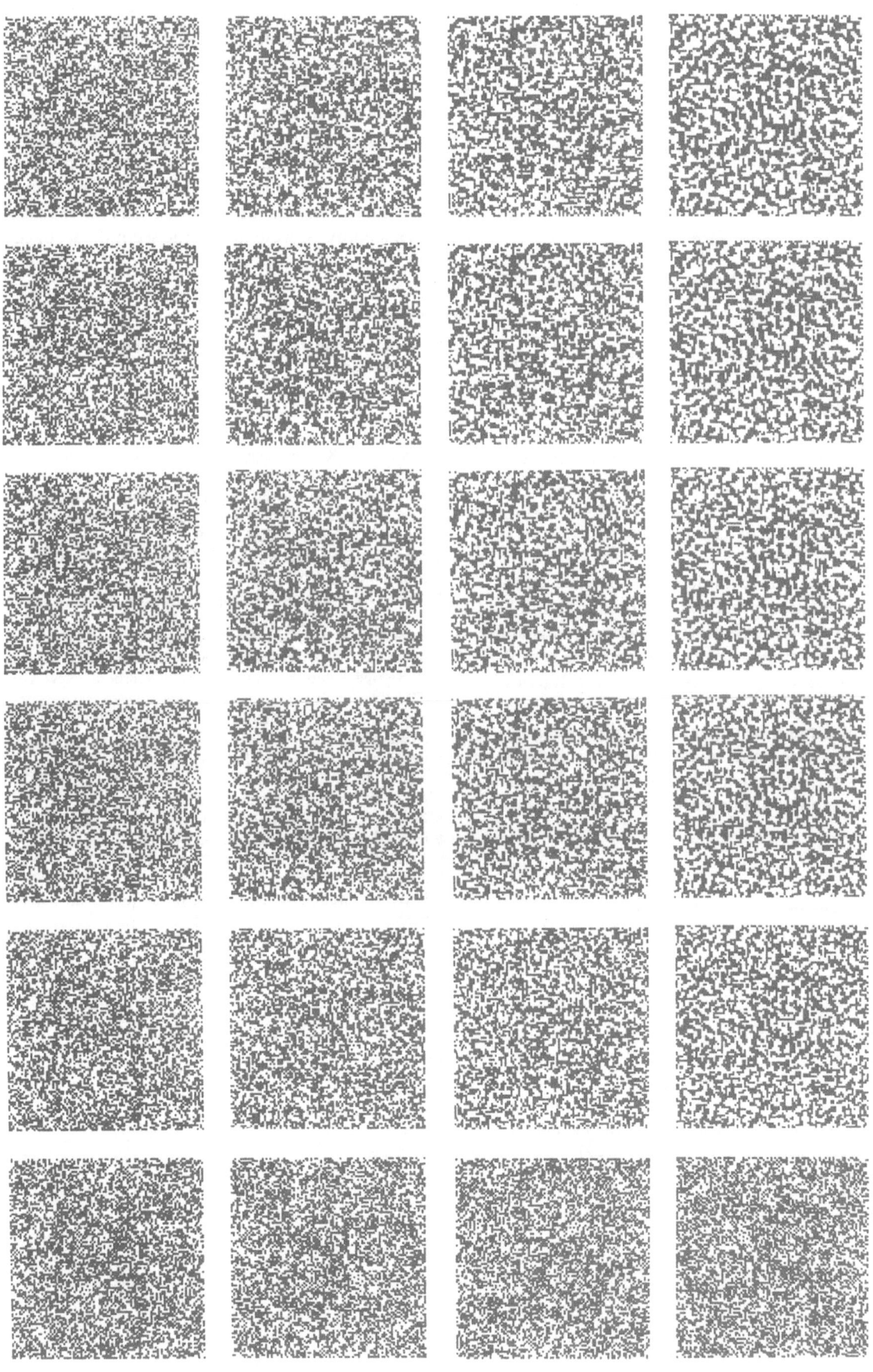

Fig. 3. Coagulation within a binary mixture of mole fraction $x = 0.3$ for different interaction energies ω; first row: $\omega = 0$, second row: $\omega = 0.1$, third row: $\omega = 1$, fourth row: $\omega = 2$. Time is increasing from left to right

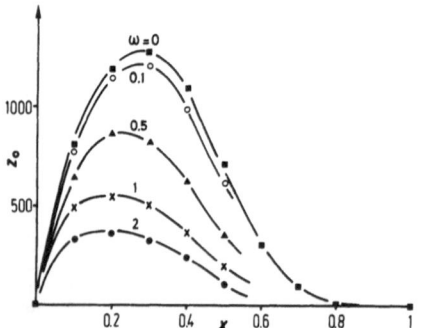

Fig. 4. Equilibrium number of clusters z_0 as a function of the mole fraction x for different interaction energies

to the accidental, uncorrelated neighborhood of particles of the same kind. If interaction energy exists, the number of clusters z_0 were calculated from averaged equilibrium values after a continuous action of the Brownian motion. With increasing interaction energies the maxima of the $z_0(x)$-plots (fig. 4) shift towards smaller x-values due to increased association; also, a simultaneous decrease of the number of clusters is observed.

An example of the complete 25 analysed time-dependent coagulation processes is shown in figure 5. The number of clusters z_0 is plotted against equidistant time intervals. The upper curve corresponds to $\omega = 0$, the lower one to $\omega = 2$ at constant mole fraction $x = 0.2$.

If these simulated time dependences of the coagulation are to be discussed quantitatively, a suitable function $z = z(t)$ is needed which describes the coagulation process analytically with appropriate parame-

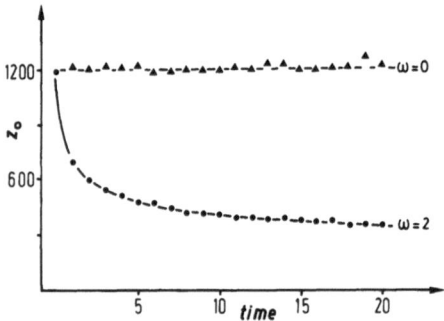

Fig. 5. Time dependence of the number of clusters for two different interaction energies at constant mole fraction ($x = 0.2$)

ters. There exist some theoretical models on the coagulation of colloids [12–14] which predict the time-dependent decrease of particle number z. The data which were simulated with the help of the Monte Carlo method can now be fitted by such theoretical models. If a whole set of measurements is available as in the present case with various parameters, then it is possible to investigate the dependence of the "experimental" parameters on those of the theoretical models. Such a study is presented in the second part of this paper.

Acknowledgement

The work is part of project No. 2.025.0.81 of the Swiss National Science Foundation.

References:

1. McDonald, I. R. and Singer, K. Q., Rev. Chem. Soc. **24**, 238 (1970); Chem. Brit. **9**, 54 (1973).
2. Christen, H. and Eicke, H. F., J. Phys. Chem. **78**, 1423 (1974).
3. Binder, K. (edit.). "Monte Carlo Methods", Topics Current Phys. Vol. 7, Springer, Berlin 1979.
4. Jensen, K. and Wirth, N., PASCAL User Manual and Report, Springer Verlag, Berlin 1975.
5. Pavlidis, T., Graphics and Image Processing, Springer Verlag, Berlin 1982.
6. Torstendahl, S., PASCAL Compiler; distrib. by DECUS.
7. Hildebrand, J. H., Prausnitz, J. M. and Scott, R.L., "Regular and Related Solutions", van Nostrand Reinhold, New York 1970.
8. Guggenheim, E. A., Thermodynamics, North Holland, Amsterdam 1949.
9. Hildebrand, J. H., Nature **168**, 868 (1951).
10. Herzfeld, K. F. and Heitler, W. Z., Elektrochm. **31**, 536 (1925); Heitler, W., Ann. Physik **80**, 630 (1926).
11. Münster, A., in: "Physik d. Hochpolymeren", Vol. II (H. A. Stuart, edit.), Springer, Berlin 1953.
12. v. Smoluchowski, M., Physik, Z. **17**, 557, 585 (1916); Z. Physik. Chem. **92**, 129 (1918).
13. Müller, H., Kolloid-Z. **38**, 1 (1926), Kolloid-Beih. **26**, 257 (1928).
14. Overbeek, J. Th. G., Colloid Science (H. R. Kruyt, edit.) Vol. I, Amsterdam 1952.

Received January 14, 1983

Authors' address:

H.-F. Eicke
Institute of Physical Chemistry
University of Basel
CH-4056 Basel

Order formation in binary mixtures of monodisperse latices

I. Observation of ordered structures*)

S. Yoshimura and S. Hachisu

Institute of Applied physics, University of Tsukuba Sakura, Ibaraki 305 Japan.

Abstract: Formation of alloy structures in binary mixtures of monodisperse latices was studied. When the particle concentration of the mixture exceeded certain extent, a phase separation took place in the system to deposit (one or two) ordered structure(s). The structures were observed by a light microscope. Several lattice types appeared depending upon the composition of the mixture; they are $NaZn_{13}$, AlB_2, $CaCu_5$, $MgCu_2$ and a hexagonal one with a composition of AB_4 which is not yet identified to any of alloys or compounds. In view of the fact that the interaction between the particles in a stably dispersed latex is effectively repulsive, this phenomenon of order formation in binary systems is of the same kind as that in single component or monodisperse latices. The phase transition in the present system would be the binary version of the Alder transition that explains the order formation in monodisperse latices.

Key words: Latex, order, binary system, alloy, phase transition.

I. Introduction

Monodisperse latex [1] is an excellent colloid for the study of concentrated systems. It is highly monodisperse and stable even at very high particle concentrations. Furthermore, the particles are almost perfect spheres not to cause unnecessary complications in theoretical treatments.

Bright iridescence [1, 2] it exhibits under certain conditions indicates that some orderly lattice structure is formed in the dispersion. This evoked a number of studies on concentrated latices. So far, liquid structure [3, 4, 5], crystalline structures [5, 6, 7], transition from liquid to crystalline state [8, 9, 10] have been studied. When the particle size is fairly large, they can be seen by light microscope [11]; liquid, and crystalline states and the coexistence of these two states were observed. Now, it has been established that there is a far reaching parallelism between monodisperse colloids and atomic systems. The studies that have been made so far are concerned only with monodisperse or single component latices. The present work is an extention of the knowledge of single component latices to binary cases.

It is well known that a monodisperse latex shows phase separation to deposit an ordered phase (that is iridescent as mentioned above) when its volume fraction exceeds a certain value [8]. As far as concerned with this general tendency, a binary latex would behave in the same way. Actually, as will be shown in the present experiment, when its volume fraction exceeds certain limit, the binary latex shows phase separation to deposit some ordered phase (or phases). However, the parallelism ends here. Upon going into further details the difference becomes significant; the condition for the phase separation to occur and the structure of deposited ordered phase (or phases) are not uniquely determined as is the case with single component systems, but vary in complicated manner depending upon the composition of the systems.

The main object of the present work (part I) is to find out as many types of the ordered structures as possible. The stability or the mechanism of the formation of the structures will be discussed in the subsequent report (part II).

The study of the order formation in binary colloids was initiated by the discovery of a strange structure in opal by Sanders and Murray [12]. They found by electron microscopy that two kinds of silica spheres of different sizes form a super lattice structures. The structure in opals must have been formed in stable dispersions of silica spheres and then been dehydrated. Therefore, the same structures would be produced in

---

*) Dedicated to the memory of Professor Dr. B. Tamamushi.

some artificial colloids. The authors succeeded in producing the structures in binary mixtures of monodisperse latices [13], and observed them under a light microscope in their stably dispersed state; the lattice types of them were determined from the observation. Some of them were the same as those found in opals.

It is remarkable that the same structures are found in atomic systems, and that most of them belongs to a special group of alloys named size factor intermetallic compounds.

II. Experiment

1. Preparation

Monodisperse polystyrene latices were used, the diameter of which ranged from 2000 A to 1μ. They were synthesized by one shot method. The control of the particle diameter was made by adjusting the amount of the emulsifier (sodium dodecyl sulphate).

The polymerized latices were dialysed and then deionized by ion exchange resin, Amberlite MB-3. When these treatments were finished, the latices had volume fractions (here after designated by ϕ) near 0.1. They were stocked at 5 °C in a refrigerator. Just before use, it was condensed up to $\phi = 0.3$ by ultra-filtration. As will be mentioned, this is to facilitate the search for the optimum condition for the ordered structure to appear.

2. Method of experiment

Two kinds of latices with desired diameters each of which is 0.3 in volume fraction, were mixed at suitable particle number ratio (here after it will be expressed by η). The mixtures were highly viscous and sticky making volumetric measurement very difficult. They were formulated by a gravimetric method by dropping latices directly into an observation cell placed on a dial scale; then the content was gently stirred with a thin glass rod for a few minutes. The cell was placed on a metallurgical microscope of inverted type (fig. 1) and observed by oil immersion method. Owing to low ionic concentration (at the conductivity water level) of the system, the particles were separated with significant distances to produce good visibility.

A mixture thus prepared had irregular structure (cf. fig. 2) and was in a frozen state because of its high particle concentration; the particles seen in the view field were bound at their individual positions without any visible Brownian motion. Only very rarely, intermittent moves of particles (four or five particles moved cooperatively at the same time) were recognized. To allow them undergo fast relaxation, the mixture was diluted with distilled water to an optimum concentration, where the movement of the particles was rather active. This condition was, in fact, the state of phase separation in this binary latex. In most cases, dilution up to 2 ~ 3 times the original volume was suitable; too much dilution made the disordered state stable, and any ordered structure did not come out. When the condition was optimum, the ordered phase appeared in 20 ~ 30 hours.

This rather delicate adjustment for producing the ordered structure is a peculiarity of binary systems. With single component systems, ordered structure is readily produced when the critical value (of particle concentration for the phase transition) is exceeded.

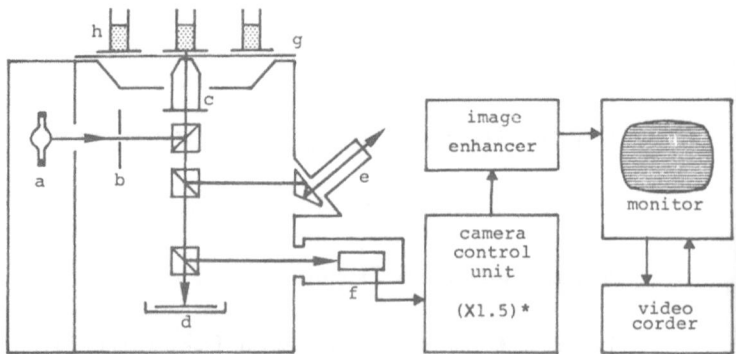

Fig. 1. Schematic diagram of microscope apparatus. a: Xenon arc lamp house, b: Pin hole, c: Objective lens, d: 35mm camera & 16mm cine camera, e: eye piece, f: TV camera, g: X-Y stage, h: observation cell

In binary systems, a great many trials and errors in the arrangement of the particles is necessary for an alloy structure to form. Therefore, a fast relaxation is necessary as the experimental condition. In order to realize it, the system is to be brought in or near the state of phase separation, where diffusional movement of the particles is active.

III. Result

3.1 General feature

The structure tended to form in the bulk (one or two layers apart from the surface) of the mixture rather than at the surface of the window, especially when the surface was contaminated by adhering particles. Therefore, the microscope focus should be moved to and fro in the course of survey for the structures.

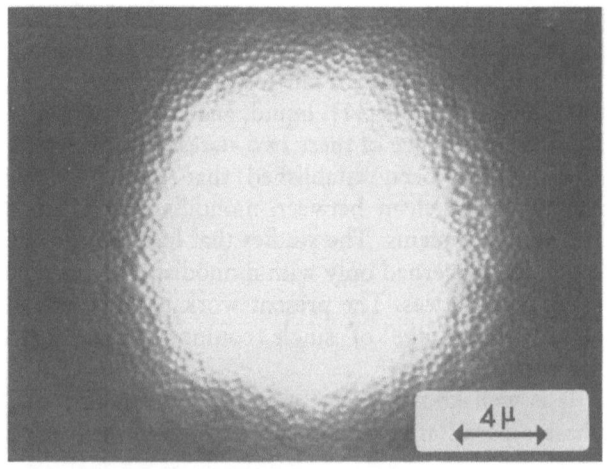

Fig. 2. Disordered state in a latex mixture

The structures appeared, in their early stage of formation, as small islands (10 ~ 15 μ in diameter) in the sea of disordered phase (cf. fig. 3, it shows a state of the phase separation). At the periphery of each island, order-disorder fluctuation was observed. The islands finally grew into large areas several hundreds of microns in diameter. The structures were easily broken by a small mechanical shock and reformed in several hours. After long period of time (several weeks), the ordered phase sedimented onto the bottom of the cell indicating that it has a higher density than the coexisting disordered phase.

In most cases, only one type of ordered structure appeared. In an exceptional case, two different alloy structures coexisted in the sea of the disordered phase. Sometimes, in addition to an alloy structure, a single component pure phase (of smaller particles in most case, since η value is usually very large) appeared. It was noticeable that in these multi-phase systems, the number of phases (including disordered phase) did not exceed 'three'. This is in accordance with the phase rule ($p = c - f + 2$, where p is the number of phases, c is that of components and f is the freedom).

One structure exposed several different net planes (that happened to be parallel to the surface of the window). From the patterns in these net planes, the lattice types were determined. Furthermore, by moving the microscope focus to and fro, the mode of superposition of the planes was known together with a rough estimation of the distances between the planes.

Depending upon the type of the structure, 3 to 13 superimposing planes were observed.

Until now, five types of structures have been found, of which, four were assigned to lattice types of $NaZn_{13}$, AlB_2, $CaCu_5$ and $MgCu_2$. Except AlB_2-type, these belong to a peculiar group of alloys named 'size factor compounds'. The fifth one is a hexagonal structure with a composition of AB_4, but is not yet assigned to any of really existing alloys or compounds. In addition to these ones, NaCl structure was also found. However, it was not in a latex system but in a mixture of silica and gold sols. It will be reported elsewhere.

When a cell containing a sedimented alloy structure was seen by the naked eye, iridescence was not seen as evident as is with single component latices. This was

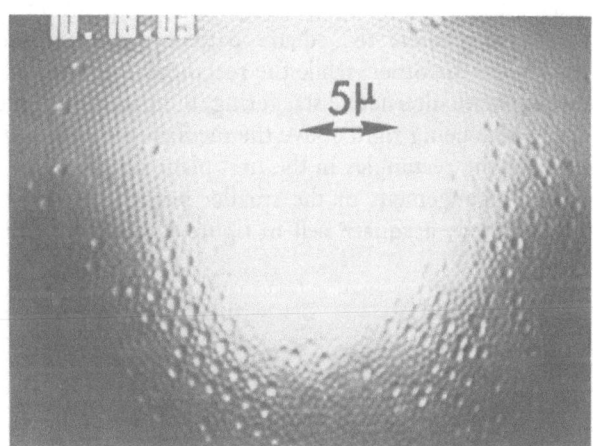

a

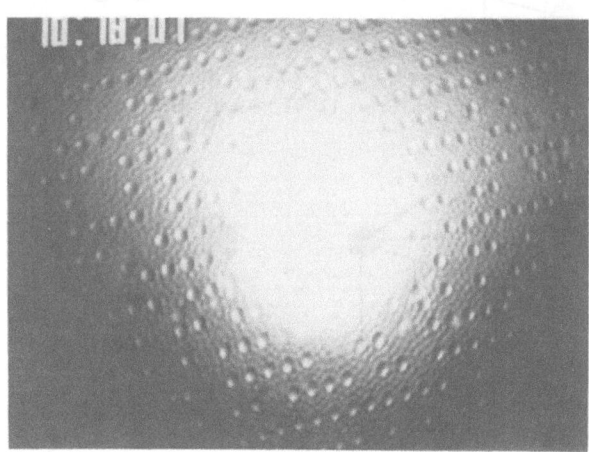

b

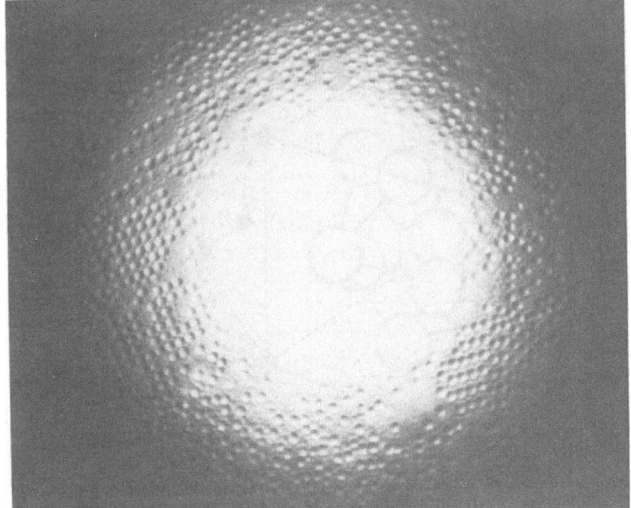

Fig. 3. A state of phase separation; small islands of ordered structure are seen in the sea of disordered phase. Some of different patterns represent the different net planes of the same structure

Fig. 4. Patterns observed in 6000A- 3100A mixture. a) (111) plane of $NaZn_{13}$ formed together with a single component phase, b) (110) plane of $NaZn_{13}$

due to rather large spacial periodicities in the structure. By immersing the cell in water, an iridescence could be recognized on the side and the bottom of the cell.

3.2 Observation of the structures

1) NaZn$_{13}$ structure: This was found in 4700A–2000A, 5500A–2500A and 6000A–3100A (in core diameter) mixtures.

The observed patterns are shown in figure 4. The pattern shown by the larger particles are remarkable. It is obvious that the side of the square pattern in figure 4a is equal to the short side of the rectangular pattern in figure 4b, and that in the rectangle, the length ratio of the shorter to the longer side is 1 : 1.4. This indicates that these patterns respectively represent (100) and (110) planes of a simple cubic lattice. This was supported by the mode of superposition of these net planes; the square pattern superimposed rightly on the other, while the rectangular pattern did not but with alternative staggering, the particles in the next plane being right above the medians of the longer sides of the rectangles in the first plane (cf. fig. 5).

The arrangement of the smaller particles is rather complicated; a square cell in figure 4a contains four

small particles which form a rhombus (strictly speaking, it may be a shallow tetrahedron), while a rectangular cell in figure 4b contains 6 particles, five of which forming a distorted pentagon and one particle in the center. Several trials using two sets of plastic spheres of different sizes led to a conclusion that each cubic cell of the larger particles contains a centered icosahedron of small particles. It is shown in figure 6. The composition of this structure is AB$_{13}$. The staggered arrangement of the rhombuses in the (100) plane (fig. 4a) shows that the neighboring icosahedrons stagger in their orientations by 90 degrees. Therefore, a single cubic cell of the larger particles is not the true unit. The true unit is a block of 8 cubes as shown in figure 6. This structure also explains the alternative mode in the arrays of the distorted pentagons in figure 4b. The unit cell contains 112 particles (8A + 104B). It is remarkable that the same structure is found in the atomic system; it is an alloy structure NaZn$_{13}$. It belongs to a class named size-factor intermetallic compound [14].

Another observation was that in the same latex mixtures, sometimes, other type of structure was found to coexist, which will be mentioned in the following.

2) AlB$_2$ structure: This was found in 4700A–2000A, 5500A–2500A, 5600A–2100A and 8000A–2800A mixtures. In the first two of them, the above mentioned NaZn$_{13}$ structure was found to coexist. Typical pattern of this new structure is shown in figure 7a, in which large particles are packed very closely. By forwarding the microscope focus the next

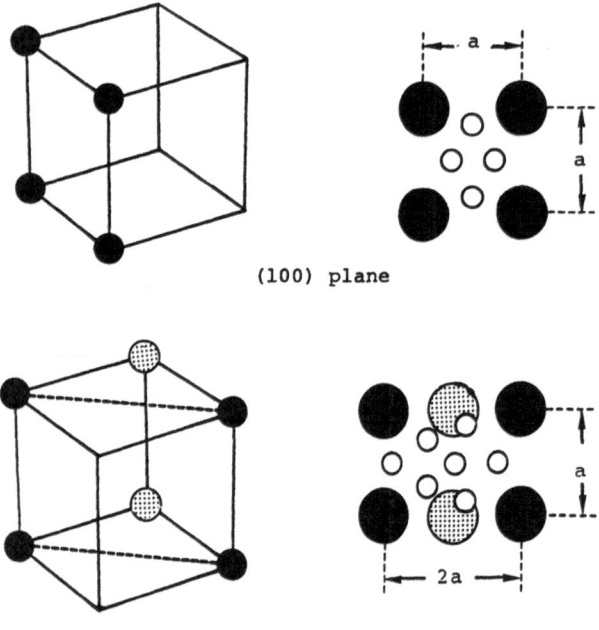

Fig. 5. Illustration of the correspondence of the observed patterns to the net planes in the lattice that is formed by the large particles in NaZn$_{13}$ structure

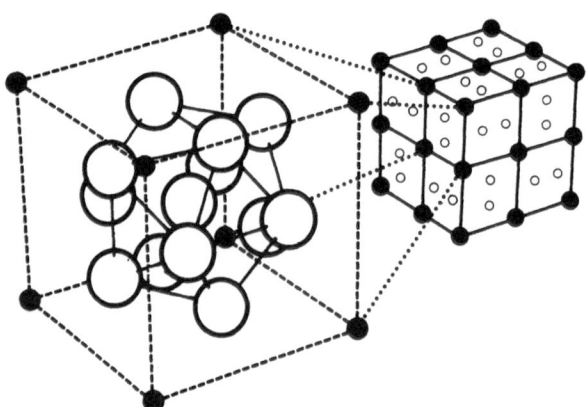

Fig. 6. Pictorial drawing of NaZn$_{13}$ structure. The large particles form simple cubic lattice, in each cell of which a cluster of small particles is accommodated. The cluster is a centered icosahedron. Two icosahedrons stagger in orientation with each other by 90 degrees. Then, the unit cell of this structure is a block consisting of 8 cells of the sigle cubic lattice

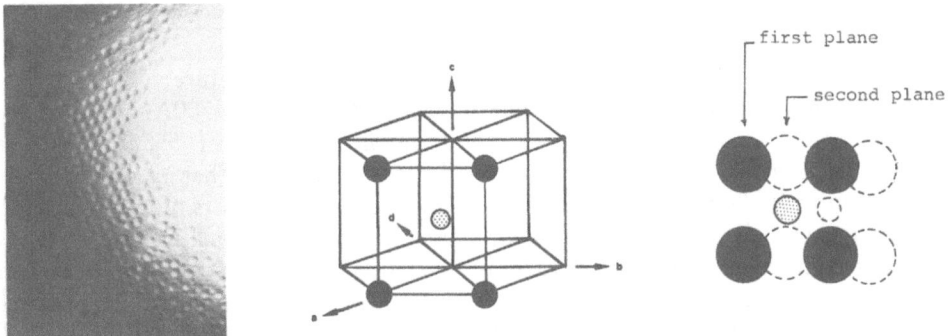

Fig. 7a. Photograph at the left shows a pattern found in a 5500 A–2500 A mixture, in coexistence with the NaZn$_{13}$ structure. As illustrated by the drawing, it represents (10$\bar{1}$0) plane of AlB$_2$ structure

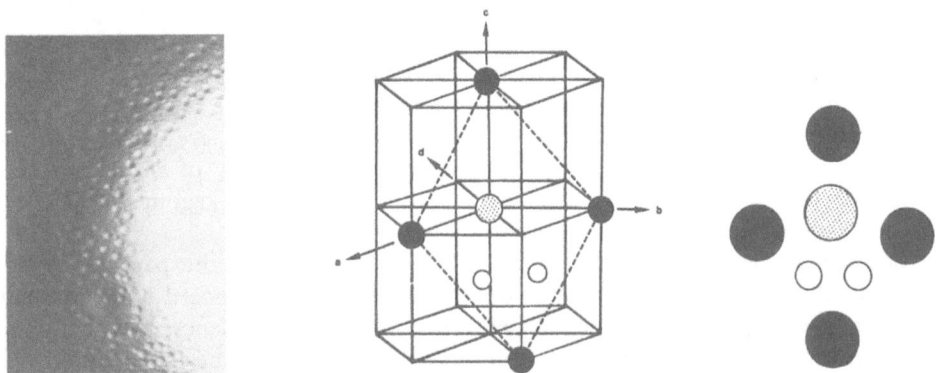

Fig. 7b. One of observed pattern of AlB$_2$ structure, which represents (11$\bar{2}$1) plane

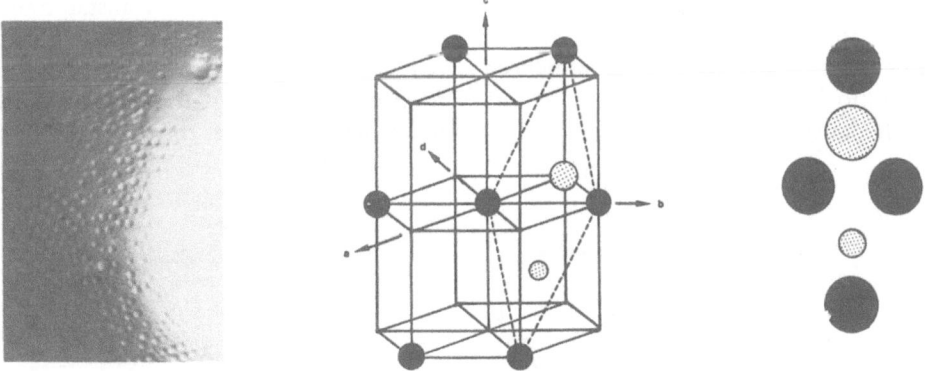

Fig. 7c. One of observed patterns of AlB$_2$ structure, which represents (10$\bar{1}$1) plane

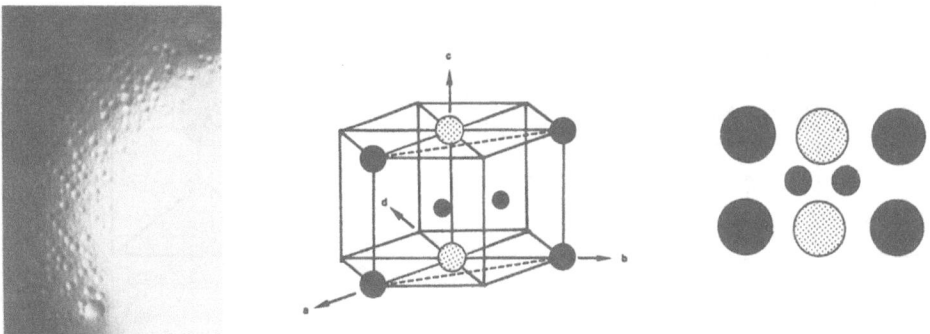

Fig. 7d. One of observed patterns of AlB$_2$ structure, which represents (11$\bar{2}$0) plane

plane was found. It was the same as the first plane, but was staggered as shown by the drawing at the right end of the figure.

In addition to this, several other patterns were found as shown in figures 7b, 7c and 7d (at their right ends). The analysis was very difficult, but fortunately Sanders had already deduced the lattice model from his observation on opal structure [15]; it is AlB_2 structures as shown in figure 8 (In figures 7a, b, c and d, the lattice structure was already shown in aid of understanding). It is a hexagonal lattice consisting of two planes, one is made of large particles and the other of small particles; each of the small particles situating in the center of a trigonal prism made of six large particles. The patterns show by the photographs in figures 7a, b, c and d are explained by this structure as shown by the illustrations attached side by side to the respective photographs.

The coexistence of the two structures AlB_2 and $NaZn_{13}$ is very interesting, because the composition differs so much between the two. It was suspected that the composition of the mixture may favour the appearance of either of them; namely, if the smaller particles exist in heavy excess, the appearance of $NaZn_{13}$ might be favored over AlB_2, and vice versa. Actually however, any appreciable effect was not recognized over a wide range of η-value (from 6 to 70). This leads to an important conclusion that the lattice type formed in a latex mixture is not influenced by the number ratio but determined solely by the combination of the latices.

3) $CaCu_5$ structure: This was found in 5500A–3100A and 4700A–3100A mixtures. The patterns in the case of 5500A–3100A are shown in figures 9a and b. That of figure 9a is of hexagonal symmetry and consists of large and small particles, the number ratio in which is 1 : 2. On moving the focus upwards (by about 0.8 μ), the next plane of the same arrangement appeared. All the particles, large and small, were rightly above those in the first plane; no staggering was found.

Then, by moving the microscope focus back from the second plane very carefully, another new plane (see fig. 9a) was found lying between the above mentioned two planes. In this new plane the small particles were arranged in a different way from those in figure 9a, and the large particles are out of focus indicating that they are not in this plane. This new plane consists solely of small particles arranged in a manner so called kagome structure as shown in figure 10.

Figure 10 shows the particle arrangement in this structure. It is composed of an alternative superposition of these two planes. Brief consideration leads to the formula AB_5 as the composition. This corresponds to the structure of an inter-metallic compound $CaCu_5$ [14].

The structure was rather unstable and in most cases disappeared in a few days, turning into a pure phase (either of large or small particles). Sometimes, it transformed into $MgCu_2$ structure. In this connection, it was noticeable that this structure did not appear as

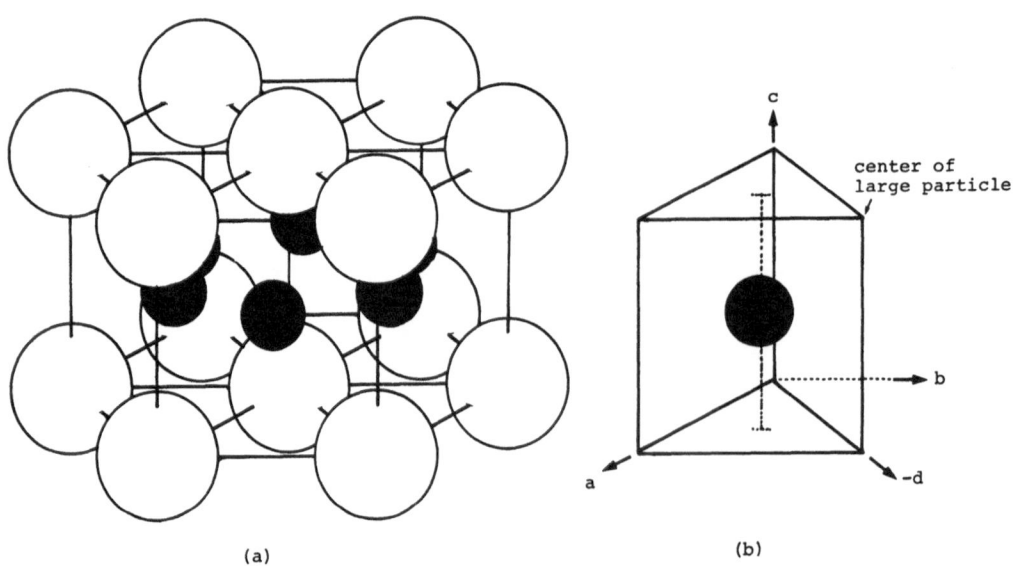

(a)   (b)

Fig. 8. AlB_2 lattice structure: A trigonal prism of large particles accomodates one small particle

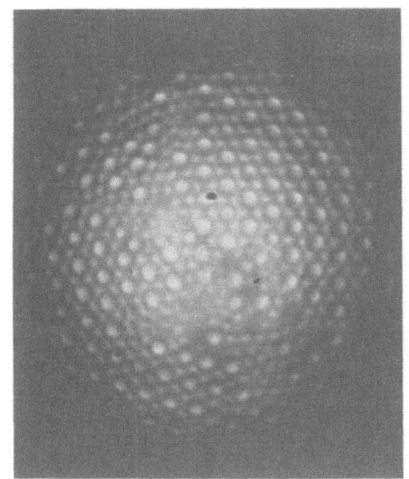 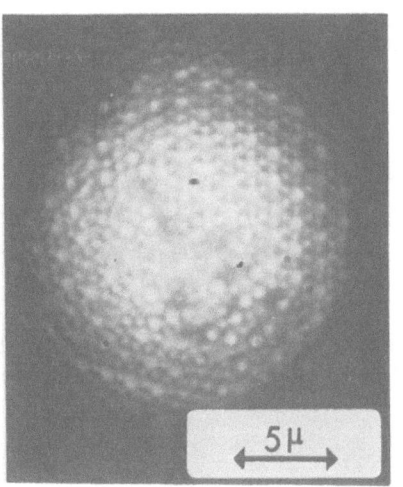

Fig. 9. Two types of hexagonal patterns observed in 5500 A–3100 A mixture. (a) shows the first plane and (b) the second plane

isolated islands but was always found at the margin of islands of the pure phase of small or large particles (cf. fig. 11). Due to rather unstable nature of this structure, observation under different η-valuse was not done over a wide range. The cases of $\eta = 6.4$ and 4 were tested; the same structure was found.

4) $MgCu_2$ structure. This was found in three mixtures, 4500A–2800A, 4500A–3100A and 8000A–5600A. All of the observed patterns were hexagonal (cf. figs. 12, 13 and 14).

In the first sight, patterns looked like the same as $CaCu_5$-type. However, close observation revealed that the large and small particles in each of the patterns (figs. 12 and 13) were not in the same plane but actually form a stack of two or three net planes, each plane consisting solely either of large or small particles

as indicated in the figures. They changed from one pattern to another in an intricate fashion. The distances between two successive net planes were small compared with the interparticle distance in a net plane. These features reflect complex nature of the lattice constitution of this structure.

At last, it was concluded that these patterns were explained by the lattice structure of $MgCu_2$, an intermetallic compound named Laves phase. This was independently found by Hasaka et al. [16], they also reported that $MgZn_2$ structure was found in the same mixture. Detailed explanation will be given in Appendix.

5) AB_4 Structure: In figure 15, the pattern that appears in a mixture of 6500A and 3100A latices is shown. Remarkable feature is the arrangement of

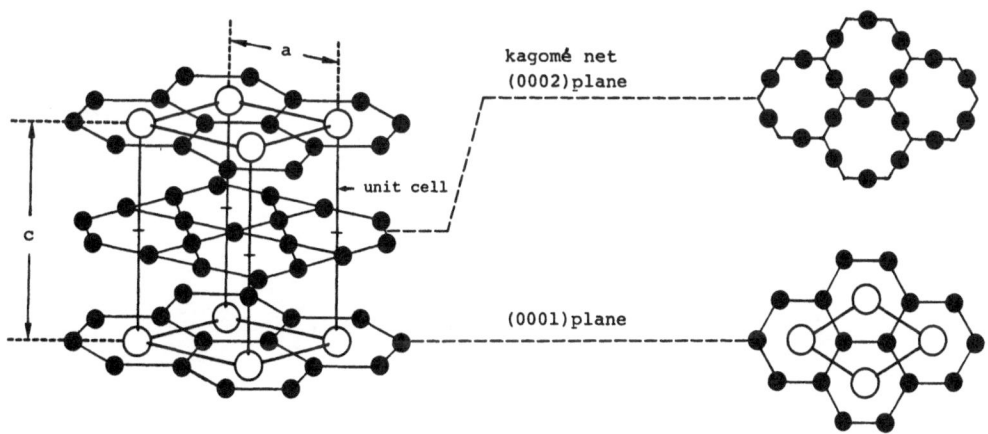

Fig. 10. Illustration of $CaCu_5$ lattice. It consists of alternative superposition of the two kinds of net planes

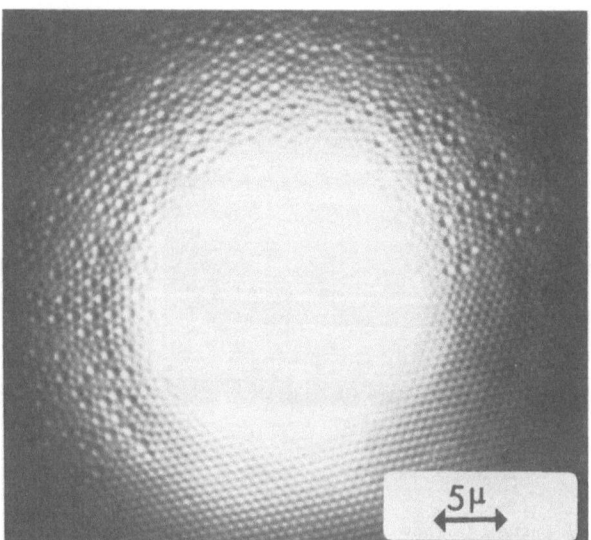

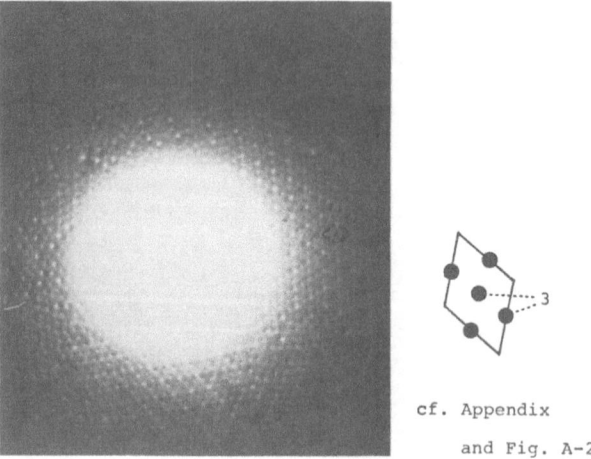

Fig. 13. A pattern found in the same mixture as that in figure 12. This was observed at the same play by moving the microscope focus

Fig. 11. An example of the coexistence of three phases. Upper right corner is a disordered phase, lower right is a pure phase and on the left side, CaCu₅ structure is seen

trigonal clusters of larger particles, with smaller particles being accomodated in areas surrounded by three of the clusters. The actual pattern shown in the photograph has several defects and irregularities. It is completed by reasonable corrections as given in figure 16. In the first place, let us examine the frame work or the sub-lattice made of the large particles. When the microscope focus was advanced, the second and third planes of the large particle appeared one after another; and an alternative recurrence was recog-

nized in the mode of superposition. In figure 17, the first and the second planes are shown; the black particles are in the first plane (which will called A-plane, hereafter), and the white ones are in the second plane (B-plane). It is obvious that B-plane is obtained by rotating A-plane by 60 degrees. The frame work of the large particles is thus known to be an alternative (in orientation, by 60 degrees) superposition of A and B planes. The symmetry type is 6 m, with C-axis perpendicular to the plans of the page.

In this frame work, there are two kinds of tunnel-like vacancies along C-axis, which accommodate the small particles; one is wide tunnels designated as W, the other is narrow ones named N (cf. fig. 17). The

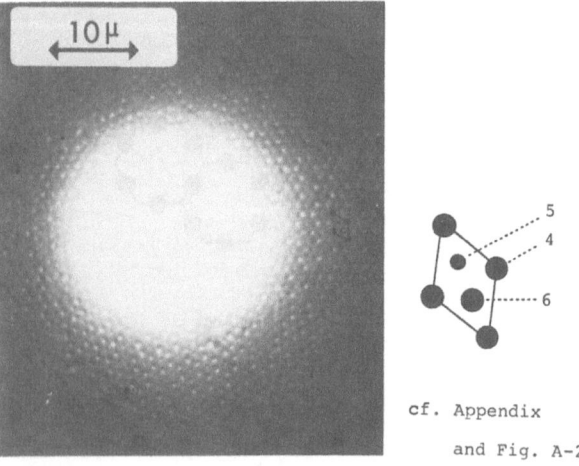

Fig. 12. A pattern observed in a 4500A-2800A mixture. This represents one of the net planes of MgCu₂ structure

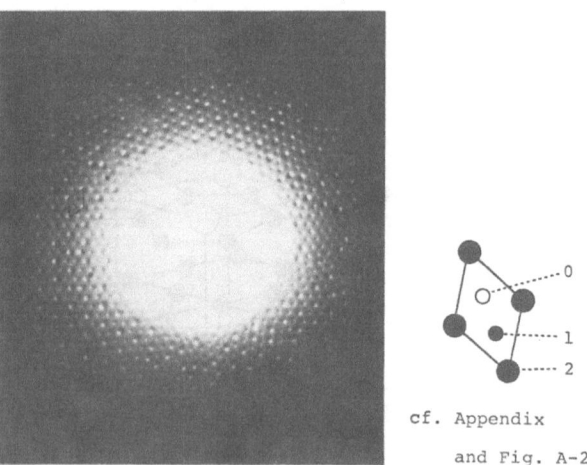

Fig. 14. Another pattern at the same place in the same mixture as that of figure 12

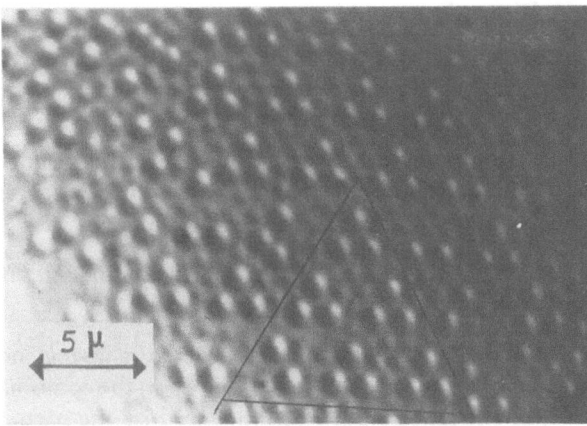

Fig. 15. A pattern observed in a 6500 A–3100 A mixture. Trigonal clusters of large particles are remarkable; the typical pattern is seen in the area surrounded by a drawn triangle. This photograph was taken from a Video picture

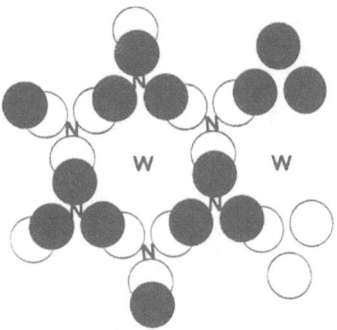

Fig. 17. Observed mode of superposition of hexagonal net planes formed by the large particles in the structure (cf, fig. 16). Black and white circles represent the first (A) and the second (B) planes respectively

latter is periodically (along C-axis) squeezed; it is rather to be said a chain of holes than a tunnel. Each of the W-tunnels contains a package of a large number of small particles, while a N-tunnel accommodates a series of small particles, one in each hole.

The fashion of the packing of small particles in a W-channel was surveyed by carefully forwarding the microscope focus. Due to poor visibility, the obtained information was limited. In these cases a knowledge of packing of hard spheres is useful, because latex particles can be approximated by hard spheres (with

an effective radius that is little larger than the actual particle radius.) From the view point of sphere packing, it was inferred that the pattern observed on the small particles in the W-tunnel must represent an icosahedron with one particle in its center. Furthermore, the whole structure must also represent a state of sphere packing; by model construction-experiment using two sets of plastic spheres, the whole structure was concluded to be as shown in figure 18a.

In the figure, a triagonal cluster of small black circles lies in A-plane and form the top facet of an icosahedron. Other circles around the black circles, which are either half toned or drawn by broken lines, form the side facets of the same iscosahedron. Circles forming the bottom facet are not seen and lie in B-plane. Other small spheres (broken line in N-tunnels) form linear chains along C-axis. Icosahedrons in a W-tunnel form a linear linkage along C-axis; one icosahedron in which shares its top and bottom facets with its neighbors above and below respectively.

The unit cell of this structure is the hexagon shown in figure 18b. It contains three planes of large spheres (A-, B- and A-planes) and two icosahedrons of small spheres sharing one facet. It consists of 6 large spheres and 24 small spheres producing a composition of AB_4. Figure 18b shows a photograph of the model constructed by two sets of plastic spheres, which consists of A- and B-planes of large spheres and small spheres between them.

The spheres large and small in figure 18 are in face contact (small spheres forming the facets of an icosahedron contact to the cetering sphere but not with each other). This is, of cause, not the actual state. The spheres represent the imaginary particles (with the effective radii) approximating the latex particles. Actually, the latex particles are smaller and keep

Fig. 16. The complete picture of pattern in the photograph in figure 15. Small particles drawn by thin lines were seen clearly when the focus was forwarded slightly; they must be situated little inside the structure (cf. fig. 18)

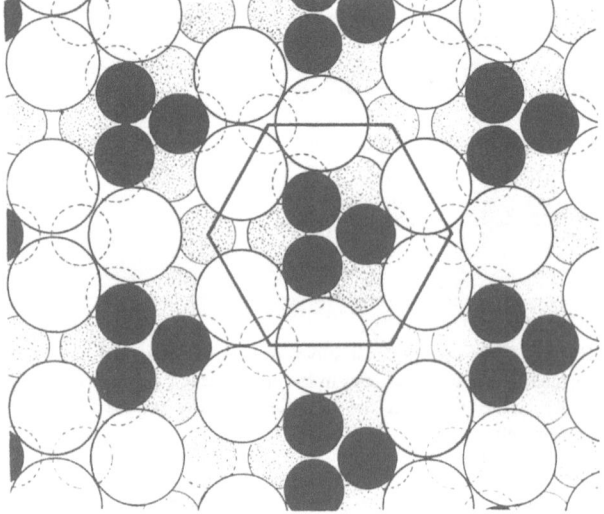

a

b

Fig. 18. (a) Picture of AB$_4$ structure, as a packing of two sets of spheres. Small particles in a *W*-tunnel form an icosahedron. (b) Model of AB$_4$ structure constructed by two sets of plastic spheres. Respective diameters of the spheres are 5 and 3 cm

significant distances between them by the repulsive interaction, and are arranged as seen in the photograph of figure 15.

The radius ratio obtained from figure 18a is near 0.6, and the two sets of plastic spheres in figure 18b have diameters of 5 cm and 3 cm respectively. This value is larger than the actual ratio 3100/6500 = 0.48. Detailed explanation will be given in part II. The exact symmetry type of this structure in terms of space group expression is *p* 6 mmc [17]. The authors have not yet found the counter part of this structure in the real atomic systems.

IV. Discussion

The result of the experiment is summarized as follows:

1) A binary mixture of monodisperse latices undergoes phase transition from disordered to ordered state, and produces a state of phase separation.

2) The ordered phase has a higher volume fraction than that of the coexisting disordered phase.

3) The lattice type of the ordered phase is not unique but multifarious depending upon the combination of the latices, but not influenced by the particle number ratio of the component latices. This strongly suggests that the lattice type is determined by the mode of packing of the two sets of hard spheres.

4) In the state of phase separation, two kinds of ordered phases can coexist together with the disordered phase. It looks like that 'three' is the maximum number of phases to coexist in accordance with the phase rule.

5) For every kind of lattice structure, except AB$_4$ which is pending, we can find its atomic counterpart in alloy system.

The result above mentioned manifests, again, the excellence of colloids as a model system for the study of statistical physics of concentrated systems. Considering the difficulty of computer study of this problem, the merit would be enormous.

As to the phase transition property, the behavior of monodisperse latices was explained by involking the concept of Kirkwood-Alder transition [10], whereby the latex particles being approximated by hard spheres. For binary systems too, we could expect the parallelism between hard sphere and latex systems.

Although we do not know exactly the behavior of a concentrated binary hard sphere system, we can presume its general tendency, as an extension of the Alder transition. The presumption is that it will undergo a phase transition from disordered to ordered state upon increase of the particle concentration, and shows a phase separation whereby the deposited ordered phase is more concentrated than the coexisting disordered phase.

This presumed behavior is in accordance with the above summarized observations on the binary latices. Employing the concept of sphere packing as the means of analysis, we can proceed further into the problem of the stability of the ordered phases, which will be mentioned in part II.

Acknowledgement

The authors would like to express their thanks to Dr. M. Iwata and Dr. T. Ishii in the National Institute for Research on Inorganic Materials for their valuable advice and kind cooperation.

Appendix

Determination of the structure formed in 4500A–2800A mixture: Because of the complexity of the patterns, the structure was difficult to deduce from the observed result. The determination was made by collating the patterns with lattice structures of some of well known alloys. An alloy that was picked up for the collation was one of the so called Laves phase. The reason for this preference was that, this alloy has an atomic radius ratio (of the component metal atoms) of 0.8, which is very near the effective radius ratio (cf. Part II) of the present mixture. Laves phase includes three classes, $MgCu_2$, $MgZn_2$, and $MgNi_2$. At last, $MgCu_2$ was found to fit with the observed patterns.

$MgCu_2$ structure is characterized by the diamond lattice formed by the large particles (cf. fig. A-1), each of its interstices accommodating four small particles (forming a tetrahedron with its basal plane parallel to (111) plane of the diamond lattice, but they are not shown in the figure). When this structure is seen from [111] direction, it is a stack of net planes of hexagonal symmetry, each consisting of one kind of particles, large or small. It is shown in figure A-2. The unit consists of 12 net planes (they are numbered for the convenience for the explanation in the following).

In order to identify the structure, observation must be made over at least 12 net planes that appear one after another as the microscope focus is forwarded. Actually, however, it was impossible to see small particles situated far from the window surface, because the optical path was disturbed by the particles lying in front. The identification, thus, was made by

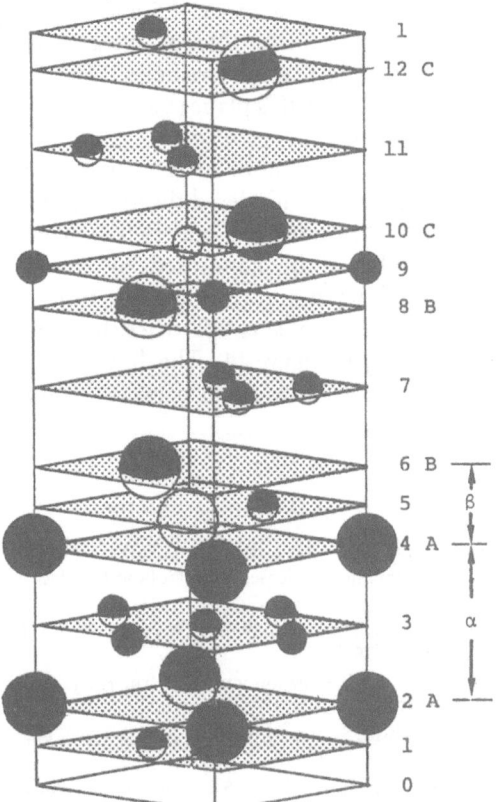

Fig. A-2. Hexagonal expression of $MgCu_2$ structure. Net planes of the large particles are marked A, B and C

observing the large particles that were supposed to form the diamond lattice.

The diamond lattice is characterized by the mode of stacking of the net planes, *AABBCCAA* . . ., as shown in figure A-2, and by the large distance between two adjacent net planes of the same mark, such as between A and A (indicated by α in the figure), which is three times as large as that between A and B (indicated by β in the figure); actual distance between A and A was measured to be from 5000 A to 8000 A depending on the condition.

The diamond structure of the large particles was ascertained by visual observation through microscope on the manner of staggering of 14 net planes of the large particles. Photograph of figure 12 is a typical one; it was obtained by focusing on β- part of figure A-2.

Three net planes are seen at the same time, because the distances between them are very small (~ 2000 A). The primitive cell is shown by the drawing and marked in the photograph; the particles that are respectively on the three net planes are

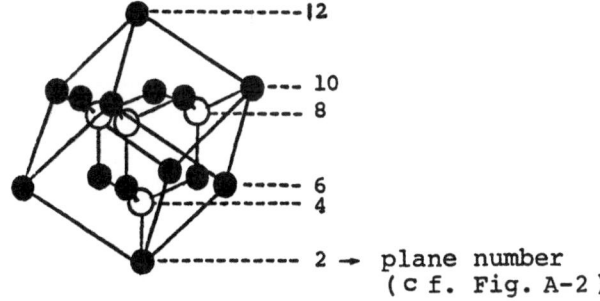

Fig. A-1. The diamond lattice fromed by the large particles in $MgCu_2$ structure. Small particles are not shown; they are packed in the interstices of the structure

marked by the 'net plane number' 4,5 and 6. Photograph in figure 13 is plane No. 3, which shows kagome pattern of the small particles. The large particles above and below this plane (No. 4 and No. 2) are seen rather clearly.

At last, it is to be mentioned that there was a complication that greately hindered the identification; net plane No. 0, the nearest plane to the window surface, which should be composed of the large particles if the whole structure is $MgCu_2$ type, was almost always consisted of small particles. It is shown by the photograph in figure 14. The large particles must have been replaced by small particles owing to some surface effect.

References

1. Onsager, L., Ann. N. Y. Acad. Sci., **51**, 627 (1949).
2. Hachisu, S., Kobayashi, Y., Kose, A., J. Colloid Interface Sci., 42. 342 (1973).
3. Brown, J. C., Pussy, P. N., Goodwin, J. W., Ottewill, R. H., J. phys. A, **8**, 604 (1975).
4. Nieuwenhuis, E. A., Pathmanohoran, C., Vrij, A., J. Colloid Interface Sci., **81**, 196 (L(83).
5. Takano, K., Hachisu, S., usw. J. Colloid Interface Sci., **66**, 124 (1978).
6. Krieger, I. M., O'Neil, F. M., J. Am. Chem. Soc., **90**, 3114 (1968).
7. Luck, V. W., Klier, M., Wesslau, H., Die Naturwissenschaften, **14**, 485 (1962).
8. Kose, A., Hachisu, S., J. Colloid Interface Sci., **46**, 460 (1974).
9. Hachisu, S., Kobayashi, Y., J. Colloid Interface Sci. **46**, 470 (1974).
10. Wadachi, M., Toda, M., J. Phys. Soc. Jpn., **32**, 1147 (1973).
11. Kose, A., Ozaki, M., Takano, K., Kobayashi, Y., Hachisu, S., J. Colloid Interface Sci., **44**, 330 (1973).
12. Sanders, J. V., Murray, M. J., Nature, **275**, 201 (1978).
13. Hachisu, S., Yoshimura, S., Nature, **283**, 188 (1980).
14. Pearson, W. B., Crystal Chemistry and Physics of Metals and Alloys, Wiley-Interscience, London, 1972.
15. Sanders, J. V., Phil. Mag., **42**, 705 (1980).
16. Hasaka, M., Nakamura, H., Oki, K., J. Japan Inst. Metals, **45**, 347 (1981).
17. Kato, K., Private Communication (1983).

Received April 6, 1983;
accepted June 6, 1983

Authors' addresses:

S. Yoshimura
Izumi High School
Takane-cho 875-1, chiba-shi
280-01 Japan

S. Hachisu
15-3, minamisawa 5
Higashikurume
Tokyo 203 Japan

Topological features of oily streak defects and parabolic focal conics in dilute lamellar lyotropic liquid crystalline phases*)

W. J. Benton and C. A. Miller

Department of Chemical Engineering, Rice University, Houston, Texas

Abstract: Conventional polarizing microscopy and Hoffman modulation contrast optics were used to investigate the defect structures of lamellar liquid crystalline phases in dilute anionic surfactant systems, i. e., those containing at least 85% water. Based on the results obtained an improved model of the so-called oily streak defects is proposed. It features two rows of edge dislocations which have wavy configurations due to buckling. Some defects in parabolic focal conic arrays are described.

Key words: liquid crystals, lyotropic, lamellar phase, oily streak, defect, dislocation, disclination.

Introduction

The nature of the defects which occur in lamellar liquid crystalline phases has been investigated both for thermotropic smectic phases and for lyotropic phases containing amphiphilic compounds and water. In this paper we are concerned with phases of the latter type which are of interest because of their relevance to certain phenomena in detergency, emulsion technology, biological systems, and enhanced oil recovery.

Many years ago, Friedel and Grandjean [1] recognized that in most systems the energy required to change the distance between layers in a lamellar phase is much greater than that required to bend layers. Hence they proposed that focal conic structures where the layers take the form of uniformly spaced Dupin

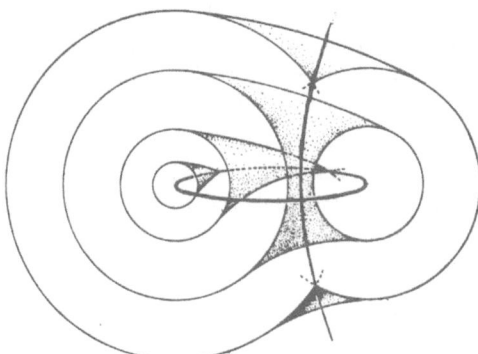

Fig. 1. Focal conic showing hyperbola passing through a focus of ellipse (after Y. Bouligand)

cyclides should be common. Figure 1 shows a view of such a focal conic structure.

It can be shown that the points of discontinuity in slope of individual layers (cyclides) in figure 1 lie on the hyperbola shown. Other discontinuities not visible in the diagram lie along the ellipse shown. The hyperbola passes through one focus of the ellipse and has one of the apices of the ellipse as its own focus. Since the hyperbola and ellipse are defect lines where discontinuities in optical properties occur, they can be observed by optical microscopy and focal conic domains thus identified. The defect lines are also sites of increased strain energy. Further details of focal conic structures may be found in the papers of Bouligand [2, 3].

Another arrangement where defects occur in such a way that the spacing between layers remains uniform is the parabolic focal conic (PFC) structure of figure 2a which was first proposed by Rosenblatt et al. [4]. Here the discontinuities in the slope of individual layers lie along interlocking parabolae as shown in figure 2b. Domains where the PFC arrangement exists can also be identified by optical microscopy, as is discussed further below. It has been seen in thermotropic smectic phases [4] and in lyotropic liquid crystals containing surfactant [5], lipid-water systems [6] and was recently found in a system containing a graft copolymer, oil, and water [7, 8].

Yet another type of defect which occurs in lamellar phases is the edge dislocation. In a simple form (fig. 3a) the spacing between adjacent layers is not everywhere uniform. This energetically unfavorable characteristic is not found in the focal conic and PFC

---

*) Dedicated to the memory of Professor Dr. B. Tamamushi.

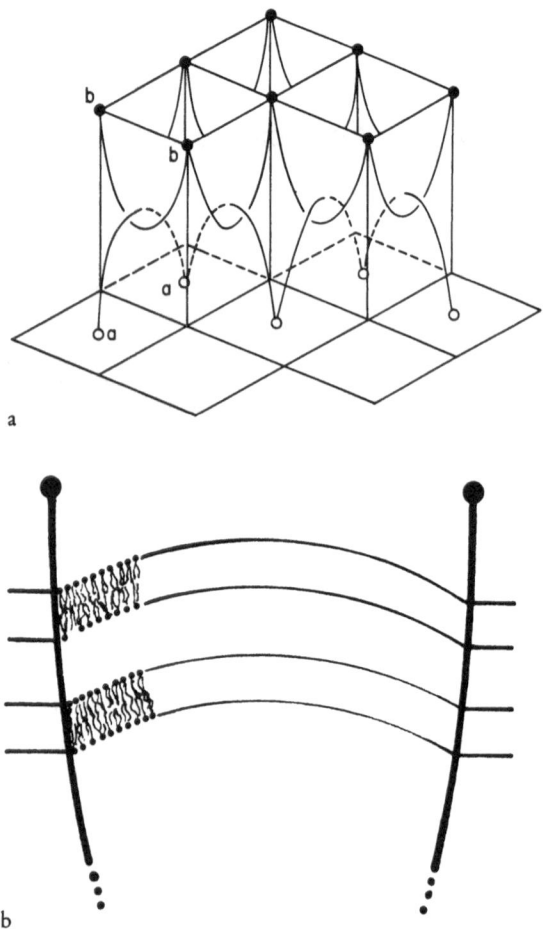

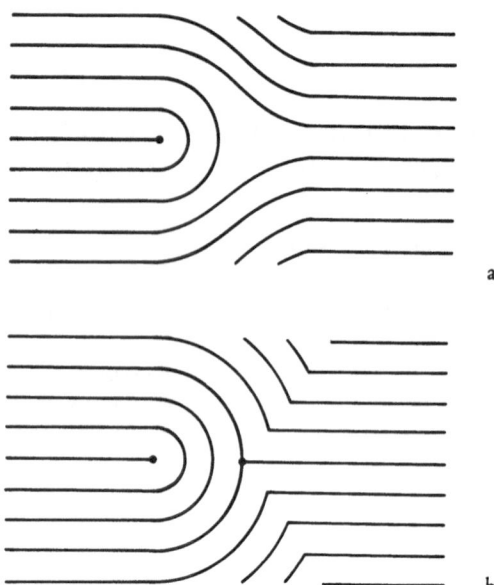

Fig. 3a). Edge dislocation in lamellar structure. b) Edge dislocation consisting of two disclinations of opposite sign (after Kleman and Friedel)

Fig. 2a). Parabolic focal conic structure (after Rosenblatt et al.). b) Curvature of bilayers in PFC

arrangements. Kleman and Friedel [9] have proposed the alternate edge dislocation structure of figure 3b where nonuniform spacing is minimized but where new defect lines have been introduced connecting points of discontinuity in layer slope. It is evident from figure 3b that the dislocation consists of two disclinations of opposite sign.

Individual dislocations can sometimes be seen in lamellar phases using optical and electron microscopy. Williams and Kleman first observed edge dislocation arrays in the thermotropic smectic *A* phase of *p*-cyano benzylidene-*p*'-octyloxyaniline by phase contrast microscopy [10]. Recently they were observed in a thermotropic smectic phase near the *A* to *C* transition of *n*-undecyl-*p*-azoxy-α-methylcinnamate by a novel technique employing oblique filtering of transmitted polarized light [11–13]. Interaction between nearby dislocations has been discussed by Pershan [14].

Of greater interest here are defects called "oily streaks" which are commonly seen with optical mi-

croscopy in lamellar phases. Their appearance is discussed extensively below. An early model proposed for oily streaks was linear twin arrays of focal conics [1]. In recent years, however, Kleman and co-workers [15, 16] and Asher and Pershan [17] have presented evidence that edge dislocations run longitudinally along oily streaks. Indeed, the former group has proposed that an oily streak is made up of an array of multiple parallel edge dislocations [18].

In this paper we report observations with optical microscopy in lamellar phases which exist in dilute anionic surfactant systems, i. e., containing at least 85% water. In most of the systems a short chain alcohol and sodium chloride are also present. Information on phase behavior in these systems including the composition range over which the lamellar phase exists is given elsewhere [19]. Because of their high water content these phases, which consist of many parallel surfactant bilayers separated by thick water layers (> 10nm), are much less viscous than the concentrated lamellar phases studied by previous workers.

Two features of our experimental technique are particularly noteworthy. In the first place, the samples are observed while in sealed cells of rectangular cross-section and some 100 μm to 200 μm thick. Since no evaporation is possible, time dependent changes and structural reorientation are readily observed. In the second place, conventional polarizing microscopy is supplemented by the use of Hoffman modulation contrast (HMC) optics. The latter amplifies gradients in optical density and thus facilitates observation of structural characteristics of the sample.

In the following sections we describe observations of both oily streaks and the PFC structure. The information obtained with HMC optics allows us to extend previous knowledge concerning the positions and configurations of the arrays of edge dislocations within oily streaks. The observations of the PFC structure show defects in the structure which are more complicated than those considered previously.

Materials

Dilute homogeneous lamellar phases were formed from ternary and quarternary mixtures consisting of pure anionic surfactants, short chain alcohols, and water (or brine). Similar phases were made with several commercial petroleum sulfonate surfactants, which are complex mixtures of anionic surfactants. Partial phase diagrams for these dilute systems (85% or more water content) have been reported elsewhere along with details of purities, solution mixing procedures and criteria for equilibrium (19). These results showed that a general pattern of phase behaviour exists for both pure and commercial systems. In brief, lamellar phases could be formed from isotropic micellar solutions by making the surfactant-alcohol mixture more hydrophobic. An increase in salinity or an increase in the concentration of an oil-soluble alcohol could effect this transformation.

Ternary systems were made to form lamellar phases with sodium octanoate-n-decanol-water [20, 21, 22] and sodium di-2-ethylhexyl-sulfosuccinate-water-sodium chloride [22]. Quaternary systems studied were sodium dodecyl sulfate-n-hexanol-water-sodium chloride [23, 24] and sodium 4-(l'-heptylnonyl) benzene-sulfonate-n-propanol-water-sodium chloride [25]. Various petroleum sulfonates, which contained considerable amounts of unreacted oils, inorganic salts and water as impurities, were mixed with short-chain alcohols and brine in appropriate amounts to obtain the lamellar phase.

The short chain alcohols and salts used in the experiments were all reagent grade. The water was deionized and triple distilled. Glassware and cells were cleaned first with chromic acid solution and then rinsed thoroughly with distilled water and dried. Further information on the sources and purity of the surfactants used and the compositions studied in the various systems may be found elsewhere [19].

The solutions were mixed as 10 gram or 10 ml. systems in teflon-capped, flat-bottomed test tubes. These were viewed macroscopically between crossed polarizers at constant temperature [19] to determine their birefringent textures and the number of phases present. In this way the composition ranges of lamellar phases in the various systems were determined.

Optical microscopy and sample techniques

Transmission optical microscopes equipped with both polarized and Hoffman modulation contrast optics were used for sample observation. The microscopes (Nikon PoH and Optiphot-Pol) were equipped with 35 mm cameras and automatic exposure control (Nikon HFM and UFX systems respectively).

Traditionally transmission polarized light microscopy has been a primary means of identifying and interpreting liquid crystalline

structures from discontinuities in the birefringent textures [26]. Hoffman modulation contrast optics (HMC) [27], on the other hand, produces a phase gradient across the sample by means of a series of carefully aligned slits in the condenser and objective parts of the microscope. With this arrangement gradients in the optical density of light passing through the samples are amplified. As a result, observation of structural features such as defects is facilitated. The three-dimensional structure within a sample can be reconstructed from the images obtained at a series of planes in register but at different depths.

Temperature control for samples was obtained with a Mettler FP-52 controller and a FP-5 microscope hot-stage attachment modified for temperatures close to ambient and below by a flow of cooled nitrogen gas [28].

Solutions were drawn into optical rectangular cells (Vitro-Dynamics Inc.) by capillary action and then sealed with a two-part epoxy adhesive which had a cure time of five minutes. The cells were then attached to microscope slides. Image intensity and contrast could sometimes be enhanced by imbibing an immersion fluid into the space between the microscope slide and cell. The cells had pathwidths ranging from 10 μm to 500 μm with lengths of 25 or 50 mm. Generally cells of 100 μm or 200 μm pathwidth were used with objectives of 20 and 10 times magnification respectively.

The sealed optical rectangular cell was found to be preferable to the usual microscope slide and cover slip technique for the weakly birefringent dilute lyotropic solutions (85% or more water content) of interest in this work. With a closed cell samples could be observed for long periods of time (several months in some cases) and as a function of temperature without evaporation. The uniform geometry of the closed cell ensured consistency for comparative purposes from sample to sample and also eliminated effects of external forces associated with the meniscus between a slide and cover slip.

Results

Figure 4 is a series of photomicrographs which illustrate development of the defect structure with time in a dilute lamellar phase containing 3.6 wt % sodium octanoate, 2.8 wt % n-decanol, and 93.6 wt % water. Taken only 0.12 hour after filling of the cell, figure 4a shows a birefringent texture which occupies nearly the entire field of view and which has a general directionality from left to right, the direction of flow during filling. The cell is aligned so that its longitudinal axis is at 45° from the polarizing axes.

Figures 4b-4e, in registration with figure 4a, illustrate in time lapse changes which occur over the next 22 hours. The birefringence becomes concentrated into well defined filaments known as oily streaks. These have fairly uniform birefringence and are separated by dark regions where the surfactant-alcohol bilayers are believed to be parallel, i. e. planar with the top and bottom surfaces of the cell. After 22 hours the relaxation process takes place more slowly (figs. 4f–4h), but it is clear that the dark regions grow in extent at the expense of the oily streaks. At still later times the planar regions between the oily streaks develop a regular cellular appearance (fig. 5). This

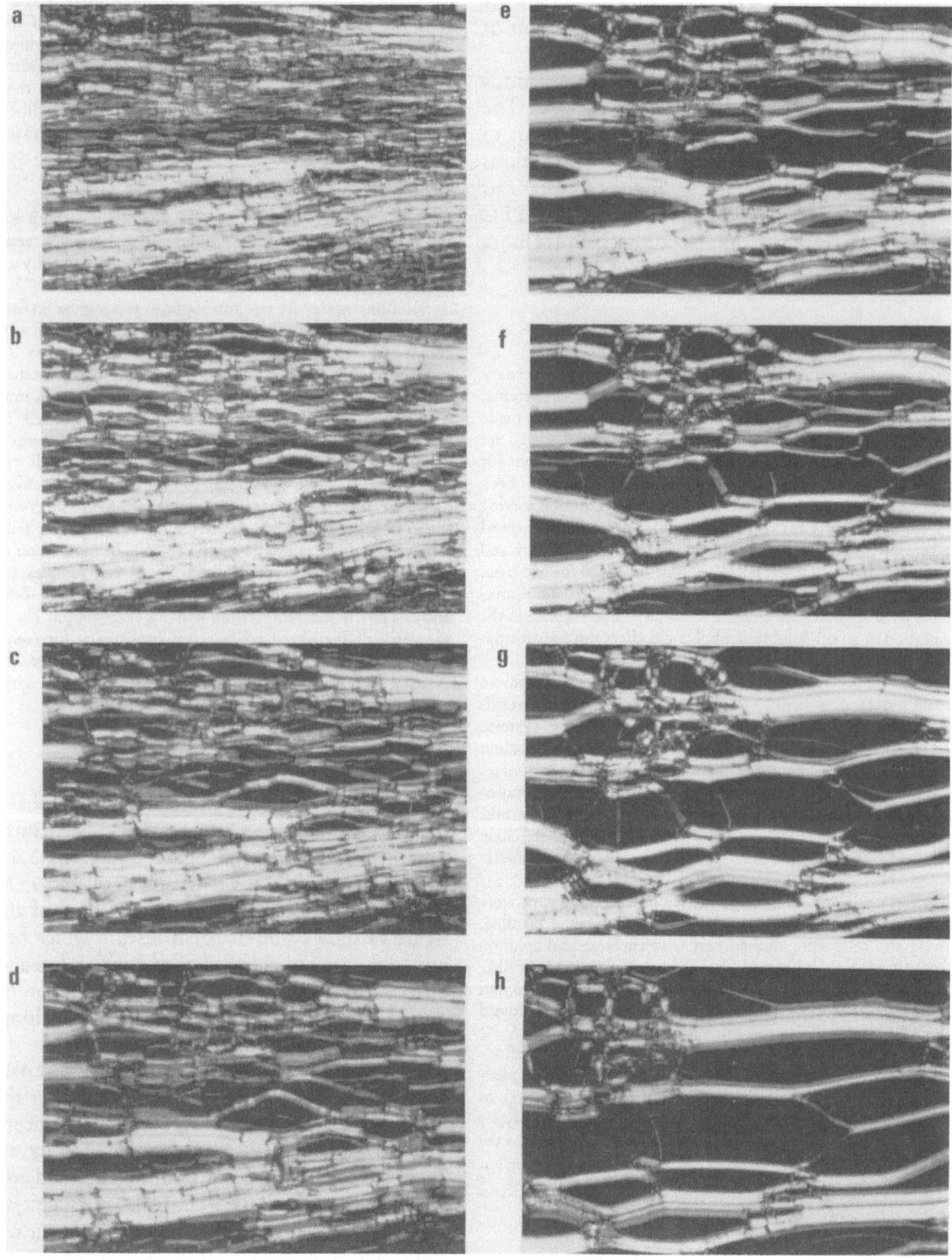

Fig. 4a). Time lapse series of lamellar phase relaxation process (a) = 0.12, (b) = 1.5, (c) = 3.0, (d) = 9.0, (e) = 22.0, (f) = 45.0, (g) = 75.0, (h) = 124.0 hours. Polarizers at 45° to longitudinal axis of optical cell

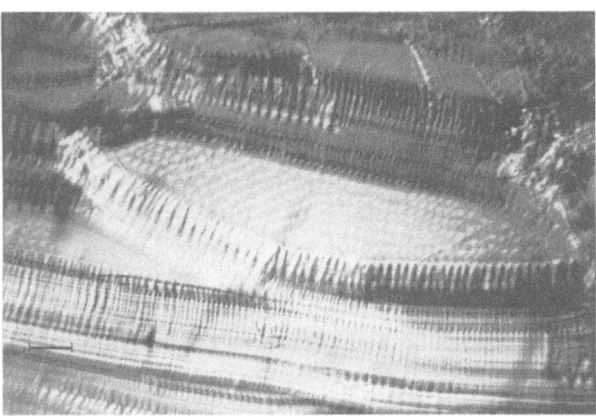

Fig. 5. Lamellar phase showing PFC texture and oily streaks. Polarized light +¼ with HMC optics superimposed

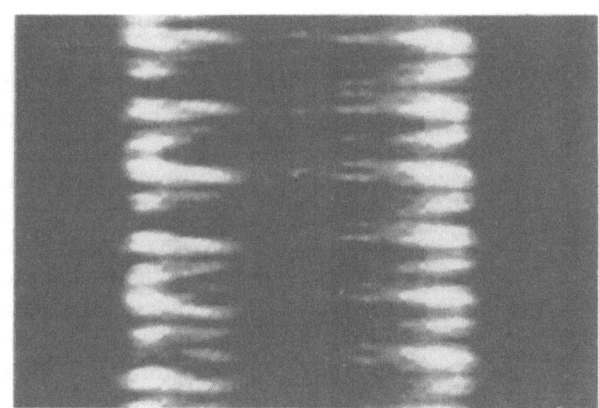

Fig. 6a

texture is characteristic of the parabolic focal conic (PFC) arrangement mentioned previously. We note that textures characteristic of the ordinary focal conic defects of figure 1 do not appear in figures 4 and 5 and indeed are only rarely seen for the dilute lamellar phases of interest here.

We have observed basically the same process in several other dilute lamellar phases containing anionic surfactants, short-chain alcohols, and water (or brine). The rate of development of the oily streaks, planar regions, and PFC structure differs from system to system, however. As might be expected, the rate is greatest in systems where viscosity is lowest. Indeed, even in the same system the rate increases when the composition is changed in a direction to lower the viscosity. We note that, owing to the low surfactant content, some of the lamellar phases we have studied have viscosities of only a few centipoise [29].

Fig. 6b

Oily Streaks

We have studied the internal structure of oily streaks using both polarizing and HMC optics. Figures 6a and 6b, taken in registration, are views of one portion of an oily streak using the two techniques. The image in figure 6b obtained with HMC optics shows a remarkable array of longitudinal defects within the oily streak, most of which appear as undulating lines. More lines occur near the edges of the oily streak than near its center. Several adjacent lines may have the same wavelength and more or less uniform spacing as in the band of seven lines on the left side of the figure. In this particular case the wavelength and spacing are about 30 μm and 10 μm respectively. Note, however, that the wavelength is

Fig. 6c

Fig. 6. a) Section of oily streak. Polarized light. In register with figure 6b. b) Section of oily streak upper array. HMC optics. Bar = 25 μm. c) Section of oily streak. Lower array. HMC optics. In register with figures 6a and 6b. Bar = 25 μm

quite different for other lines in other portions of the oily streak. As discussed below, we consider that individual lines are edge dislocations in the lamellar structure and that they buckle, probably due to compressive stresses.

In the image taken with polarized light, figure 6a, the polarizers are positioned perpendicular and parallel to the longitudinal axis of the oily streak. The wavelength of the periodic birefringent pattern at the left is exactly twice that of the seven lines in the band seen in figure 6a. Indeed, portions of the lines within this band are visible in figure 6b as thin dark lines crossing the bright regions. Little birefringence is seen in the central portion of the oily streak, but another birefringent pattern appears at the right. Its wavelength differs slightly from that of the pattern on the left, at least in the lower portion of the figure. Note that the outermost lines in figure 6b on both sides of the oily streak are outside the corresponding regions of birefringence observed in figure 6a.

Figure 6c is still another view of the same oily streak but with the focus about 15 μm above the bottom surface of the rectangular cell, which is 100 μm thick. In contrast, the focus in figures 6a and 6b is about 15 μm below the top surface of the cell. The pattern of lines in figure 6c is much like that of figure 6a. It is not identical, however, in the sense that individual lines at the top are not necessarily in the same positions as lines at the bottom. Even for pairs of lines which do nearly coincide, the wavelengths in the longitudinal direction generally differ. We note that only occasional defects were seen with Hoffman optics in the central portion of the cell and none very near the top and bottom surfaces. Thus, any explanation of the oily streak must account for the existence of two arrays of undulating linear defects, one somewhat below the top of the cell, the other somewhat above the bottom.

When the lines in figures 6b and 6c are closely examined, a variety of configurations is found. Some are nearly straight, others sinusoidal, still others straight along some portions of their length and undulating elsewhere. Fairly common is deviation from sinusoidal behaviour with the "crests" becoming narrow and the "troughs" broad. The crests even develop into cusps in some cases with the points of the cusps prominent in the image. In other oily streaks we have seen some individual lines terminate as well as more complex patterns of undulation with a periodicity of short wavelength superimposed on one of long wavelength along a given line.

As noted previously, patterns in which several lines are parallel and equally spaced are frequently seen. We note that such patterns e. g., the one on the left side of figure 6b, may be described in terms of the symmetry of bands [30].

Based on observations of oily streaks in several anionic surfactant systems with both HMC and polarizing microscopy, we believe that periodic longitudinal structure similar to that shown in figures 6a–c is a general property of well developed oily streaks in lamellar phases. Figure 7, which shows several oily streaks, illustrates this point. Here the polarizing axes are perpendicular and parallel to the longitudinal axis of the optical cell. The observations of previous workers seem also to be consistent with this conclusion [17, 18].

A further observation in agreement with previous workers and clearly illustrated in figure 7 is the conservation law of Kirchoff where the width of an oily streak leaving a node is the sum of the widths of the oily streaks entering the node [15, 17].

A particularly intriguing phenomena is seen in figure 8. Here it appears that a transverse defect is advancing along the oily streak to the left, removing several longitudinal defects (edge dislocations) along part of its length in such a way that it induces a twisting and leaves behind a single or possibly a pair of zig-zag lines. Although further study of such behavior is needed, we consider it likely that the zig-zag represents a linear array of PFC's but entirely within the oily streak. The parabolic defects characteristic of the PFC arrangement are seen as lines connecting adjacent points along the zig-zag line. As figure 8 is focused near the top surface of the cell, these parabolae face upward and terminate near the top surface. Faintly visible is an interlocking zig-zag line which apparently represents downward facing parabolae

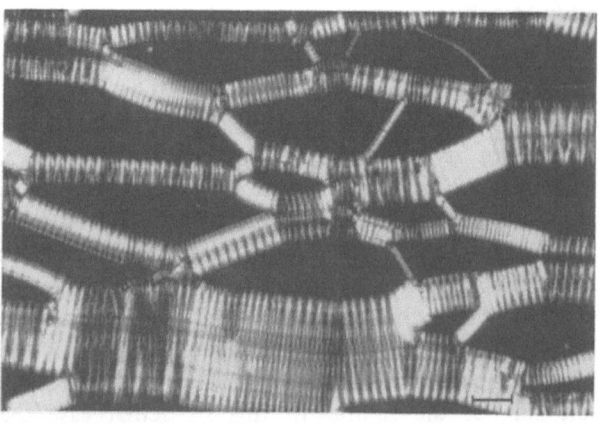

Fig. 7. Oily streak network and planar regions. Polarized light. Bar = 117.57 μm

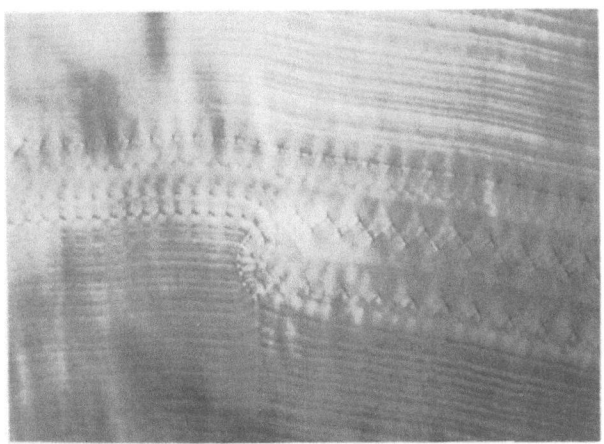

Fig. 8. Transverse defect in an oily streak with linear array of PFC's. HMC optics

which terminate near the bottom surface. For clarity a diagram of the two zig-zag lines is given in figure 9a. We emphasize that this behaviour appears to be a special but not unique case. Other lines we have seen with something of a zig-zag appearance seem not to be associated with the PFC structure but instead are edge dislocations with troughs joined together at cusps as shown diagrammatically in figure 9b and observed in figures 6b and 6c. Below the node of the transverse defect in figure 8 the edge dislocations are not removed but may be seen to flare inward towards the node.

Parabolic focal conic arrays

As indicated previously and as illustrated in figure 5, the planar regions which develop between the oily streaks during the annealing process ultimately transform into parabolic focal conic (PFC) arrays. The process by which these arrays form is not entirely clear. Figures 10a and 10b, taken in register with focus near the top and bottom surfaces of the cell respectively, show the boundary between an oily streak and a

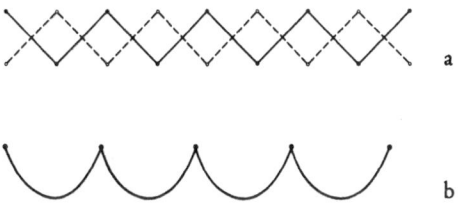

Fig. 9a). Schematic of linear array of PFC's. ●—● = upward parabolae ○——○ = downward parabolae. b) Zig-zag of pointed cusps and troughs, possible catenary band

developing PFC array. It appears that the PFC structure begins to form at the edge of the oily streak and grows outward into the planar region. The parabolae of the PFC structure do not extend all the way to the top and bottom cell surfaces but terminate at the same depths as those where the two arrays of edge dislocations are seen in the oily streaks.

Two features of figures 10a and 10b are striking. One is that the PFC cell size in the row adjacent to the oily streak is different from the characteristic longitudinal wavelength in the oily streak itself. A possible explanation of this behavior is that both longitudinal and transverse periodicity in the oily streak influence cell size in the PFC array. The second is the rapid decrease in cell size with distance away from the oily streak. Associated with the change in cell size are many defects in the cellular array. Figure 11 is a diagram of the interlocking cellular patterns of PFC texture for regions near the top and bottom surfaces.

As we have discussed elsewhere [31], the node of an isolated edge dislocation in a PFC array, i. e., a position where an additional row of cells begins, is characterized by two disclinations of opposite sign. Points in a regular array at both the top and the bottom of the cell have four-fold rotational symmetry, as figures 10a and 10b show. At an isolated dislocation site there is a single point with three-fold symmetry at the top (bottom) surface and a single point with five-fold symmetry at the bottom (top) surface. As a result, the cellular pattern exhibits a single pentagon in the otherwise square array at the top (bottom) surface and a single triangle at the bottom (top) surface.

In figures 10a–b and 11 so many extra rows are introduced so rapidly that more complex patterns are seen, e. g. two pentagons with a common side (fig. 10a) or two triangles with a common vertex (fig. 10b). We especially call attention to the arrangement near the middle of figure 10a and at two nearby locations where a pentagon and a triangle have a common vertex with the result that two new rows of cells are initiated simultaneously. The cell in the corresponding position of figure 10b is also very interesting. It has four sides but bears little resemblance to an ordinary square cell. Its top portion is like the triangle in the left portion of the same figure but its bottom portion is like the pentagons of figure 10a. In the lower right portion of figure 10a are three pentagons with a common vertex. A triangle is seen at the same location on the bottom surface (fig. 10b).

A well developed PFC structure is shown in figure 12. Quite striking is a secondary network of curves which seem to be grain boundaries in the PFC array and which may be associated with edge disloca-

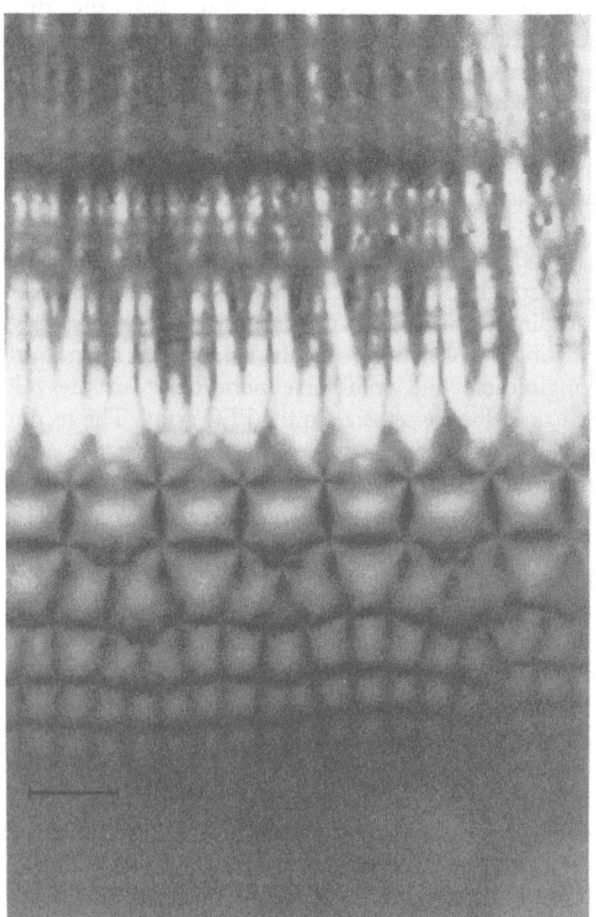 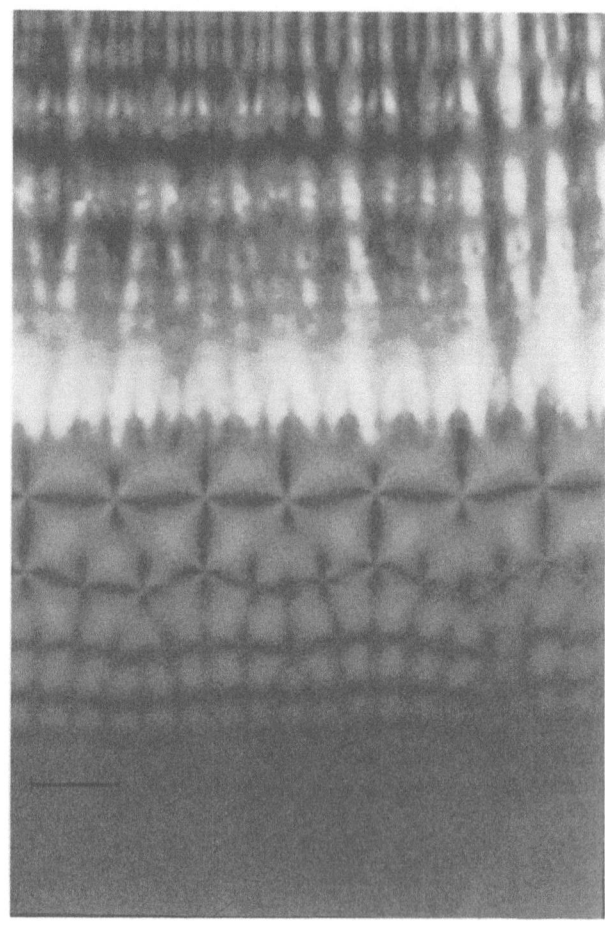

Fig. 10a). Oily streak/PFC interface, upper boundary polarized light, Bar = 35.28 μm. b) Oily streak/PFC interface, lower boundary polarized light. Bar = 35.28 μm

tions. Here too there are interesting defects within the array. The circled region shows two pentagons with a common side arranged such that two new rows in perpendicular directions are introduced at the same location. Figure 13 is a diagram of the interlocking cellular pattern shown near this defect along with the pattern which has two triangular cells required on the other surface of the cell.

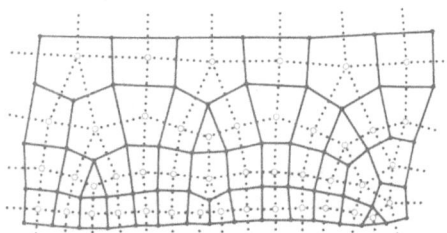

Fig. 11. Tracing of PFC arrays of figures 10a and 10b showing two-dimensional tessalation and defects ●——● = upward facing parabolae ○— — —○ = downward facing parabolae

A birefringent pattern is found along the edges of the optical rectangular cell as seen at the top of figure 12. It resembles the pattern seen near the edge of an oily streak (of fig. 6a). Two rows of PFC cells can be seen adjacent to part of this edge texture. In contrast to figures 10a and 10b, it appears that cell size does correspond rather closely to edge periodicity with two cells having approximately the same width as one birefringent lobe.

Discussion

The photomicrographs discussed above provide considerable information about the annealing process for lamellar liquid crystalline phases although, as we shall see, they also raise questions for which answers are not yet available. As the flow is basically parallel during entry of the lamellar phase into the rectangular cells, the bilayers immediately after filling should be

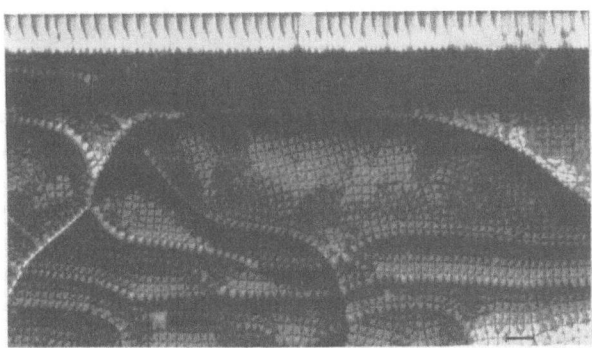

Fig. 12. Well developed PFC array. Polarized light. Circle indicates region shown schematically in figure 13. Bar = 100 μm

roughly parallel to the top and bottom surfaces of the cell. Many structural defects can be expected, however, including edge dislocations of the type shown in figure 3 or possibly long, cylindrical defects consisting of a single lamella wound into a spiral, as observed by Papahadjopoulos et al. [32].

Lateral and vertical motion of dislocations (glide and climb) continues after the flow ceases as the solution relaxes. The motion of dislocations takes place in such a way that strain energy is reduced. We note that glide presumably occurs more easily than climb because the latter requires that some layers be broken and reformed [18].

At some locations in the cell where the initial density of dislocations is high, a reduction in strain energy can be achieved if the dislocations move to take up positions shown by the short dashes in figure 14. Here one row of dislocations has formed in the upper part of the cell with another in the lower part of the cell. Each short dash actually represents two disclina-

tions of opposite sign. Except for the regions near the top and bottom surfaces of the cell the bilayers undulate, forming tilt walls where dislocations can fit in with less distortion than that caused by a single isolated dislocation. This arrangement provides a coupling between the two dislocation arrays.

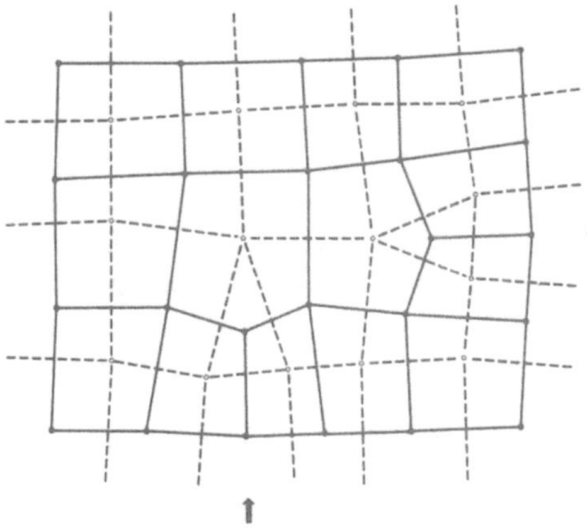

The arrangement of figure 14 is consistent with our observations described above. It explains why the dislocations observed are almost all relatively near the top and bottom surfaces with very few near the center of the cell. It further explains why dislocations near the bottom surface are not directly underneath dislocations near the top surface although the spacing between adjacent lines is much the same at both the top and the bottom. We note that the total birefringence exhibited by an oily streak is, according to this model, due partly to the local curvature of bilayers in the individual dislocations of both arrays, partly to deformation of the bilayers throughout the central portion of the cell, and partly to the longitudinal undulations of the dislocation arrays.

As many dislocations with their extra rows of bilayers move into an incipient oily streak, the constant overall cell thickness dictates some local decrease in thickness of the water layers separating the bilayers. Fluid incompressibility then requires a simultaneous increase in thickness of the water layers in nearby regions where the gliding away of the dislocations leaves the bilayers almost perfectly aligned parallel to the top and bottom surfaces of the cell. In order to prevent water flow from these planar regions into the oily streak, pressure within the oily streak must rise. The long and relatively thin edge dislocations thus

Fig. 13. Schematic of circled area of figure 12 including corresponding downward facing parabolae

Fig. 14. Model of oily streak. See text for details

become subjected to compression. Since the nodes joining individual oily streaks are relatively immobile and may even be anchored to the cell surfaces [17], the amount of longitudinal expansion within any one streak is limited and axial compressive forces develop. These could ultimately lead to buckling, which could produce the observed sinusoidal dislocation lines within the oily streak (figs. 6b and 6c) and the associated periodic pattern of birefringence seen along the longitudinal direction (fig. 6a).

Eventually, the planar regions transform into the parabolic focal conic (PFC) texture. As we have seen, this texture appears to nucleate along the oily streaks forming the boundary of a region and to grow inward until it occupies the entire region. Possibly the dilation within the planar layers mentioned above eventually reaches the point that buckling occurs and the PFC forms. We note that Asher and Pershan [6] attributed formation of the PFC texture in their samples to dilation of the planar regions. Ribotta and Durand [33] and Scudieri et al. [34] also found textures which, in the light of present knowledge, are clearly those of the PFC upon dilation of thermotropic smectic phases.

The proposed model of an oily streak is a development of previous ideas concerning their structure. Indeed, as indicated above, Kleman [18] proposed that an oily streak was made up of parallel arrays of edge dislocations. Our proposed concept of two rows of dislocations near the top and bottom surfaces of the cell is novel, however, and stems from our observations described above. We recognize that in some cases there may be dislocations present in the central part of the cell which have not yet managed to climb to one of the two rows. Such a situation is especially likely in lamellar phases having higher viscosities and more rigid bilayers than those studied here since, as indicated previously, the process of climb involves breaking and reforming some bilayers. We further recognize that the diagram of figure 14 is idealized in that there may be some variation in size among the dislocations present and hence also some variation in the distance between adjacent dislocations in a row.

Also novel is the concept that the dislocations may buckle due to compression although such behaviour seems a plausible conclusion to draw from the undulating lines of figures 6 and 8. We note that while some longitudinal structure within oily streaks has been detected by previous workers, the detailed information provided by HMC optics in views such as figures 6b and 6c was not available to them.

Kleman et al. [18] and Rault [35] suggested that the periodic structure seen along oily streaks was caused

by local transformation into focal conic regions. Such transformation was considered favorable because it would result in more uniform spacing between layers. If it does occur, periodic arrays of ellipses should be visible along an oily streak. As we see individual undulating dislocations instead of ellipses, we propose the alternate mechanism of dislocation deformation by buckling due to compression. We do see cusps, however, where the dislocations appear to break into short segments. Perhaps this breakage does involve some discontinuities in bilayer slope and some increase in uniformity of bilayer spacing. Finally, as indicated above (fig. 8), we occasionally see what appear to be linear parabolic focal conic arrays within an oily streak. These presumably form for the same reason, i. e., an increase in uniformity of spacing.

Our concept of rows of regularly spaced dislocations positioned in undulating layers in such a way as to relieve strain energy is not novel, however. Clark and Meyer [36] and Delaye et al. [37] have shown that such a situation can arise when the layers of a lamellar phase became unstable and buckle upon being subjected to dilation.

We again emphasize that our observations have been restricted to dilute lamellar phases of relatively low viscosity. Further work is required to determine whether our proposed model of an oily streak is valid for more concentrated systems or for smectic thermotropic liquid crystals. The basic mechanism of oily streak formation, including the reduced spacing in the oily streak relative to nearby planar regions produced by dislocation migration, would not appear to be restricted to our systems, however. What does remain to be seen is whether dislocations can climb sufficiently in more viscous systems to form arrays as well defined as those we have observed.

Summary

Observations of lamellar liquid crystalline phases in dilute anionic surfactant systems, especially observations using HMC optics, have revealed additional features concerning oily streak defects, including the existence of two arrays of multiple, parallel edge dislocations. Previous models of oily streaks have been adapted and extended to conform to this new information.

Some complex defects in parabolic focal conic arrays are also described involving two or more edge dislocations at a given position. The basic pattern seen previously for an isolated edge dislocation is maintained in the complex defects. This pattern consists of two disclinations of opposite sign on the top and bottom surfaces of the array, one appearing as a point of three-fold symmetry and the other as a point of five-fold symmetry.

Acknowledgement

This research was supported partly by a grant from the U. S. Department of Energy and partly by grants from Amoco Production Company, Gulf Research and Development Company, Exxon Production Research Company, and Shell Development Company.

References

1. Friedel G. and Grandjean, F., Bull. Soc. Fr. Mineral **33**, 409 (1910).
2. Bouligand, Y., J. Physique **33**, 525 (1972).
3. Bouligand, Y., in: "Dislocations in Solids" ed. F. R. N. Nabarro **5**, 301 North-Holland (1980).
4. Rosenblatt, C. S., Pindak, R., Clark, N. A., and Meyer, R. B., J. Physique **38**, 1105 (1977).
5. Benton, W. J., Fort, T., Jr. and Miller, C. A., SPE Preprint 7579, Pres. at 53rd. Ann. Fall. Tech. Conf. and Exh. of Soc. Pet. Eng. of AIME, Houston, Texas, October (1978).
6. Asher, S. A. and Pershan, P. S., J. Physique **40**, 161 (1979).
7. Candau, F., Ballet, F., Debeauvais, F., and Wittman, J-C., J. Colloid Interface Sci. **87** (2), 356 (1982).
8. Candau, F. and Ballet, F., in: "Microemulsions" ed. I. D. Robb, Plenum Press, New York, (1982).
9. Kleman, M. and Friedel, J., J. Physique 30 C4–48 (1969).
10. Williams, C. E. and Kleman, M., J. Physique **35**, L–33 (1974).
11. Meyer, R. B., Stebler, B., and Lagerwall, S. T., Phys. Rev. Lett. **41**, (20) 1393 (1978).
12. Lagerwall, S. T., Meyer, R. B., and Stebler, B., Ann. Phys. 3, 249 (1978).
13. Lagerwall, S. T. and Stebler, B., in: "Liquid Crystals" ed. Chandrasekhar, S., Heyden Publ. London (1980).
14. Pershan, P. S., J. Appl. Phys. **45**(4), 1590 (1974).
15. Kleman, M., Colliex, C., and Veyssie, M., "Lyotropic Liquid Crystals" ed. S. Friberg, Adv. in Chem. Ser. **52**, 71 (1976).
16. Kleman, M., Williams, C. E., Costello, M. J., and Gulik-Krzywicki, T., Phil. Mag. **35**(1) 33 (1977).
17. Asher, S. A. and Pershan, P. S., Biophys. J. **27**, 393 (1979).
18. Kleman, M., "Points, Lines and Walls, in: Liquid Crystals, Magnetic Systems and Various Ordered Media." Wiley Inter. Sci. Publ. New York (1983).
19. Benton, W. J. and Miller, C. A., in press J. Phys. Chem.
20. Mandell, L. and Ekwall, P., Acta Polytechn. Scand. Chem. Met. Series **74**, 1, 1968.
21. Friman, R., Danielsson, I., and Stenius, P., J. Colloid Interface Sci. **86**(2), 501 (1982).
22. Benton, W. J. and Miller, C. A., Proc. of the Int. Sym. on Surfactants in Solution held in Lund, Sweden, June (1982).
23. Natoli, J. N., Ph. D. Thesis, Carnegie-Mellon Univ. (1980).
24. Natoli, J. N., Benton, W. J., Miller, C. A., and Fort, T., Jr. Submitted to J. Disp. Sci. Tech. (1983).
25. Benton, W. J., Natoli, J. N., Qutubuddin, S., Mukherjee, S., Miller, C. A., and Fort, T., Jr. Soc. Pet. Eng. J. **22**, 53 (1982).
26. Demus, D. and Richter, L., "Textures of Liquid Crystals", VEB Deutscher Verlag, Leipzig (1978).
27. Hoffman, R. and Gross, L., Applied Optics **14**, 1169 (1975).
28. Neubert, M. E., Norton, P., and Fishel, D. L., Mol. Cryst. Liq. Cryst. **81**, 253 (1975).
29. Miller, C. A., Mukherjee, S., Benton, W. J., Natoli, J. N., Qutubuddin, S., and Fort, T., Jr. AIChE Symp. Ser. **212**(78) 28 (1982).
30. Shubnikov, A. V. and Kopstik, V. A., "Symmetry in Science and Art," Plenum Publ. New York (1974).
31. Benton, W. J., Toor, E. W., Miller, C. A., and Fort, T., Jr. J. Physique **40**, 107 (1979).
32. Papahadjopoulos, D., Vail, W. J., Jacobson, K., and Poste, G., Biochim. Biophys. Acta **394**, 483 (1975).
33. Ribotta, R. and Durand, G., J. Physique **38**, 179 (1977).
34. Scudieri, F., Ferrari, A., and Gunduz, E., J. Physique 40 C3-90 (1979).
35. Rault, J., Phil. Mag. **34**(5), 753 (1976).
36. Clark, N. A. and Meyer, R. B., Appl. Phys. Lett. **22**(10), 493 (1973).
37. Delaye, M., Ribotta, R., and Durand, G., Phys. Lett. **44A** (2), 139 (1973).

Received May 19, 1983

Authors' address:

W. J. Benton
Department of Chemical Engineering
Rice University
P. O. Box 1892
Houston, Texas 77251, USA

Progress in Colloid & Polymer Science

Progr. Colloid & Polymer Sci. **68**, 82–89 (1983)

An experimental study of liquid/liquid displacement phenomena in periodically shaped capillaries*)

E. Wolfram, É. Kiss, and J. Pintér

Department of Colloid Science, Loránd Eötvös University, Budapest, Hungary

Abstract: The displacement of n-dodecane by aqueous solutions of sodium dodecyl sulfate in glass capillaries with periodically varying cross-section has been studied. By measuring the dynamic contact angles of the moving fluid interface upon variation of surfactant concentration c, flow rate v and viscosity ratio $\bar{\eta}$, of oil to water according to a three-level quadratic experimental design, significant pair-interdependences have been found between the influence of c and v as well as between that of c and $\bar{\eta}$ on the displacement. The results indicate that in addition to these parameters which are generally used to characterize the efficacy of oil displacement, other factors such as surface viscosity and rigidity of the liquid/liquid interfacial film as well as the stability of the oil film adhered to the capillary wall have to be taken into account, too.

Key words: Liquid/liquid displacement, dynamic wetting, contact angle, capillary flow, tertiary oil recovery.

Introduction

Displacement of a liquid (or, more generally, a fluid) phase by another one being immiscible with it from the surface of a solid is a common phenomenon. As a matter of fact, even spreading or contact angle formation involves displacement of that fluid phase, *viz.* gas or vapour, with which the solid had been in contact prior to be wetted. However, in common usage the word "displacement" is restricted for systems consisting of two liquids, one of them being polar (water), and the other one non-polar ("oil"). For this phenomenon, the name *"immiscible liquid/liquid displacement"* (LLD) has been coined in general, and if it occurs in a tube or in a porous (capillary) system, it is often called *"two-liquid flow"* (TLF).

TLF is governed both by bulk properties such as (volumetric) flow rate, bulk viscosities of the two liquid phases, pressure (gradient) and by surface properties such as solid/liquid interaction (wettability, contact angle, adherence of liquid film to the solid wall), interfacial viscosity, adsorption of solutes at the solid/liquid and the liquid/liquid interface.

For a long time, TLF has been a domain of classical fluid mechanics with the main emphasis on bulk properties, taking surface properties mostly not at all or only in an oversimplified manner into account. For instance, complete wettability of the solid by one of the liquids, or, at best, a constant contact angle independent of rate and direction of flow ("static contact angle with no hysteresis") is often assumed.

For real systems, however, this assumption is generally not valid, and both the dynamic contact angle and its hysteresis should be considered. Experimental investigations of this kind have been carried out since long modelling different practical processes as technical adhesion, flotation, surface coating, liquid distribution in soils or powdered systems [1–5] and from a pure academic point of view [6–10]. The main impetus to study LLD in cylindrical tubes [11–14] was given by the necessity of learning the role of these dynamic parameters in enhanced oil recovery [15–17].

In the following, we will restrict ourselves to TLF. As a continuation of previous work, of which a preliminary paper was published in this Journal recently [18], we will report in detail results of measurements made in capillaries whose diameter is a periodic function of the axial distance. In such tubes, the effect of surfactant concentration (in the aqueous phase), of the flow rate and of the viscosity ratio on the mutual displacement of the two liquid phases has been studied.

---

*) To the memory of Professor Bun-ichi Tamamushi.

Preliminary considerations

Movement of the three-phase line of contact (TPL)

Macroscopically, liquid-liquid displacement consists in the movement of the TPL along the solid surface. This, however, is apparently at variance with principles of classical hydrodynamics according to which a more or less thin layer of liquid should remain in an immobilized state, i. e., adhering to the surface. On the other hand, it is obvious that in reality fluid flow has to occur at the geometrical surface as well.

There are several approaches to circumvent this dilemma. For instance, Blake and Haynes [14] consider the mechanism of an advancing TPL as a sequence of adsorption and desorption steps. Other authors [19–21] suppose the existence of a narrow region near the TPL where the no-slip condition becomes invalid, i. e., the flow disobeys the classical hydrodynamic equations. It has even been assumed [22] that the real microscopic (intrinsic) contact angle is in all cases of the TLF equal to 180° and any deviation from this in the measured values should be attributed to the inaccuracy of the experimental determination. However, there is no experimental evidence for this assumption which also contradicts surface thermodynamics.

Displacement in capillaries

The qualitative features of the LLD in a capillary are depicted in figure 1 in which the series from *b* to *f* shows the flow which is expected to occur under normal conditions, i. e., if none of the rates of flow and wetting, resp., is too high relative to the other one. In this, say normal, case the dynamic contact angle, θ_{dyn}, is characteristic for the process, and the deformation of the liquid/liquid interface as influenced by the movement of the TPL is relatively small, so that its shape is not very different from the spherical one. If, however, wetting is too rapid even at a relatively low flow rate, then the displacing (wetting) liquid runs ahead (*film-imbibition*, fig. 1a). In the other extremity, if the flow rate is high, the displacing liquid protrudes into the liquid to be displaced leading to what is called *fingering* or *finger-forming* (fig. 1g).

According to Elliott and Riddiford's interpretation [23] the film-imbibition is a "relaxing", in other words spontaneous, flow and the finger-forming is a "forced" one.

From our point of view it is important to note that both imbibition and finger-forming may, and in fact generally do, result in the dispersion of one liquid in the other one during the flow.

Let us now consider both the imbibition and the finger-forming configuration as the initial states. As figure 2 illustrates, two kinds of liquid distribution (as noted by I and II in fig. 2) may form from either of the initial states *a* (imbibition) and *b* (finger-forming). It is to be noted here that even although the configurations *a* and *b* are seemingly geometrical mirror images of each other, they represent basically different situations since the flow direction is the same in both cases.

As a semiquantitative operational approach to analyse these processes, one can make use of a simple surface energy treatment. Clearly, the condition for imbibition to occur can be taken as being identical with that of spreading, i. e.,

$$\gamma_{SL_1} > \gamma_{L_1L_2} + \gamma_{SL_2} . \tag{1}$$

For finger-forming the deformation of the liquid/liquid interface will be predominant over the wettability of the solid.

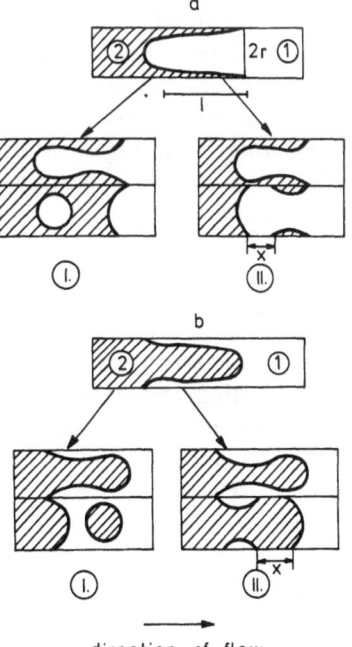

Fig. 2. Possibilities for dispersion formation during immiscible liquid displacement in cylindrical capillary tubes

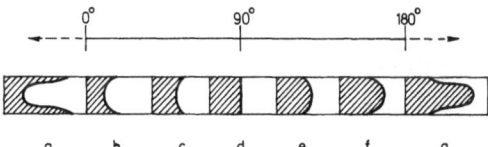

Fig. 1. Shapes of moving interfaces in cylindrical capillary tubes

Obviously, the driving force for droplet formation to occur from either liquid phase (process I in fig. 2) is the decrease in the area of the liquid/liquid interface (with the solid/liquid interfacial area remaining unchanged). This is, of course, the more promoted the larger the liquid/liquid interfacial tension.

Similarly, it can be easily shown that the conditions for process II (entrapment) to occur are

$$2r\pi x\gamma_{SL_2} + 2r\pi(l - x)\gamma_{SL_1} + 2r\pi(l - x)\gamma_{L_1L_2}$$
$$< 2r\pi l\gamma_{L_1L_2} + 2r\pi l\gamma_{SL_1} \qquad (2)$$

which in case b gives

$$\gamma_{SL_2} < \gamma_{L_1L_2} + \gamma_{SL_1} \qquad (3)$$

and

$$2r\pi x\gamma_{SL_1} + 2r\pi(l - x)\gamma_{SL_2} + 2r\pi(l - x)\gamma_{L_1L_2}$$
$$< 2r\pi l\gamma_{L_1L_2} + 2r\pi l\gamma_{SL_2} \qquad (4)$$

leading to

$$\gamma_{SL_1} < \gamma_{L_1L_2} + \gamma_{SL_2} \qquad (5)$$

in case a.

Since this latter condition is at variance with the afore-mentioned spreading condition, process II starting from situation a in figure 2 is unlikely to occur. It is to be noted that throughout this consideration we have tacitly assumed that all interfacial tensions involved are constant during the flow. Also the geometry of the interface has been simplified.

On the other hand, if experiments reveal that the latter process still occurs this means that the assumptions are oversimplified, and other parameters as surfactant adsorption resulting in changes of the interfacial energies and/or a high interfacial viscosity, may also influence the flow patterns.

Flow in non-uniform tubes

There has been very little published on this type of flow. A general thermodynamic analysis was given by Everett [24] dealing with wetting and adhesion in different solid/fluid/fluid systems. The stability of equilibrium states of a fluid/fluid interface and contact angle hysteresis in capillaries are also discussed. Oh and Slattery [25] have presented a theoretical treatment concerning equilibrium positions, movement of a residual oil segment in a supposed capillary with periodically varying cross-section and how it could be displaced from it. The movement of an oil bubble driven at a constant pressure difference is not uniform in this type of capillary. Since the curvature

and so the capillary pressure of the oil-water interface depends on the position of the oil bubble in the tube, sections of accelerating motion (Haines-jump) and creeping motion occur alternatively. Furthermore, the movement of an oil segment may be discontinuous since the profile of the capillary and the shape of the oil-water interface determined by the contact angle cannot be described by a smooth common curve. This results in the entrapment of the residual liquid phase in the pore waist.

According to Oh et Slattery an as complete as possible oil displacement from a chain of pores is promoted if the wettability by the displacing liquid of the solid wall is neither too high nor too low.

Capillary pressure of a curved oil-water interface moving through a capillary system consisting of narrow and wide tube sections was measured by Brown et al. [16] using a sensitive manometer. Besides other conclusions, they found that the interfacial viscosity was of great importance in displacement processes. This was supported theoretically also by Giordano et Slattery [26].

Experimental

Materials

Capillaries with periodically varying cross-section were made of pyrex glass by a HP capillary drawer apparatus. Their shapes and sizes are shown in figure 3.

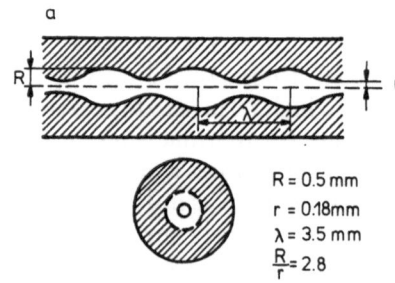

$R = 0.5$ mm
$r = 0.18$ mm
$\lambda = 3.5$ mm
$\frac{R}{r} = 2.8$

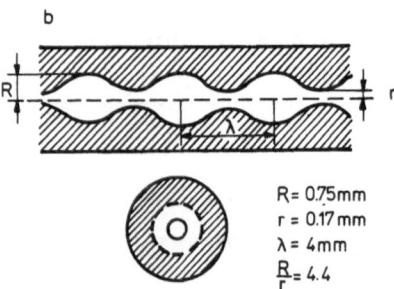

$R = 0.75$ mm
$r = 0.17$ mm
$\lambda = 4$ mm
$\frac{R}{r} = 4.4$

Fig. 3. Geometries of periodically shaped capillary tubes used in the experiments

As a non-polar (oil) phase *n-dodecane,* and as a polar (water) phase *aqueous sodium dodecyl sulfate* (SDS) solutions with different surfactant concentrations were used. In order to regulate the viscosity ratio of the two liquid phases small amounts of either paraffin oil or glycerol were added to the oil and, respectively, the water phase.

n-dodecane: (Fluka, pract.) $c = 95\%$, $Mw = 170.3$, density $= 0.748 \cdot 10^3$ kg/m^3, viscosity $= 1.35$ mPas (at 293 K).

High purity *paraffin oil:* density $= 0.86 \cdot 10^3$ kg/m^3, viscosity $= 164$ mPas (at 293 K).

Sodium dodecyl sulfate (SDS): (Merck, lab. pur.) recrystallized from 1:1 ethanol: benzene mixture, $Mw = 288.3$, $c_M = 8$ mol/m^3.

Glycerol: (Reanal, anal. pur.) $Mw = 92.1$, density $= 1.26 \cdot 10^3$ kg/m^3, viscosity $= 1499$ mPas.

Equipment

The capillary was placed in a glass cell and surrounded by an oil mixture with the same refractive index as the glass capillary. The oil and water phases were injected in a given sequence directly into the tube using a piston-like liquid driver (Infucont, Kutesz. Hung.) allowing for a flow velocity range from 1 to 10^3 nl/s. Since the liquids were forced to move at a constant volumetric flow rate, the linear flow velocity changes periodically in the capillary along its axis.

The displacement of the fluid phases in the tube was observed with a microscope. The shape of the moving interface and the advancing dynamic contact angle measured in the water phase, $\theta_{w,o}$, was studied in enlarged photos.

Experimental design

The experimental conditions for the displacement measurements were determined by using a Box-Behnken type (descriptive and non-optimizing) three-level quadratic experimental design [27]. The three parameters affecting the LLD are (see table 1):
– surfactant concentration in the aqueous phase, c_s,
– flow velocity of the liquids, v, and
– viscosity ratio of the liquids, $\bar{\eta} = \eta_o/\eta_w$.
The points corresponding to the experimental conditions can also be represented on a cube with axes of the three transformed variables (fig. 4).

The advancing oil-water contact angles or the amounts of film-imbibition on the capillary wall were measured as a dependent variable characterizing the dynamic wetting. Quadratic polynomials were fitted to these values. The relative magnitudes of the coefficients in the polynomials show both the effects of the individual parameters and of their interdependences.

Having three independent variables the response-surface can be visualized by taking one of the variables (e. g. v) constant. Then the response-surfaces can be represented over the plane determined by the two other parameters (c_s and $\bar{\eta}$). Therefore, the shapes of the response-surfaces show the influence of c_s and $\bar{\eta}$ upon the dynamic wetting ($\theta_{w/o}$) while comparing several surfaces corresponding to different flow velocities indicates the effect of flow velocity.

In order to interpret our results in this system, we classified the contact angles measured at different diameters in the capillary into three groups:
– contact angles at pore constrictions: $d < 400$ μm (C)
– contact angles at pore waists: $d > 800$ μm (W)
– contact angles at intermediate sections: 400 μm $< d <$ 800 μm (I).

This arrangement is possible since the error caused by it is not higher than the variations of contact angles ($\pm 5°$) measured in the same position in the capillary (even though there is a diameter dependence of contact angles).

This displacement sequences were:
– air displacement by oil phase;
– oil displacement by water after pretreating (60 min) of capillary with oil;
– water displacement by oil again.

Results

Air displacement

The shape of the advancing oil-air interface is similar to the equilibrium one ($\theta_e = \theta°$). Significant deviation from the equilibrium shape occurs only if the flow velocity in the tube constrictions is as high as 2 mm/s.

Table 1. The three-level Box-Behnken experimental design applied for three variables and the corresponding experimental conditions ($c_M =$ c.m.c.)

Variables of experimental design			Experimental conditions		
x_1	x_2	x_3	c_s/c_M	v nl/s	$\bar{\eta} = \eta_o/\eta_w$
0	0	0	1.0	196	1.5
+1	+1	0	1.9	368	1.5
+1	−1	0	1.9	24	1.5
−1	+1	0	0.1	368	1.5
−1	−1	0	0.1	24	1.5
+1	0	+1	1.9	196	2.5
+1	0	−1	1.9	196	0.5
−1	0	+1	0.1	196	2.5
−1	0	−1	0.1	196	0.5
0	+1	+1	1.0	368	2.5
0	+1	−1	1.0	368	0.5
0	−1	+1	1.0	24	2.5
0	−1	−1	1.0	24	0.5

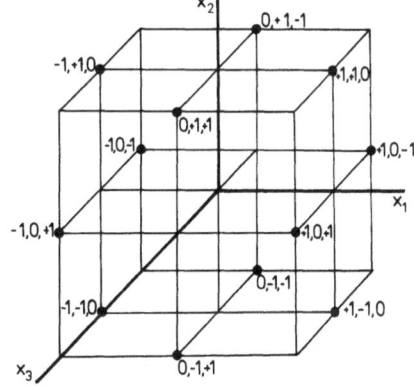

Fig. 4. Points of three-level Box-Behnken experimental design applied for three variables representing experimental conditions

Oil displacement

a) Wetting in the capillary as of figure 3a

As expected, SDS in the aqueous phase promotes the local displacement of oil on the wall independently of the capillary diameter. The effect of flow velocity is the opposite, i. e. cos $\theta_{w/o}$ decreases and so $\theta_{w/o}$ increases at higher flow velocity. Accordingly, the dynamic water/oil contact angle is at maximum at high flow velocity and at low SDS concentration in every section of the capillaries (fig. 5, 6, 7).

Comparing the three results obtained in the three parts (C, W, I) of capillaries, we find that the influence of flow rate resulting in deformation of the water/oil interface is the greater the smaller the tube diameter (fig. 7). The reason for this is that the same volumetric flow velocity gives a higher linear flow rate at pore constriction than at the waists. Then $\theta_{w/o}$ is an obtuse angle almost in the whole range of variables at pore constrictions, while $\theta_{w/o}$ is less than 90° at pore waists.

The coefficient of the third parameter, x_3 (the viscosity ratio) has a small value in the polynomial, so its role is negligible. This is also valid for the interdependent influence of it and of the flow velocity.

The smaller the diameter of that capillary tube section where the displacement occurs, the more definite is the shape characterizing the response-surfaces. It clearly appears that the effect of surfactant concentration on the wettability is not independent of

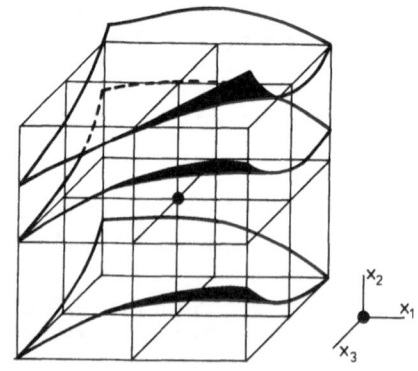

$$y = 0.317 + 0.087x_1 - 0.139x_2 - 0.058x_3 - 0.181x_1^2 - 0.224x_2^2 +$$
$$+ 0.182x_3^2 + 0.051x_1 x_2 - 0.065x_2 x_3 + 0.196x_1 x_3 =$$
$$= 0.641 + 0.152c + 2.398v - 0.748\bar\eta - 0.223c^2 - 7.573v^2 +$$
$$+ 0.182\bar\eta^2 + 0.329cv - 0.378v\bar\eta + 0.218c\bar\eta$$

Fig. 6. Response-surfaces and the polynomial fitted to the cosine of dynamic contact angles measured in intermediate parts of capillary tubes of type figure 3a. x_1, x_2 and x_3 as in figure 5

that of the two other parameters. The coefficients of both pair-interdependences of c_s (x_1x_2 and x_1x_3) are high in the polynomials.

In other words, this result means that the increase of c_s is favourable for wetting ($\theta_{w/o}$ small) in the following circumstances:

– in the whole range of v, for high $\bar\eta$;
– at high and middle v, for middle $\bar\eta$;
– at high v, for low $\bar\eta$.

The most surprising is the reversal of the concentration effect when going from high velocity and viscosi-

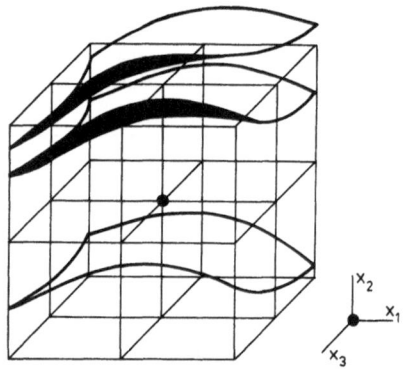

$$y = 0.698 + 0.136x_1 - 0.116x_2 + 0.097x_3 - 0.258x_1^2 - 0.356x_2^2 +$$
$$+ 0.110x_3^2 + 0.127x_1 x_2 - 0.013x_2 x_3 + 0.084x_1 x_3 =$$
$$= 0.280 + 0.487c + 3.337v - 0.311\bar\eta - 0.318c^2 - 12.03v^2 +$$
$$+ 0.110\bar\eta^2 + 0.820cv - 0.076v\bar\eta + 0.093c\bar\eta$$

Fig. 5. Response-surfaces and the polynomial fitted to the cosine of dynamic contact angles measured in pore waists in capillary tubes of type figure 3a. x_1: transformed variable of SDS concentration; x_2: transformed variable of flow velocity; x_3: transformed variable of viscosity ratio of liquids

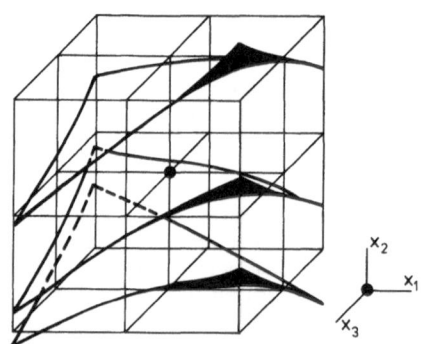

$$y = -0.259 + 0.167x_1 - 0.223x_2 + 0.032x_3 - 0.085x_1^2 + 0.152x_2^2 +$$
$$+ 0.045x_3^2 + 0.230x_1 x_2 - 0.022x_2 x_3 + 0.338x_1 x_3 =$$
$$= 0.772 - 0.459c - 4.604v - 0.453\bar\eta - 0.105c^2 + 5.139v^2 +$$
$$+ 0.045\bar\eta^2 + 1.486cv - 0.128v\bar\eta + 0.376c\bar\eta$$

Fig. 7. Response-surfaces and the polynomial fitted to the cosine of dynamic contact angles measured in constrictions of capillary tubes of type figure 3a. x_1, x_2 and x_3 as in figure 5

ty ratio to lower ones. The reason for this will be discussed in the last paragraph.

b) Wetting in the capillary as of figure 3b

In this type of capillary dynamic wetting seems to be more sophisticated.

The relative influence of the parameters upon the dynamic contact angle is similar to that obtained for the previously discussed tube. Namely, the pair-interdependences of c_s and v as well as of $\bar{\eta}$ become the most significant.

In addition, however, dynamic water/oil interfaces form that cannot be characterized by contact angles (fig. 1a), and oil entrapment also occur. Since the dynamic contact angle was chosen as a dependent variable of polynoms these phenomena do not fit into this interpretation.

c) Film-imbibition and oil entrapment in the capillary as of figure 3b

Water/oil interface advancing in the pore waist of this type of capillaries is similar to that shown in figure 2a. A water film imbibing on the capillary wall overtakes the bulk of water in the middle of the tube. The film approaching the pore constriction coalesces if the bulk water did not reach that place of the tube before. So a droplet of the oil phase is formed and separated from the displaced bulk oil. This drop will be trapped if its diameter is bigger than that of the pore constriction.

If this entrapment process repeatedly occurs in a series of pore constrictions, this leads to the dispersion of oil and to a very low efficacy of the oil displacement.

The change in the areas of the dynamic water/oil interface before and during the entrapment process is shown in figure 8. As the TPL is moving towards the constriction, the area of the interface increases to a value highly exceeding that of the spherical interface and, just in front of the constriction, becomes very small followed by the drop rupture.

How frequently oil entrapment occurs is determined by c_s and by v for a given pore geometry. The simultaneous influence of these two parameters can be seen in figure 9.

At a medium flow velocity, oil entrapment often occurs at any surfactant concentration. The probability of the process is small when both of the parameters are either sufficiently small or high. In these cases there is no essential difference between the velocity of wetting and that of flow.

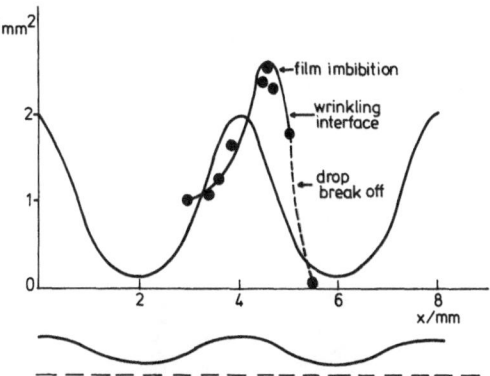

Fig. 8. The area of water/oil interface advancing continuously in the periodic tube as compared to the area of a similar interface upon wrinkling and subsequent rupture

It is obvious that the good wettability at low flow rates promotes the oil entrapment. This behaviour is similar to the spontaneous capillary rise.

It is, however, to be mentioned that repeated entrapment may take place by another mechanism at low c_s and high v. The surface viscosity of the water/oil interface becomes important in this process. A rigid interfacial film which is unable to contract rapidly has formed, resulting in a "wrinkling" of the surface layer that can clearly be seen by the microscope (see fig. 10).

Water displacement

The last process in the TLF is the displacement of water by injecting *n*-dodecane. This oil/water interface shows finger-forming in all systems. The water film covering the capillary wall is stable and remains continuous even during the oil flow.

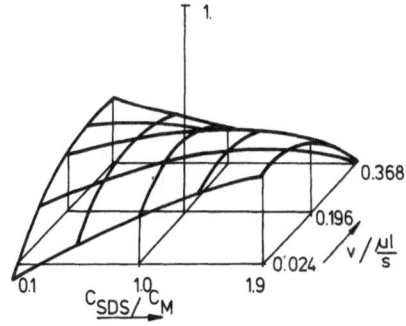

Fig. 9. Probability of oil entrapment as a function of flow velocity and SDS concentration

Fig. 10. Rigid interfacial film showing wrinkling

Discussion

Let us summarize what can be said about the factors affecting oil displacement by surfactant solution in a capillary system containing pore waists and constrictions.

Among the chosen controlled variables influencing the dynamic wetting, the effect of surfactant concentration and of flow velocity met our expectation. A higher surfactant concentration promotes the oil removal both from the capillary tube and from the wall. The dynamic water/oil contact angle increases with increasing flow velocity.

It was not possible to predict an interdependent influence of SDS concentration and of flow velocity as well as of the viscosity ratio of liquids. Experimentally, the latter was found to become appreciable during the displacement in pore constrictions (see fig. 7).

If the viscosity ratio is higher than 1.5, the surfactant promotes the wetting but, contrarily, solutions of low SDS concentration enhance wetting if the viscosity ratio is less than 2.5. The probable reason for this is

Table 2. Stability (measured by the life-time t_{film}) of oil-films adhered to glass surfaces contacting with a drop of SDS solution of concentration c_s (c_M = c.m.c.)

c_s/c_M	t_{film}/s
0.1	0.98 ± 0.17
1.0	1.07 ± 0.13
1.9	1.50 ± 0.30

the stability of the oil film adhered to the glass surface. According to previous experience, the film stability may play a great role in advancing of the TPL. The data in table 2 were obtained under similar experimental conditions but on plane glass surfaces. These results can be correlated to those obtained in pore constrictions. At a low flow velocity, the wetting by the "water" phase is not hindered by the stability of an oil film with a life-time of 1 s or so but the film becomes a hindrance against the movement of the TPL at higher flow velocities. Accordingly, the interface turns inside out, and the contact angle increases.

As a consequence of the common influence of SDS concentration and flow velocity there is a water-film imbibition on the capillary wall and, in certain cases, the oil drop can be trapped in front of pore constrictions. At small flow velocities the good wettability by the surfactant solution is favourable for the film-imbibition and coalescence in constrictions.

At low SDS concentration, the reason for oil entrapment is the formation of a rigid interfacial film due to lauryl-alcohol which could not be eliminated even with careful purication. The film formation occurs at low SDS concentration, since above the c_M (8 mol/m³) lauryl-alcohol gets solubilized in the SDS micelles [28].

The entrapment is, of course, also a function of the actual pore geometry. It does not occur at all in a capillary with narrower waists.

Another phenomenon in connection with the shape of the capillary is the diameter dependence of dynamic contact angle. We may compare contact angles forming at different volumetric flow velocities in different sections of the tubes, but at the same linear flow velocity. We have found that the contact angles are higher in pore constrictions than in waists [18]. This interesting difference raises the question whether it is justified to calculate the linear flow velocity from the volumetric one and to connect this value to the contact angle. It would be useful to consider the flow distributions in both liquid phases in the vicinity of the interface.

We can conclude that dynamic wetting cannot be regarded as a sequence of equilibrium wetting positions following each other, since fluid interfaces of very different shape can be formed. This leads to oil dispersion in pore systems of certain geometry.

The most favourable oil displacement is not obtained with a surfactant solutions showing the highest wetting power. It is necessary to meet simultaneously the requirements of both oil displacement from the tube and from its wall.

In addition to the three chosen variables (SDS concentration, flow velocity, viscosity ratio of liquids) other parameters as stability of oil film adhering to the glass surface and the surface viscosity of the oil/water interface are also of importance.

Acknowledgment

A part of this work was supported by the Hungarian Academy of Sciences.

References

1. Schonhorn, H., Frisch, H. L., Kwei, T. K., J. Appl. Phys. 37, 4967 (1966).
2. Stechemesser, H., Geidel, Th., Weber, K., Colloid Polymer Sci. 258, 109 (1980), 258, 1206 (1980), 259, 767 (1981).
3. Summ, B. D., Gorunov, U. V., Fiziko-khimicheskiye osnovi smatchivaniya i rastekaniya Moskva, Izdat. Khimiya 1976, p. 110.
4. Haines, W. B., J. Agric. Sci. 20, 97 (1930).
5. Taber, J. J., Soc. Petr. Eng. J. 9, 3 (1969).
6. Ablett, R., Phil. Mag. 46, 244 (1923).
7. Schwartz, A. M., Tejada, S. B., J. Colloid Interface Sci. 38, 359 (1972).
8. Coney, T. A., Masica, W. J., NASA Techn. Note N 69-20794 (1969).
9. Radigan, W., Ghiradella, H. L., Frisch, H. L., Schonhorn, H., Kwei, T. K., J. Colloid Interface Sci. 49, 241 (1974).
10. Elliott, G. E. P., Riddiford, A. C., Nature 195, 795 (1962).
11. Goldsmith, H. L., Mason, S. G., J. Colloid Interface Sci. 18, 237 (1963).
12. Chittenden, D. H., Spinney, D. U., J. Colloid Interface Sci. 22, 250 (1966).
13. Blake, T. D., Everett, D. H., Haynes, J. M., in: "Wetting", SCI Mon. 25., p. 164, London 1967.
14. Blake, T. D., Haynes, J. M., J. Colloid Interface Sci. 30, 421 (1969).
15. Brown, C. E., Neustadter, E. L., J. Can. Petr. Techn. 19, 100 (1980).
16. Brown, C. E., Jones, T. J., Neustadter, E. L., In: "Surface Phenomena in Enhanced Oil Recovery" (Shah, D. O., ed.) Plenum Pr. N. Y. 1981, p. 495.
17. Pintér, J., Wolfram, E., ibid. p. 479.
18. Kiss, É., Pintér, J., Wolfram, E., Colloid Polymer Sci. 260, 808 (1982).
19. Huh, C., Mason, S. G., J. Fluid. Mech. 81, 401 (1977).
20. Hansen, R. J., Toong, T. Y., J. Colloid Interface Sci. 37, 196 (1971).
21. Huh, C., Scriven, L. E., J. Colloid Interface Sci. 35, 85 (1971).
22. Oliver, J. F., Mason, S. G., J. Colloid Interface Sci. 60, 480 (1977).
23. Elliott, G. E. P., Riddiford, A. C., Rec. Progr. Surf. Sci. 2, 111 (1964).
24. Everett, D. H., Pure Appl. Chem. 52, 1279 (1980).
25. Oh, S. G., Slattery, J. C., Soc. Petr. Eng. J. Apr. 19, 83 (1979).
26. Giordano, R. M., Slattery, J. C., AIChE J. Symp. Ser. 78, 212 (1982).
27. Box, G. E. P., Behnken, D. W., Technometrics 2, 455 (1960).
28. Gupta, L., Wasan, D. T., Ind. Eng. Chem., Fundam. 13, 26 (1974).

Received April 20, 1983

Authors' address:

Prof. Dr. E. Wolfram
Dr. É. Kiss (Mrs.)
Dr. J. Pintér
Department of Colloid Science, Loránd Eötvös University, H-1445 Budapest 8, POB 328, Puskin u. 11/13, Hungary

Progress in Colloid & Polymer Science   Progr. Colloid & Polymer Sci. **68**, 90–96 (1983)

The study of the state of adsorption layers of long-chain one-one valent electrolytes at liquid interface*)

E. D. Shchukin, Z. N. Markina, and N. M. Zadymova

Institute of Physical Chemistry, Academy of Sciences, Moscow, USSR

Abstract: A state equation for an ionized monolayer and its corresponding adsorption isotherm for one-one valent long-chain electrolytes taking account of the interaction between amphiphilic ions are proposed. The validity of the equations is demonstrated using adsorbed layers of cetyl trimethyl ammonium bromide (CTAB) on the basis of high precision ($\pm 0,02$ mJ$\cdot$m^{-2}) tensiometric measurements at the water-air and water-octane interfaces at 303–343 K. Analysis of the values measured for interfacial tension and ionisation potential has shown for both interfaces the existence of three regions characterized by different extents to which the interactions between adsorbed amphiphilic ions are manifested. The region of low surface occupancy, $\theta \gtreqless \sim 0,1$ (ideal behaviour of monolayer), the region $\sim 0,1 < \theta < \sim 0,7$ where interactions are weak (attraction constant $a_0 = \text{const} = \sim 1$) and the region $\theta > \sim 0,7$ where the layer approaches the condensed state (a_0 increases monotonously) were examined. An important role is shown to be played by the hydrophobic interactions between the segments of hydrocarbon chains of adsorbed cetyl trimethyl ammonium (CTA$^+$) ions immersed into water.

Introduction

Studies of the state of adsorbed layers of surface active stabilizers are of a great current interest in connection with the general problem of stability of disperse systems [1, 14, 27].

Because of the lack of direct and sufficiently sensitive methods for studying the structure of adsorbed layers of surfactants (especially ionized) at mobile interfaces, very little work has been done in this field. Therefore, considerable importance becomes attached to indirect methods such as interfacial tension measurements, for example plotting the isotherm π (surface pressure) against A (area per molecule) on the basis of the Gibbs equation and then by choice of an appropriate state equation describing the behaviour of a given layer.

The effect of electric charge in the surfactant monolayer on its behaviour was studied previously by many authors [1–9] who proposed various state equations. Davies [1, 2], Guastalla [3], and Phillips and Rideal [4, 5] estimated the electrostatic component of the surface pressure on the basis of the Gouy-

Chapman theory of the double layer. However Pal et al. [11] experimentally found that the Gouy-Chapmann theory overestimates the contribution of electrostatic interactions to the π value. These authors compared the π vs. A isotherms for monolayers of two-basic non-ionized acids (sebacic, suberic, pimelic) and their ionized salts at the water-oil interface and found experimentally the electrostatic component of the surface pressure (π_{exp}) which was always lower than predicted by the Gouy-Chapman scheme. This discordance was attributed by the authors [11, 12] to the effect of counterions binding by the charged monolayers.

In contrast to the conventional assumption about the absence of cohesive interactions between hydrocarbon chains of surface active molecules (ions) at the water-hydrocarbon interface due to their solvation by the non-polar phase, Ghosh [6–10] has shown that for adsorbed layers of long-chain ions without electrolyte additives the cohesion component of surface pressure at this interface is not equal to zero because of the attraction between chain segments submerged into water of the adsorbed amphiphilic ions. Ghosh [6] gave a state equation for adsorbed layers of long-chain one-one valent electrolytes at the solution-air

---

*) Dedicated to the memory of Professor Dr. B. Tamamushi.

and solution-hydrocarbon interfaces. The verification of this equation by the same author has supported its validity only for sufficiently low θ values (where V degree of surface occupancy). This fact can be ascribed to a possible interdependence between electrostatic and cohesion components of surface pressure [1] especially marked for intermediate and high layer occupancy degrees.

Thus nowadays there are practically no equations accounting for the behaviour of ionized surfactant layers up to high occupancy degrees corresponding to the most pronounced stabilizing effect of such layers.

In this paper an attempt is made to suggest a state equation for fully ionized adsorbed layers of long-chain one-one valent electrolytes at the water-air and water-hydrocarbon interfaces as well as the corresponding adsorption isotherm, valid over a wide range of surface occupancy degrees.

Generally, the surface pressure π of an ionized surfactant adsorption layer may be presented as the sum of three components [1, 13].

$$\pi = \pi_K + \pi_e - \pi_s, \tag{1}$$

where π_k is the kinetic term accounting for the thermal motion of ions; π_e is the electrostatic term; π_s is the cohesion term accounting for attraction between the hydrocarbon chains of the adjacent adsorbed ions. It is known [4–6], for the electrolyte-free solutions of ionic surfactants that the surface pressure of adsorption layers is caused not only by surface active ions but also by their counterions. Hence the state equation of an ionized adsorption layer must include as a kinetic component the term $2\pi_k$ which can be written using the van Laar equation:

$$\frac{2kT}{A_0} \ln \frac{A}{A - A_0} \text{ or } 2\Gamma_{\max}RT\ln(1 - \theta), \tag{2}$$

where A_0 is the "own" area of the amphiphilic ion; A is the area per surfactant ion in the monolayer; $\Gamma_{\max}$ is the maximum adsorption; k, R, T have the usual meanings.

It is further necessary to estimate electrostatic and cohesion components of the surface pressure. Taking into account the Davies' idea [14] that it is not always correct to separate these components as well as the fact that for long-chain (above C_8) ions attraction prevails upon repulsion [15], we have written an expression for "effective" cohesion term (π_s') in line with Frumkin's work [16]:

$$\pi_s' = \pi_e - \pi_s = - \frac{\alpha}{A^2}, \tag{3}$$

where α is a constant characterizing the attractive forces between hydrocarbon chains of adsorbed ions (per ion).

In the case of adsorption of ionic surfactants at liquid interfaces a decrease in concentration of inorganic electrolyte in solution results in increasing thickness of the 2nd electric layer near the surface (up to several tens of Å in the absence of electrolyte). Repulsion between polar heads of amphiphilic ions also rises. An increase in horizontal distances between adsorbed surfactant ions is not favoured by thermodynamic factors because it might cause a decrease in θ and consequently an increase in interfacial tension (σ) of the solution. Hence one may expect the repulsion between polar heads of amphiphilic ions to depend on the different extents to which they are immersed into the aqueous phase. This gives rise to π_s^6.

Summarizing the components of the surface pressure and taking into account (2) and (3), one can derive the following state equation for adsorbed layers of long-chain ions:

$$\begin{aligned}
\pi &= \frac{2kT}{A_0} \ln \frac{A}{A - A_0} - \frac{\alpha}{A_2} \\
&= - 2\Gamma_{\max}RT\ln(1 - \theta) - a'\Gamma^2 \\
&= - \Gamma_{\max}RT\{2\ln(1 - \theta) + a_0\theta^2\}
\end{aligned} \tag{4}$$

where a_0 is, as in Frumkin's work, [16] the attraction constant equal to $\dfrac{\Gamma_{\max}a'}{RT} = \dfrac{N^2\Gamma_{\max}\alpha}{RT}$

(N is Avogadro number, a' is the cohesion constant per mole of surfactant).

Our expression (4) differs by a factor 2 before the logarithmic term from the known Frumkin's state equation for non-ionized adsorption layers of long-chain surfactants.

It is possible to show by expanding $\ln(1 - \theta)$ into power series that for low occupancy degrees of the adsorption layers where ($\theta \to 0$) equation (4) becomes $\pi A = 2kT$.

In the case of adsorption of anionic surfactant from aqueous solution in the presence of an inactive electrolyte having a common cation, the Gibbs equation $\Gamma = - \dfrac{1}{RT} \dfrac{\partial \sigma}{\partial \ln a}$ (a is the molar activity) can be written [17, 18] as follows:

$$\Gamma = - \frac{1}{iRT} \frac{\partial \sigma}{\partial \ln c} \Bigg/ \left(1 + \frac{v\partial \ln \gamma_\pm}{i\partial \ln c}\right) \tag{5}$$

where $i = v_+ \left(\dfrac{v_+ c}{v_+ c + v'_+ c_a} \right) + v_-$; $\gamma_\pm$ is the average

molar activity coefficient of the surface-active electrolyte; v_+, v_- and v are numbers of cations, anions and the total number of ions resulting from dissociation of one amphiphilic molecule, respectively; v'_+ is the number of cations resulting from dissociation of a molecule of the inactive electrolyte; c and c_a are surfactant and inactive electrolyte concentrations in solution, respectively.

For aqueous solutions of one-one valent surface active eletrolytes without inorganic electrolyte added the equation (5) takes the following form:

$$\Gamma = - \frac{1}{2RT} \frac{\partial \sigma}{\partial \ln c \gamma_\pm} = - \frac{1}{RT} \frac{\partial \sigma}{\partial \ln a_\pm} , \quad (6)$$

where $a_\pm$ is the average ionic activity of the surfactant ($a = a_\pm^v = a_\pm^2$).

Combination of equations (4) and (6) yields the adsorption isotherm:

$$b_1 a_\pm = \frac{\theta}{1 - \theta} e^{-a_0 \theta} , \quad (7)$$

where b_1 is the adsorption equilibrium constant. Expression (7) differs from Frumkin's adsorption isotherm [16] by the lack of factor 2 in the exponent.

Validity of the derived equations (4) and (7) has been verified in the present work using as an example aqueous solutions (without electrolytes) of a cationic micelle-forming stabilizer: cetyl trimethyl ammonium bromide (CTAB) $CH_3(CH_2)_{15}\overset{+}{N}(CH_3)_3\overset{-}{Br}$ at the solution-air and solution-octane interfaces.

Experimental

Materials and methods

In the present investigation we have used CTAB from Schuchardt, FRG (purity 97%), tripli-recrystallized from absolute ethanol and subsequently dried in vacuum at 308 K to a constant weight. The purity after recrystallization (from elementary microanalysis data) was of 99,8%.

High purity water was used to make up the solutions. For cleaning it from surfactant traces a specially treated activated charcoal was used which did not alter practically the electrical conductivity of water [19] ($\varkappa_{298K} = 1$ to $1,5 \cdot 10^{-4}$ S·m¹). Polargraphically tested surfactant concentration in water was no more than $5 \cdot 10^{-8}$ mole·l⁻¹. Octane was chromatographically pure.

Tensiometric measurements at the interfaces solution-air and solution-octane were carried out using the maximum bubble (droplet) pressure method, the accuracy being of $\pm 0,02$ mJ·m⁻². The interfacial tension was calculated [20] from:

$$\sigma = 0,5\, g \varrho\, hX , \quad (8)$$

where g is the acceleration due to gravity; ϱ is the density of the liquid or density difference between two liquids in the case of σ measurements at the interface solution-air or solution-hydrocarbon respectively; h is the distance between the capillary rim and the lowest point of the interfacial boundary in the manometric vessel (fig. 1), as measured using a cathetometer KM-10 with an accuracy of $\pm 10^{-6}$ m; X is the effective capillary tube radius (fig. 1) calculated using Sugden's tables [21, 22] by sequential approximations from the value of the inner capillary tube radius measured with an accuracy of $\pm 10^{-7}$ m by means of a microscope UIM-21 and from the capillary constant.

Densities of surfactant solutions were determined with a two-capillary picnometer with an accuracy of $\pm 0,2$ kg·m⁻³ [23]. Temperature variations were held within $\pm 0,1$ K by using an air thermostatting mantle. Bubble (doplet) formation time was from 1 to 60 min which allowed surfactant solution to attain equilibrium values of interfacial tension (fig. 2).

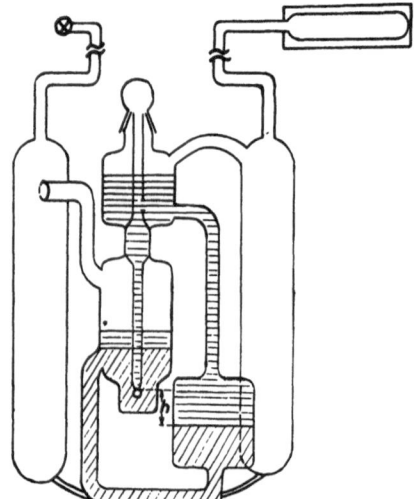

Fig. 1. Perfected device for measuring interfacial tension

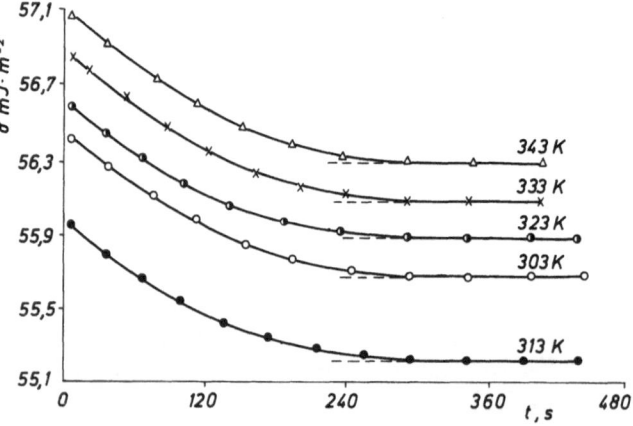

Fig. 2. Variation of surface tension for CTAB aqueous solution at concentration of $2,5 \cdot 10^{-4}$M with time at various temperatures

Tensiometric measurements at the liquid-liquid interface were carried out after holding the phases in contact during two days until mutual saturation of the phases was verified by sensitive conductometric tests.

Ionisation potential (ΔV) of CTAB adsorption layers at the solution-air interface was measured by the ionisation method using a standard scheme with an Am[241]-sonde [24]. The measuring device used was a VA-1-51 electrometer with a dynamic condensor from RFT (GDR) having the input resistance of 10^{15} Ohm. The accuracy of measurements was of ± 1 OmV.

Results and discussion

High precision tensiometric measurements under equilibrium conditions were made for CTAB aqueous solutions at concentrations varying from 0,03 to $19,04 \cdot 10^{-4}$ mole $\cdot l^{-1}$ at the solution-air and solution-octane interfaces in the temperature range 303–343 K. The results are represented in figure 3.

Figure 4 shows an example of π vs. A isotherms for CTAB adsorbed layers at 303 K at the solution-air and

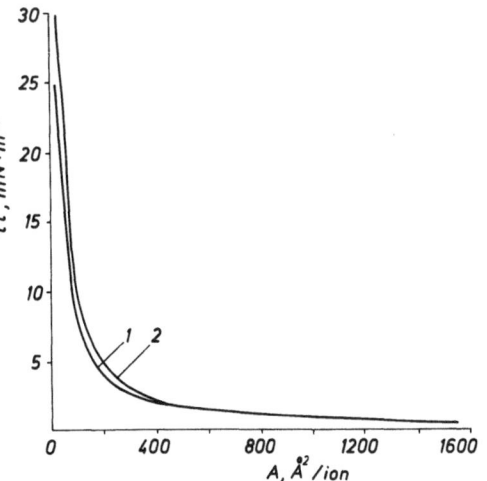

Fig. 4. Surface pressure of CTAB adsorption layers at the water-air (1) and water-octane (2) interfaces as a function of area per surfactant ion at 303 K

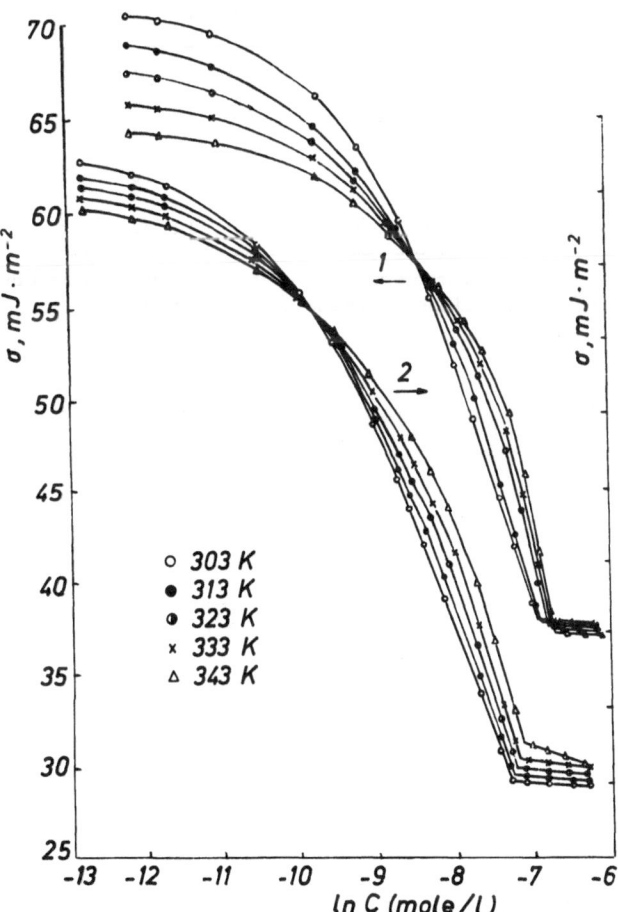

Fig. 3. Equilibrium isotherms of interfacial tension for CTAB aqueous solutions at the interfaces solution-air (1) and solution-octane (1) at 303–343 K

solution-octane interfaces. The π value was estimated from the difference between interfacial tensions for the solvent and the surfactant solution at the boundary with the same phase: $\pi = \sigma_0 - \sigma$. The A values were calculated by using the Gibbs equation (eq. 6) on the basis of the dependence $\sigma = f(a_+)$. Average ionic activity coefficients were taken from of Chand and Malik the work [25].

The analysis of the data obtained for the spread CTAB adsorption layers ($\theta < \sim 0,1$) shows that for both the interface studied at 303–343 K the product πA is constant and equals $\sim 2kT$. This means that in the spread CTAB layers at the water-air and water-octane interfaces not only amphiphilic ions but also counterions are present, participating in the thermal motion without mutual interaction. It is to be noted that the experimental iostherm ΔV vs. ln c for aqueous CTAB solutions at the solution-air interface at 303 K (fig. 5) has a linear part in the region of low surface occupancy degrees ($\theta \sim 0,1$; $A \sim 430$ Å²; $c \sim 0,27 \cdot 10^{-4}$ mole $\cdot l^{-1}$) which evidently correspond to an ideal two-dimensional gas.

On the basis of the θ values found by using the Gibbs equation and taking account of the linearity of equation (7); we find,

$$- \ln \frac{a_\pm(1 - \theta)}{\theta} = \ln b_1 + a_0\theta. \qquad (9)$$

We have attempted to estimate values for the attraction constant a_0 and the adsorption equilibrium constant b_1 in the systems studied.

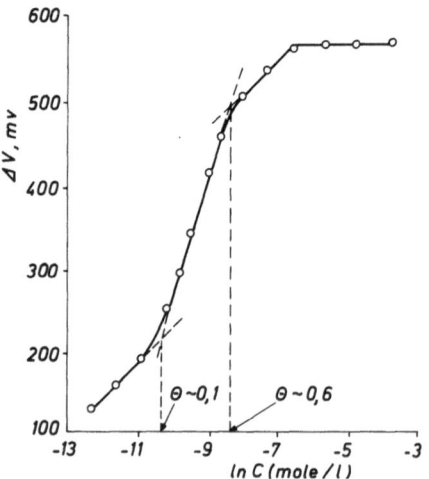

Fig. 5. Equilibrium isotherm of ionisation potential for CTAB aqueous solutions at the solution-air interface at 303 K

region of low and intermediate surface occupancy degrees (under a_0 = const. condition). The values of a_0 and b_1 obtained are listed in the table 1.

To elucidate the nature of interaction between long-chain CTA^+ ions at high occupancy degrees in adsorption layers having therefore the greatest stabilizing power in disperse systems, we have used the equation (7) to evaluate changes in a_0 as the layer tends to saturation (similarly to the work [26] concerned with adsorption of camphor and adamantanol from water-alcohol mixtures on the surface of a mercury electrode). The a_0 values were found using equation (7) by means of the dependence $\theta (a_{\pm})$ obtained from the experimental data using the Gibbs equation. Dependences $a_0 (\theta)$ thus found are shown in figure 7.

As one can see from figure 6 where the variation of $- \ln \dfrac{a_{\pm}(1 - \theta)}{\theta}$ as a function of θ is exemplified by the case of adsorption of ions from aqueous solutions at the solution-air and solution-octane interfaces at 303 K, the linearity holds upt to values $\theta \sim 0,65$ and $\theta \sim 0,75$ for two interfaces respectively, as well as for higher temperatures (313–343 K). It means that the equation describes satisfactory the adsorption of CTA^+ ions from aqueous solutions at the solution-air and solution-octane interfaces in the

Table 1. Attraction constant[a]) a_0 and adsorption equilibrium constant b_1 for aqueous solutions of CTAB at the solution-air and solution-octane interfaces at various temperatures

		303	313	323	333	343
	T, K					
Solution-air	a_0	1,1	1,1	1,1	1,1	1,1
	$b_1 \cdot 10^6$, $m^3 \cdot mole^{-1}$	8,4	8,6	3,1	2,8	2,5
Solution-octane	a_0	0,9	0,9	0,9	0,9	0,9
	$b_1 \cdot 10^6$, $m^3 \cdot mole^{-1}$	15,5	13,1	12,3	11,3	9,7

[a]) The a_0 values are calculated up to $\theta \sim 0,65$ and $\sim 0,75$ for the solution-air and solution-octane interfaces respectively.

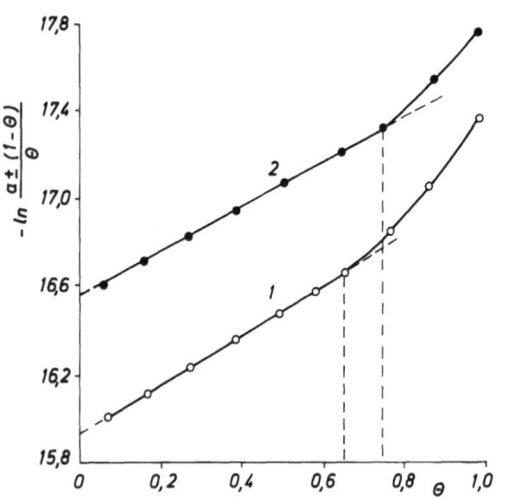

Fig. 6. Variation of $\dfrac{a_{\pm} (1 - \theta)}{\theta}$ with θ for CTAB aqueous solutions

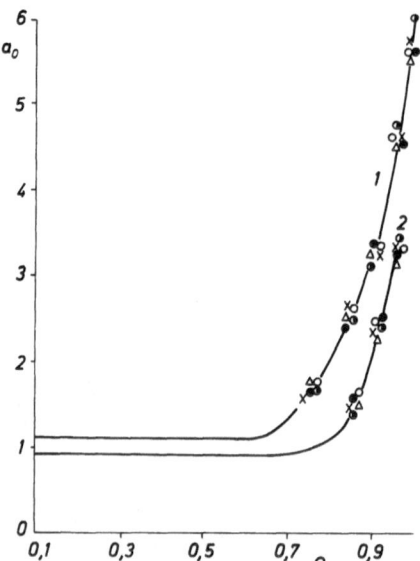

Fig. 7. Attraction constant a_0 for CTAB adsorption layers at the water-air (1) and water-octane (2) interfaces as a function of the surface occupancy at various temperatures

These dependences were used to verify the validity of our state equation (4). As table 2 shows, where examples are given for surface pressure values experimentally found (π) and calculated (π_c) from the equation (4) for the case of CTAB adsorption layers at the water-air and water-octane interfaces at 303 K, one can see a fair agreement between π and π_c over the whole range of surface occupancy degrees.

Thus the calculations show that our state equation (4) for the adsorbed layer and the adsorption isotherm corresponding to it. (7) satisfactorily describe the adsorption behaviour of CTA^+ ions in aqueous solutions in the absence of electrolyte additions at the interfaces solution-air and solution-octane over a wide range of temperatures and surface occupancy degrees.

It can be seen from figure 7 for monolayers studied in the region $\sim 0,1 < \theta < \sim 0,7$ that the attraction constant does not vary and equals 1,10 and 0,94 for the water-air and water-octane interfaces respectively.

Table 2. Experimental (π) and calculated (π_c) values from equation (4), for the surface pressure of CTAB adsorption layers at the water-air and water-octane interfaces at 303 K (bracketed numbers refer to a_0 values)

Water-air

$c \cdot 10^4$, M	θ	π, mN·m^{-1}	π_c, mN·m^{-1}	$\Delta\pi$, mN·m^{-1}
0,0544	0,027	0,53	0,53	0
0,0811	0,042	0,83	0,82	0,01
0,1632	0,085	1,69	1,67	0,02
0,2721	0,140	2,66	2,75	0,09
0,5985	0,250	5,15	4,98	0,17
1,0066	0,370	7,72	7,60	0,12
1,6868	0,520	11,55	11,50	0,05
2,0405	0,560	12,90	12,74	0,16
2,5030	0,640	15,47	15,65	0,18
3,4008	0,780	18,04	17,81 (2,0)	0,23
4,3531	0,900	20,95	20,59 (3,1)	0,36

$\Gamma_{max} = (3,9 \pm 0,2) \cdot 10^{-6}$ mole·m^{-2}

Water-octane

$c \cdot 10^4$, M	θ	π, mN·m^{-1}	π_c, mN·m^{-1}	$\Delta\pi$, mN·m^{-1}
0,0272	0,046	0,86	0,86	0
0,0544	0,093	1,73	1,74	0,01
0,0816	0,139	2,58	2,62	0,04
0,1632	0,232	4,23	4,45	0,22
0,2448	0,310	6,10	6,08	0,02
0,6802	0,580	13,12	13,23	0,11
1,1427	0,710	18,51	18,67	0,09
1,5508	0,780	22,42	22,58 (1,0)	0,16
1,8500	0,820	24,81	25,08 (1,1)	0,27
2,2037	0,870	27,04	26,76 (1,6)	0,28
2,8295	0,960	30,62	30,81 (3,4)	0,19

$\Gamma_{max} = (3,7 \pm 0,2) \cdot 10^{-6}$ mole·m^{-2}

Such a close resemblance of the a_0 values for different interfaces can be evidence for hydrophobic interactions between hydrocarbon segments submerged into water of the ions CTA^+, thus making practically negligible the effect of solvation of hydrocarbon radical parts situated in the oil phase on the resulting cohesion pressure.

It is noteworthy that an additional contribution to the cohesion interaction between adsorbed long-chain CTA^+ ions is provided by attraction forces between CH_3 groups of charged heads of adjacent ions. This correlates with the data obtained by Ghosh [6] who has pointed out that in adsorbed layers of octadecyl trimethyl ammonium bromide at the water-hydrocarbon interface π_s cannot be eliminated even at rather high (about 1 M) concentrations of an inorganic electrolyte when the whole hydrocarbon chain of adsorbed ions is "salted out" into the non-polar phase.

The contribution of the dispersion interactions between hydrocarbon chains of adsorbed ions at the water-air interface to the cohesion component of surface pressure is expressed as follows: [14]

$$\pi_s^* = \pi_{12} - \pi, \qquad (10)$$

where π_{12}, π are surface pressures of the surfactant monolayer for a given value of θ at the solution-hydrocarbon and solution-air interfaces respectively. Lower values of surface pressure at the water-air interface originate from the cohesion between hydrocarbon chain parts of adsorbed amphiphilic ions surrounded by the non-polar phase; this is absent at the water-oil interface because of solvation.

The relationship between π_s^* and A for CTAB monolayers at 303 K shown in figure 8 provides evidence for a surprisingly low contribution of disper-

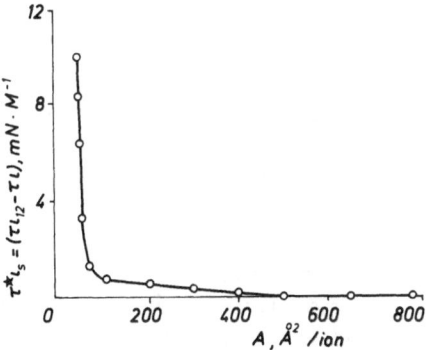

Fig. 8. The component of cohesion pressure π_s^* due to dispersion interactions between the segments of hydrocarbon chains of CTA^+ ions situated in air as a function of area per ion at 303 K

sion interactions between the segments of hydrocarbon chains of adsorbed amphiphilic ions situated in air to the cohesion component of surface pressure for a given interface at low occupancy degrees ($\sim 0,1 < \theta < \sim 0,7$). In other terms, in this region the behaviour of CTAB adsorption layers is similar for both the interfaces studied. This region corresponds to the second linear part of the isotherm ΔV vs. $\ln c$ (fig. 5).

The increase in a_0 values up to $\sim 3,0$ observed for CTAB monolayers at the water-octane interface as the layer tends to saturation (see fig. 7) results probably from an increase of the role of hydrophobic interactions between the segments of adjacent aliphatic chains submerged into water when they are brought into the neighbourhood of others.

A more pronounced increase (up to $\sim 6,0$) in the attraction constant for CTAB monolayers at the water-air interface, as compared with the water-octane interface, can be explained by the fact that in this case besides hydrophobic interactions at high surface occupancy degrees ($\theta > \sim 0,65$) bringing together the adsorbed ions favours a more considerable manifestation of dispersion interactions between their hydrocarbon radicals situated in air. Indeed, it can be seen in figure 9 that the difference π_{12}-π corresponding, as noted above, to the extent to which dispersion interactions are pronounced in adsorption layers at the solution-air interface, starts sharply increasing (from several tenths to about $10 \ mN \cdot m^{-1}$) in the region $\theta > 0,65$ ($A < \sim 65 \ Å^2$) as θ rises.

Thus the analysis of relationships between π and A, a_0 and θ, π_s^* and A provides evidence for existence in CTAB adsorption layers at the water-air and water-octane interfaces of three regions characterized by different extents to which interactions between the adsorbed amphiphilic ions are pronounced. The region $\theta < \sim 0,1$ is that of the ideal behaviour of the CTAB monolayer where a_0 and π_s^* both are equal to zero. For the region of intermediate surface occupancy degrees $\sim 0,1 < \theta < \sim 0,7$ an insignificant manifestation of interactions between adsorbed CTA$^+$ ions is characteristic, a_0 being constant and equal to ~ 1. As the layers tend towards saturation ($\theta > \sim 0,7$) they are found to approach the condensed state, which is witnessed by a monotonous rise of the a_0 and π_s^* values.

The existence of three characteristic regions follows also from independent measurements of ionisation potentials for CTAB monolayers at the solution-air interface at 303 K. Indeed, the ΔV vs. $\ln c$ isotherm

(fig. 5) exhibits an inflexion point at approximately the same θ values. Analyzing the a_0 values obtained and the character of their changes as a function of θ and of the nature of the second phase has enabled us to reveal the existence of hydrophobic interactions between the segments of aliphatic chains of adsorbed amphiphilic ions submerged into water.

References

1. Davies, J., J. Colloid Sci., 11, 377 (1956).
2. Davies, J., Proc. Royal Soc., Ser. A, **208**, 224 (1951).
3. Bauer, E., Guastalla, J., Guastalla, L., Lize, A., J. Chim. Phys., **65**, 99 (1968).
4. Phillips, J., Rideal, E., Proc. Royal Soc., Ser. A, **232**, 159 (1955).
5. Haydon, D., Phillips, J., Trans. Faraday Soc. 54, 698 (1958).
6. Ghosh, B., J. Indian Chem. Soc. **45**, 1120 (1968).
7. Ghosh, B., J. Indian Chem. Soc. 43, 19 (1966).
8. Ghosh, B., J. Indian Chem. Soc. 47, 557 (1970).
9. Ghosh, B., J. Indian Chem. Soc. 48, 561 (1971).
10. Chatterjee, B., Ghosh, K., Kolloid Z. und Z. Polymer **250**, 615 (1972).
11. Pal, R., Chattoraj, D., J. Colloid Int. Sci. 52, 56 (1975).
12. Mingins, J., Taylor, J., Owens, N., Brooks, J., in coll. Monolayers, Washington; p. 28–43 (1975).
13. Shinoda, K., Nakagawa, T., Tamamushi, B., Isemura, T., Colloidal Surfactants, Academic Press, Lodnon (1963).
14. Davies, J., Rideal, E., Interfacial Phenomena, New York and London, Acad. Press, p. 268 (1963).
15. De Boer, J., The Dynamical Character of Adsorption, Oxford University Press, London (1953).
16. Frumkin, A., Z. Phys. Chem. 116, 566 (1925).
17. Matijevic, E., Pethica, B., Trans. Faraday Soc. 54, 1382 (1958).
18. Nakagaki, M., Handa, T., Shimabayashi, S., J. Colloid Int. Sci., **43**, 521 (1973).
19. Berezina, N. P., Nikolaeva-Fedorovich, N. V., Elektrokhimiya, 3, 3 (1967).
20. Pugachevich, P. P., Zh. fizich. khim. **36**, 1107 (1962).
21. Sugden, S., J. Chem. Soc. 25, 1177 (1924).
22. Volkov, B. N., Volak, L. D., Zh. fizich. khim. 46, 1025 (1972).
23. Zatkovetski, V. M., Pugachevich, P. P., Zavodskaya, laboratoriya 33, 837 (1967).
24. Folkhardt, D., Viustnek, R., Kolloidny, zh. 36, 1116 (1974).
25. Chand, P., Malik, W., J. Electroanalyt. Chem. and Ind. Electrochem. 47, 172 (1973).
26. Stenina, E. V., Damaskin, B. B., Fedorovich, N. V., Dyatkina, S. L., Dokl. AN SSSR 236, 400 (1977).
27. Tamamushi, B., Bull. Chem. Soc. Japan 8, 120 (1933); 2, 363; (1934), Kolloid-Z. 71, 150 (1935).

Received March 29, 1983

Authors' address:

Prof. Dr. E. Shchukin
Institute of Physical Chemistry
Academy of Sciences
Moscow, USSR

Progress in Colloid & Polymer Science   Progr. Colloid & Polymer Sci. **68**, 97–100 (1983)

The effect of surfactants on aqueous dispersions of iron oxides*)

K. Meguro, S. Tomioka, N. Kawashima, and K. Esumi

Department of Applied Chemistry, Institute of Colloid and Interface Science, Science University of Tokyo, Tokyo, Japan

Abstract: The effects of an anionic and a nonionic surfactant on aqueous dispersions of three types of iron oxides were investigated by- particle size and zeta potential measurements. Iron oxide particles were suspended in an aqueous iron (III) chloride solution to render their surface positively charged. The so prepared sols were then flocculated by the addition of a small amount of sodium lauryl sulfate. The flocs of iron oxide could be redispersed by further addition of sodium lauryl sulfate or by nonionic surfactants. The flocculation and the redispersion processes can be explained in terms of a "two-fold adsorption layer" mechanism of the surfactants. The concentrations of the nonionic surfactants needed to achieve redispersion, decreased with an increase in the ethylene oxide chain length. The sols of needle-like iron oxide particles required lower concentrations of surfactants for flocculation and redispersion than those containing cubic particles.

Key words: Iron oxide, dispersion, surfactant.

Introduction

The preparation of stable pigment dispersions for emulsion and water type paints requires the addition of dispersants. Many kinds of surfactants have been utilized for this purpose. For this reason, a number of studies [1–6] have been reported on the effects of surfactants on aqueous metal oxide suspensions, particularly with respect to flocculation and peptization phenomena [7–13]. Thus, the mechanisms established for such sols are applicable to pigments dispersed in aqueous surfactant solutions.

This work describes dispersion, flocculation, and redispersion processes of three types of iron oxides suspended in aqueous solutions of anionic and nonionic surfactants. The degree of dispersion and flocculation of these suspensions was estimated by measuring the mean particle sizes and the corresponding zeta potentials.

Experimental

Three types of iron oxides used in this study (R-110-A, R-516-L and Special Yellow) were supplied by Titan Kogyo Co., Ltd.; their pertinent properties are shown in table 1. High pure grade iron (III) chloride supplied by Kokusan Sangyo Co., Ltd. served as the initial dispersing agent. The surfactants used were sodium lauryl sulfate (SLS) and polyoxyethylene cetyl ether supplied by Nikko Chemi-

cals Co., Ltd. SLS was purified by recrystallization from ethyl alcohol. The polyoxyethylene cetyl ether is abbreviated as BC-n, where n indicates the average mole number of ethylene oxide.

The suspensions of iron oxides were prepared as follows: 0.01 g of solids was added to 150 cm^3 of a desired aqueous solution and treated for 5 min in a Bransonic ultrasonic bath. The particle size distribution was determined by a centrifugal particle size analyzer CP-50 (Shimadzu Seisakusho Co., Ltd.). The zeta potential of the samples was measured with a Laser Zee 500 (Pen Kem, Inc.) electrophoresis apparatus. The specific surface area of the samples was evaluated with a Sorptograph Model ADS-1B (Shimadzu Seisakusho Co., Ltd.).

Results and discussion

X-ray diffraction patterns of three iron oxides are presented in figure 1, which show that R-110-A and R-516-L samples are α-Fe$_2$O$_3$, and that Special Yellow is α-FeOOH.

To elucidate the effect of surfactants on the stability of iron oxide sols, it is necessary to obtain a well dispersed system. For this purpose, iron oxide solids were suspended in aqueous iron (III), chloride solutions and figure 2 give their mean particle sizes as a function of the concentration of the latter. Figure 3 shows that with increasing concentration of iron (III) chloride the zeta potential of the three iron oxides becomes more positive and reaches a maximum value at about 0.5 mM iron (III) chloride. The above results indicate that the iron oxide dispersion were most stable at this iron (III) chloride concentration; there-

---

*) Dedicated to the memory of Professor Dr. B. Tamamushi.

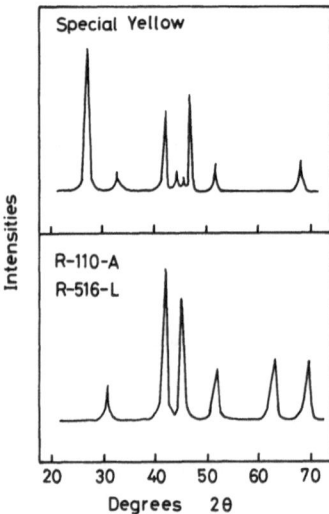

Fig. 1. X-ray diffraction patterns of the three iron oxides

Table 1. The properties of iron oxides

	Density g/cm³	Oil absorption g/100 g	Shape	Specific surface area m²/g
R-110-A	5.01	19 ± 3	cubic	6.4
R-516-L	4.42	52 ± 5	needle-like	18.3
Special Yellow	3.95	30 ± 3	needle-like	15.9

fore, the stock sol in 0.5 mM iron (III) chloride was used in further studies.

On the addition of a small amount of SLS to aqueous suspensions of iron oxide particles in 0.5 mM iron (III) chloride, the sol flocculated, but the flocs redispersed on further addition of the same surfactant. The mean particle size of R-516-L and Special Yellow increased with rising concentration of SLS, reached a maximum value at 0.3 mM of SLS in both cases, and then decreased by further addition of SLS; the mean particle size of R-110-A had a maximum value at 1 mM of SLS as shown in figure 4. The positively charged iron oxide particles in an aqueous iron (III) chloride solution were discharged in the presence of sufficient amount of SLS and then the charge was reversed to negative by further addition of this

surfactant (fig. 5). The point of zero charge of Special Yellow was at about 0.3 mM SLS and that of R-110-A at about 1 mM of SLS; both values correspond to the concentration of SLS at which the maximum mean particle size was established. It is noteworthy that the needle-like iron oxide is flocculated by a lower concentration of SLS than the sol containing cubic particles. The flocs formed by the addition of this surfactant could be easily suspended in toluene by shaking, which indicates that the surface of the solids became hydrophobic by adsorption of SLS.

As the next step, the effect of nonionic surfactants on the redispersion of the iron oxide flocs was investigated. Figure 6 a shows that the mean particle size of sample R-110-A decreased with the concentration of the nonionic surfactants (BC-n), added to the flocculated system. The complete redispersion occurred at about 5×10^{-5} M of BC-n. Figures 6b and c give analogous data for R-516-L and Special Yellow. In both cases, the required concentration of the nonionic surfactants for the complete redispersion was about 6×10^{-7} M, which is a very low value as compared to that for R-110-A. As the needle-like particles have a larger specific surface area and a higher oil absorption value than the cubic particles, the difference in the required concentration of the nonionic surfactant to redisperse the sols might be interpreted by assuming

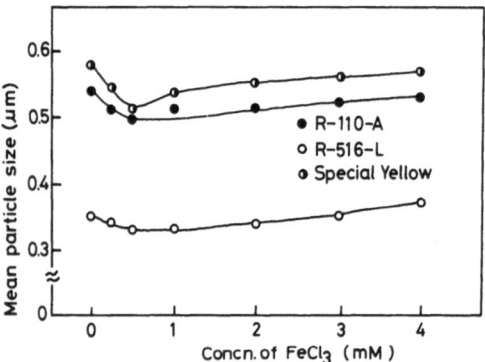

Fig. 2. The variation of mean particle size of the iron oxides with the concentration of iron (III) chloride in aqueous solution

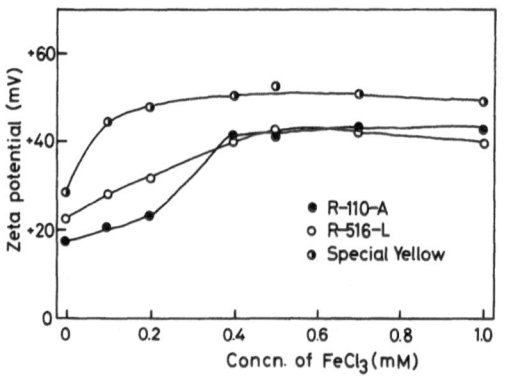

Fig. 3. The variation of the zeta potential of iron oxides with the concentration of iron (III) chloride in aqueous solution

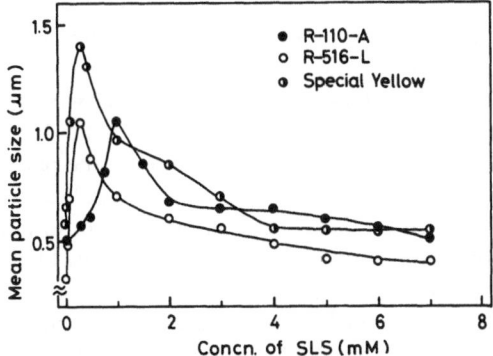

Fig. 4. The variation of the mean particle size of iron oxides in the presence of 0.5 mM iron (III) chloride as a function of added SLS

that the surfactant molecules adsorb preferentially at the edges of needle-like particles which already flocculate by the edge to edge interaction.

Figure 6 also demonstrates that the concentration of nonionic surfactants required for the redispersion is shifted to the lower concentration with an increase of ethylene oxide chain length. This finding indicates that the redispersion power of the nonionic surfactant is enhanced with an increase in the ethylene oxide chain length. Figure 7 shows that the flocs of iron oxide particles (samples R-110-A) in the presence of 1 mM of SLS were essentially uncharged and that the addition of the nonionic surfactant had no effect on the zeta potential. The above results lead to the conclusion that in this case the redispersion is not due to the electric repulsion, but to the steric hindrance caused by the adsorption of nonionic surfactant molecules.

As the conclusion, the mechanisms of flocculation and redispersion, which are schematically shown in figure 8, can be explained in terms of the "two-fold adsorption layer" similar to that suggested earlier [7,

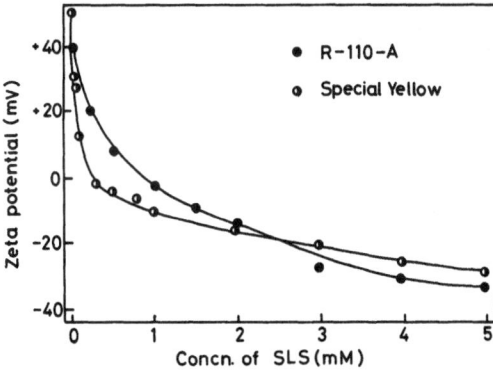

Fig. 5. The variation of the zeta potential of iron oxides in the presence of 0.5 mM iron (III) chloride as a function of added SLS

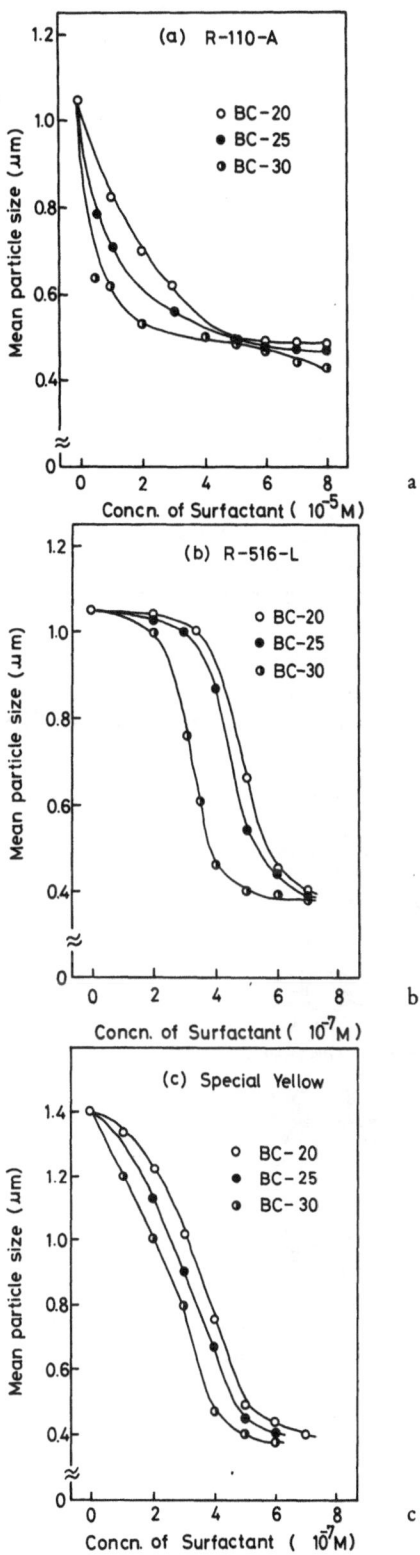

Fig. 6. The variation of the mean particle size of iron oxide in the presence of 0.5 mM iron (III) chloride and of SLS by the addition of the nonionic surfactants with different ethylene oxide chain length: a) R-110-A, in the presence of 1 mM SLS; b) R-516-L, 0.3 mM SLS; c) Special Yellow, 0.3 mM SLS

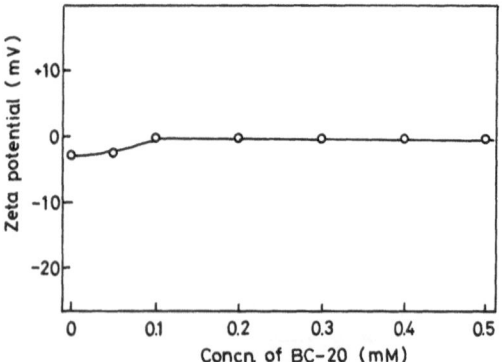

Fig. 7. The variation of the zeta potential of R-110-A with the concentration of *BC*-20 in the presence of 0.5 mM iron (III) chloride and of 1 mM SLS

8, 10]. That is, the iron oxide is positively charged in an aqueous iron (III) chloride solution providing a well dispersed state characterized by a minimum in the mean particle size and a maximum in the zeta potential. On addition of low concentrations of SLS, the particles are discharged and the size of flocculated particles increases. As the hydrocarbon chains of SLS

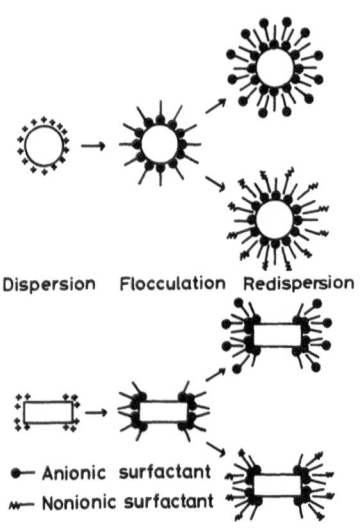

Fig. 8. A model for the mechanism of the flocculation and redispersion of iron oxide particles by surfactants

are oriented outward, the particles become hydrophobic and flocculation takes place. On further addition of SLS, the adsorption of this surfactant occurs by a chain to chain interaction resulting in a two-fold layer with the ionized groups of the second layer oriented towards the solution. Accordingly, the particles are recharged and the redispersion achieved due to the electric repulsion. Redispersion by the addition of nonionic surfactants, on the other hand, is most likely due to the steric hindrance between the adsorbed nonionic surfactant molecules. Further, from a standpoint of particle shape of iron oxide, it seems that the edge to edge interaction of needle-like particles plays an important role on the flocculation and redispersion processes shown in figure 8.

References

1. Ogihara, K., Tomioka, S., Esumi, K., Meguro, K., Shikizai Kyokaishi **55**, 546 1982).
2. Kuno, H., Abe, R., Tahara, S., Kolloid-Z. **198**, 77 (1964).
3. Fukushima, S., Kumagai S., J. Colloid Interface Sci. **42**, 539 (1973).
4. Moriyama, N., J. Colloid Interface Sci. **50**, 80 (1975).
5. Carr, W., J. Oil Col. Chem. Assoc. **54**, 155 (1971).
6. Corkill, J. M., Goodman J. F., Tate, J. R., Trans. Faraday Soc. **62**, 979 (1966).
7. Meguro, K., Kondo, T., Nippon Kagaku Zasshi, **76**, 642 (1955).
8. Meguro K., Kogyo Kagaku Zasshi, **58**, 905 (1955).
9. Mathai, D. G., Ottewill, R. H., Trans. Faraday Soc. **62**, 750 (1966).
10. Matijević, E., Ottewill, R. H., J. Colloid Sci. **13**, 242 (1958).
11. Ottewill, R. H., Rastogi, M. C., Trans. Faraday Soc. **56**, 866 (1960).
12. Ottewill, R. H., Watanabe, A., Kolloid-Z. **170**, 132 (1960).
13. Ottewill, R. H., Rastogi, M. C., Watanabe, A., Trans. Faraday Soc. **56**, 854 (1960).

Received December 13, 1982;
accepted January 17, 1983

Authors' address:

Kenjiro Meguro
Department of Applied Chemistry
Institute of Colloid and Interface Science
Science University of Tokyo
Kagurazaka 1–3
Shinjuku-ku, Tokyo 162, Japan

Direct measurement of the pressure of electrical double layer interaction between charged surfaces*)

D. B. Hough**) and R. H. Ottewill

School of Chemistry, University of Bristol, Bristol, U.K.

Abstract: The electrostatic force of repulsion between two charged solid surfaces, and acting normally between them, has been determined by direct measurement. Simultaneously the thickness of the aqueous film between the surfaces was measured optically using an interference method. One solid surface was an optically polished glass prism and the other an optically smooth spherical cap of transparent poly(-isoprene) rubber. The charges on the surfaces were effected by the use of anionic surface active agents which formed adsorbed layers on the surfaces. Both sodium dodecyl sulphate and sodium hexadecyltrioxyethylene glycol sulphate were used. The ionic strength of the solution was varied by the addition of sodium chloride. The experimental results were compared with those expected theoretically for interaction between two diffuse electrical double layers.

Key words: Surface forces, surface active agents.

Introduction

The stability of many colloidal dispersions has its origin in the electrostatic repulsion which occurs as the particles approach and their diffuse electrical double layers overlap. This fact was clearly recognised by Langmuir in 1938 who calculated the electrostatic pressure of repulsion arising from the interaction of two planar diffuse double layers [1]. Subsequently, the development of the theory of stability of lyophobic colloids as propounded by Derjaguin and Landau [2] and Verwey and Overbeek [3] utilised the electrical double layer repulsion and, in addition, incorporated an attractive interaction depending on the dispersion forces between the particles. Since the inception of the so-called DLVO (Derjaguin-Landau-Verwey-Overbeek) theory there have been some direct attempts made to measure forces of electrostatic repulsion but on very few systems. Most of the evidence on the nature of electrostatic effects has come from measurements of coagulation kinetics; this approach is very indirect and only applicable over a limited range of electrolyte concentration. Direct experimental measurement of van der Waals forces, however, has been very extensive. This work has been reviewed by Israelachvili and Tabor [4].

One of the first direct demonstrations of the dependence of electrostatic repulsive forces on potential and electrolyte concentration was that of Derjaguin et al. [5] using polarized crossed metal-wires. In 1974 Peschel, Aldfinger and Schwarz [6] measured the interaction pressure between two fused silica plates in electrolyte solutions.

Roberts and Tabor [7] introduced the idea of using an optically smooth spherical cap of transparent rubber to provide one charged surface and an optically flat glass surface to provide the other. When these surfaces were pushed together electrostatic repulsion occurred leading to the formation of a thin aqueous film. The rubber deformed easily over local undulations on the glass so that over the compression region the surfaces were essentially parallel and hence the separation distance could be obtained by light reflectance measurements. Their apparatus was so designed that a compressive force could be applied, by means of an added weight, normal to the parallel surfaces to balance the electrical double-layer force acting between the plates. However, it was found that this type of apparatus could not easily be used to make measurements at very low pressures [8] and it could not easily be adapted to study the kinetics of film drainage between solid surfaces.

---

*) Dedicated to the memory of Professor Dr. B. Tamamushi.
**) Present address: Unilever Research, Port Sunlight Laboratory, Bebington, Wirral, L63 3JW.

In 1976 we reported work [9] on the forces of electrostatic repulsion between glass and rubber surfaces coated with adsorbed layers of dodecyl sulphate ions as a function of electrolyte concentration. These measurements were made on a more sensitive apparatus than that used by Roberts and Tabor. With this new apparatus it proved possible to carry out kinetic measurements and to make equilibrium measurements over a wide range of pressures. We now report an extensive investigation of the interaction between electrically charged glass and rubber surfaces in aqueous solutions of anionic surface active agents, both in the absence and presence of added electrolytes. These measurements provide some insight into the behaviour of glass and polymer surfaces in an aqueous environment. The substrates present an alternative model surface to the mica studied by Israelachvili and Adams [10, 11], Israelachvili [12], Pashley and Israelachvili [13] and Pashley [14]. Moreover, they also give a direct method of examining the effect that surface active agents have on surfaces and their interaction properties. This was a topic in which Professor Bun-ichi Tamamushi always took a great interest [15] and we dedicate this paper to his memory.

Experimental

Materials

All the water used was twice distilled and used direct from the still.

The sodium chloride was Analar material which was roasted at 800 °C before use to remove greasy impurities.

The sodium dodecyl sulphate (SDS) was purchased from B.D.H. It was purified by continuous liquid-liquid extraction of a 10% solution in an ethanol-water mixture (70:30) with 60–80° petroleum ether (Analar grade). The extracted material was then recrystallized from alcoholic solution (10% water + 90% ethanol) at −10 °C. A freshly prepared solution of the purified SDS did not exhibit a minimum in the surface tension against concentration curve. The critical micelle concentration was found to be 8×10^{-3} mol dm^{-3}.

The sodium hexadecyltrioxyethylene glycol sulphate ($C_{16}E_3S$) was a pure sample kindly supplied by Dr. R. G. Laughlin of the Procter and Gamble Company, Cincinnati, Ohio, USA. The literature values [16] for the c.m.c. of this material are 7×10^{-5} mol dm^{-3} (surface tension) and 1.23×10^{-4} mol dm^{-3} (conductance) at 25 °C.

Apparatus for measuring repulsive forces

The central part of the apparatus used for the direct measurement of repulsive forces is shown in figure 1. The rubber surface was in the form of a cylinder (A) of polyisoprene rubber, one end of which was moulded into the form of an optically smooth spherical cap. The flat end of the cylinder was coated with a matt-black paint and was firmly supported in a blackened metal cup (B). The latter was rigidly mounted on the drive movement (C) of a micrometer unit

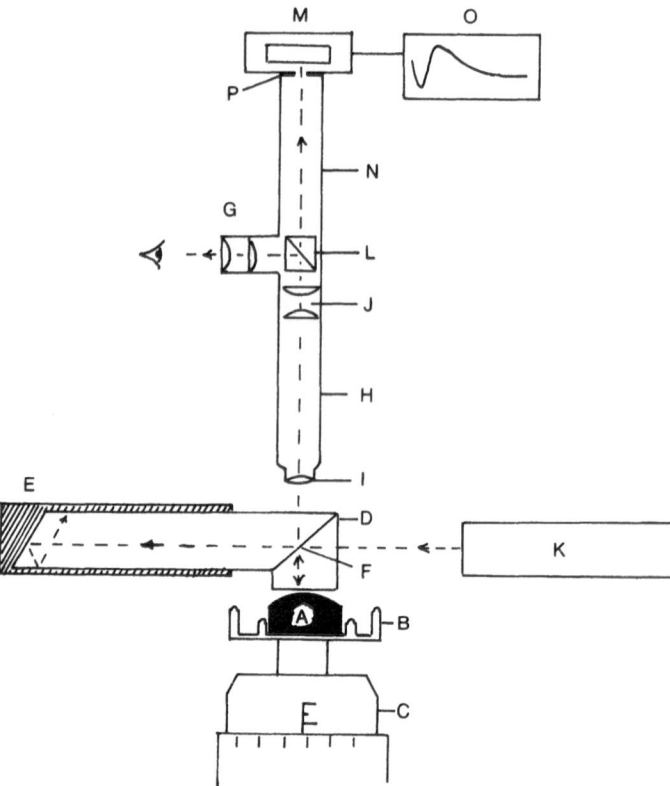

Fig. 1. Schematic diagram of apparatus for measuring interaction pressures. A = poly-(isoprene) rubber cap; B = support cup for rubber; C = micrometer drive; D = glass prism; E = clamp for prism; F = beam splitter; G = viewing device; H = microscope; I = microscope objective; J = microscope eyepiece; K = laser; L = beam splitter; M = photomultiplier tube; N = extension tube; O = chart recorder; P = variable aperture

(L.S. Starrett Company Limited; Type 468 MP) which could be read directly in 2 μm increments; this unit enabled movement of the rubber in the vertical direction to be obtained in a very precisely controlled manner. Lockable joints within the drive assembly allowed the spherical surface of the rubber to the both rotated around the drive unit and also to be tilted (shown as vertical in fig. 1). For measurements the cylinder was always set in a tilted position in order to eliminate any stray reflections from its matt-black base. These features are illustrated in the photograph, figure 2.

A glass prism (D), the planar surface of which had been polished to within a quarter of a wavelength for the sodium D line, was mounted rigidly relative to the micrometer movement, in a blackened recess (E). The latter also acted as a Rayleigh Horn and absorbed, after multiple reflections, all light transmitted horizontally through the beam-splitting cemented interface (F).

In order to carry out an experiment, the aqueous solution was injected between the lower surface of the glass prism and the spherical surface of the rubber. The reservoir of the cup (B) was filled with solution to ensure saturated vapour conditions, and the entry of dust into the region was prevented by enclosing it in very thin plastic sheeting. Direct observation of the film-area was made through the external viewing eyepiece (G), via the microscope (H) which was fitted with objective (I) and eyepiece (J). A wide range of microscope objectives and eyepieces was used to examine visually the film region according to the area desired. For example, for the

Fig. 2. Photograph showing mounted rubber cap and glass prism

whole area $a \times 1.25$ eyepiece was employed whereas to examine a selected area containing the film region, $a \times 10$ long working distance objective was needed. Illumination was provided by a 10 mW helium-neon laser (K) (Scientifica and Cooke) which emitted constant intensity, monochromatic light of wavelength 632.8 nm. This source illuminated the whole area in the field of view. A beam-splitting device (L) situated above the microscope eyepiece was used to divide the light between the viewing eyepiece and the photomultiplier tube (M). The latter, powered by a stabilised high voltage supply, was situated in a housing at the end of an extension tube (N) of adjustable length but usually in the order of 12 cm. The purpose of this tube was to allow further magnification of the image from the microscope eyepiece to occur. The electric signal from the photomultiplier tube was fed directly to a fast response flat bed chart recorder O (Oxford 1000). Although the whole film area was continuously observed through the viewing eyepiece, only light reflected from the centre of the film was allowed to be incident on the photomultiplier tube, by suitable adjustment of a variable aperture (P) situated just in front of the photomultiplier tube. The intensity of light reflected from the centre of the film was thus monitored continuously, in arbitrary units, by the chart recorder, as a function of time. The recorder was modified to run at chart speeds of up to 100 cm per min.

Preparation of rubber surfaces

The rubber used was a synthetic, 95%-cis-poly(isoprene) containing 2.5% *w/w* dicumyl peroxide. About 4 g of the uncured rubber was cut into the shape of a cone and placed apex down against the surface of a glass lens which was situated at the bottom of a cylindrical stainless steel mould, 21.5 mm in diameter. Both lens and mould were pre-heated to 150° in a hot press and then a stainless steel plunger was pressed gently downwards to give an applied pressure of 0.3 atmosphere for a period of 3 min in order to mould the rubber. Curing of the rubber took place in about 40 min in the absence of applied pressure.

The material prepared in this manner in addition to being an elastic body was optically transparent and had a refractive index of 1.520 at $\lambda_o = 632.8$ nm.

Just prior to use the lens was slowly peeled from the poly(isoprene) under propan-2-ol to minimize the adherence of dust particles to the spherical surface; the surface was then allowed to dry in a dust-free atmosphere. A drop of water placed on the rubber showed a high contact angle, thus demonstrating the hydrophobic nature of the surface.

The lens used to mould the surface was plano-concave with a diameter of 20.5 mm and a radius of curvature of −2.14 cm for the concave face. The radius of curvature of the rubber was hence taken as 2.14 cm.

Transmission electron microscopy of carbon replicas shadowed at low angles showed some asperities which were very small compared to the wavelength of light. Thus the surfaces were not molecularly smooth but because of the low Young's modulus, they easily deformed to follow the contours of the glass surface. The quality of the surfaces is demonstrated by the fringes shown in figure 3.

The glass prims

The glass prism was made from Chance Pilkington Hard Crown Glass, type 519604, and had a refractive index of 1.520 at $\lambda_o = 632.8$ nm. The Young's modulus of the glass was in excess of 10^{11} N m^{-2}.

The lower face of the prism was rectangular with dimensions 7.6 $\times$ 1.3 cm. All faces of the prism not used for light transmission were blackened. Before mounting the prism in the recess (E, fig. 1) the lower face was cleaned by successive washings in propan-2-ol and 50/50 nitric acid followed by distilled water. Zero contact angle with a water drop was taken as an indication of a clean surface.

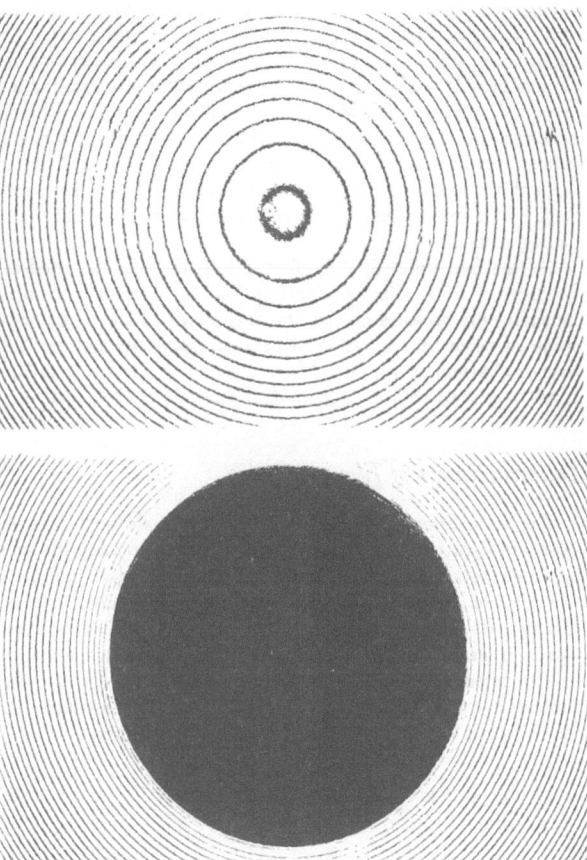

Fig. 3. a) Interference fringes, "no-contact" situation, b) interference fringes, "contact" situation showing the presence of the "black" film

Procedure for observations on thin films

The spherical surface of rubber was placed ca. 1 μm below the lower face of the glass prism and then adjusted so that the Newton's rings, so formed, were concentric with the centre of the field of view of the microscope (H). A photograph of the fringes is shown in figure 3a. The clarity of the fringes demonstrates the high quality of the surfaces. In fact, the fringe quality was used to judge the quality of the rubber surfaces and those surfaces with imperfections or dust particles were immediately rejected. At this point, the rubber was slowly advanced towards the glass surface until contact, in the form of a circular area, was made between the two surfaces. Only a very small amount of light was reflected from this contact area, for high quality rubbers, and this was recorded as the background intensity. The latter remained constant as the applied pressure was varied and was "backed-off" on the recorder.

Approximately 0.5 cm³ of solution of surface active agent was applied to the space around the rubber-glass contact region where it was held by capillary suction. The rubber surface was then slowly retracted until contact was broken and the rubber surface returned to its spherical shape About 15 minutes were allowed to elapse in order for the surface active ions to adsorb at the glass-solution and rubber-solution interfaces. The rubber was then advanced until the separation of the surfaces was of the order of 200 nm, this being a distance between the last dark and bright fringes (i.e. between $\lambda_o/2\bar{n}$ and $\lambda_o/4\bar{n}$ respectively, where $\bar{n}$ is a composite refractive index of the film, comprising adsorbed layers and aqueous core). Fringe quality was checked for this situation. At this point the thin film was quickly formed by rapidly advancing the rubber by a pre-determined amount in order to produce a small plane-parallel circular area. It was found from experience that it was inadvisable to exceed a radius of 7.5×10^{-3} cm otherwise dimple formation occurred; i.e. a small concave cavity was formed in the surface of the rubber. Figure 3b shows a photograph of the film area as seen through the microscope after equilibrium had been attained at this stage of the experiment. The black circular area of the thin film is clearly visible.

As the film thinned, the reflected intensity was recorded as a function of time. After equilibrium had been attained (constant intensity) the pressure on the film was increased and a new set of readings taken. A succession of pressure applications hence gave new equilibrium positions and from these it was possible to obtain a curve of applied pressure against equilibrium film thickness.

Determination of film thickness

The basic structure of the interfacial film was assumed to be that shown in figure 4a corresponding to the optical situation shown in figure 4b; that is, an optical sandwich bounded by glass and rubber which were chosen to have identical refractive indices n_2, and enclosing an aqueous core of refractive index n and thickness h, and two hydrocarbon layers of refractive index n_1 and thickness x.

The intensity of the light reflected from the film, I, when compared to that incident upon it, I_o, at near normal incidence was found by the Cabellero [17] method of analysis to be:

$$\frac{I}{I_o} = P \cdot P* \tag{1}$$

where,

$$P = \frac{-A + X \exp{(i\alpha)}}{1 - A X \exp{(i\alpha)}} \tag{2}$$

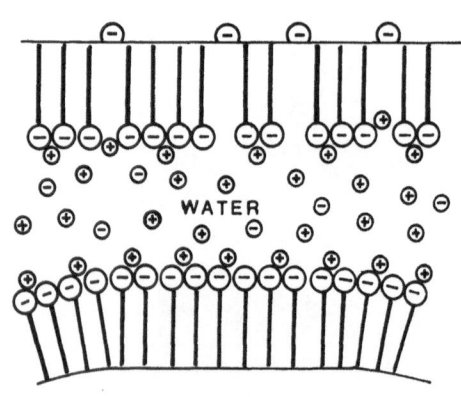

GLASS

WATER

RUBBER

a

Fig. 4. a) Schematic reperesentation of the structure of the liquid film; b) optical conditions corresponding to film structure

and

$$A = (n_1 - n_2) / (n_1 + n_2)$$

$$X = \frac{-B + ((B + A \exp{(i\alpha)}) / ((1 + AB \exp{(i\alpha)}) \text{ ex}}{1 - B((B + A \exp{(i\alpha)}) / ((1 + AB \exp{(i\alpha)}) \text{ ex}}$$

$$B = (n - n_1) / (n + n_1)$$

$$\alpha = \frac{4\pi n_1}{\lambda_o} \cdot x$$

$$\beta = \frac{4\pi n}{\lambda_o} \cdot h$$

$P* = $ the complex conjugate of P.

The magnitude of the intensity at the first bright fringe at $\lambda_o/4\bar{n}$, i.e. I_{max}, was obtained from the chart recorder. This was compared directly with the intensity of light reflected from the film, I, also obtained from the chart recorder to give the ratio I/I_{max}; I_o was maintained constant throughout the experiment. The ratio I/I_{max} was computed as a function of water film thickness h and a function of $1/h^2$. The values of x, determined from Catalin models, were 1 nm for SDS and 1.5 nm for $C_{16}E_3S$. n_1 was taken as 1.42 for SDS and 1.43 for $C_{16}E_3S$.

Determination of the average applied pressure on the film

The procedure adopted was to determine the pressure by an independent calibration procedure. The apparatus used was similar to that used by Roberts and Tabor [7] and Lewis [8]. The glass prism was securely clamped at one end of a beam of about 50 cm which was pivoted around a point close to the other end which was loaded with a suitable counter-weight. The beam was balanced and a spherical cap of the polyisoprene rubber was fixed just below the bottom (polished) face of the prism. A solution of SDS containing sufficient electrolyte to provide a stable but very thin film was inserted between the rubber and glass surfaces. Increments of weight, initially in steps of 0.1 g were then placed on the prism so depressing the rubber against the glass; the total added weight reached was usually ca. 150 g. The flattened area of the rubber was observed microscopically at each load and the radius of the flattened area, a, determined using a bifilar eyepiece.

According to the Hertz theory of elasticity [18], the radius of the flattened area is related to the radius of curvature of the spherical surface of the rubber, R, the Young's modulus of the rubber, Y_R, and the Young's modulus of the glass, Y_G, by the relationship,

$$a^3 = 0.75 \text{ mg } R \left[\frac{1 - \sigma_R^2}{Y_R} + \frac{1 - \sigma_G^2}{Y_G} \right] \qquad (3)$$

where m = the mass of applied load, g = the acceleration due to gravity, σ_R = Poisson's ratio for rubber (taken as 0.5) and σ_G = Poisson's ratio for glass (taken as 0.2). The experimental data provided a good linear plot of a^3 against m. Deviations from linearity at low applied loads were not observed, presumably because van der Waals attraction between the surfaces was prevented by the intervening stable film. Since the Young's modulus of the glass was always very much greater than that of the rubber, i.e. $Y_G >> Y_R$, then equation (3) for these conditions reduces to,

$$a^3 = 0.563 \text{ mg } R/Y_R \qquad (4)$$

thus enabling an estimate to be made of Y_R. The value of Y_R obtained by this method was $8.47 \times 10^5 \pm 0.1 \times 10^5$ N m^{-2}.

In all the experiments reported the area of the film was measured using the apparatus shown in figure 1. Hence, since the relationship between πa^2 and m was known from the calibration graph, the pressure applied to the film was calculated as the central applied pressure, P_{ap}, as given by the equation,

$$P_{ap} = 3 \text{ mg}/2\pi a^2 \qquad (5)$$

It has been shown by Hughes and White [19] that under equilibrium conditions, for low pressure and low electrolyte conditions a correction is needed to the applied pressure, i.e.

$$P_{ap}(\text{corr}) = C(\lambda) \cdot P_{ap} \qquad (6)$$

where $C(\lambda)$ is a correction to the central pressure as calculated using the Hertz theory of elastic deformation. The value of λ was given by the expression,

$$\lambda = \frac{2 P_R (x R_e)^{1/2} (1 - \sigma_R^2)}{Y_R} \qquad (7)$$

where P_R = the electrostatic force of repulsion (see later) and R_e = the effective radius of the deformable rubber cap for the situation of interaction between a planar glass plate and a spherical rubber cap. As pointed out by Hughes and White in this case $R_e = 2R$ and so in the present work R_e was taken as 4.28 cm. The values of σ_R and Y_R were those given above. From the computed curve given by Hughes and White it is clear that as soon as λ becomes greater than 5, $C(\lambda)$ tends to unity, and P_{ap} is correctly given by Hertz theory.

Values of λ were calculated for all the conditions used in the present work. As shown in figure 6 the correction makes a small difference to the results obtained with 4×10^{-3} and 6×10^{-3} mol dm^{-3} SDS at the distances greater than 26 nm. In all other cases the value of λ was substantially greater than 5 and hence correction of the data was not considered to be worthwhile.

Results

Structure of the thin aqueous film

For most experiments, the surface active agents were used in the region of the c.m.c. For hydrophobic substrates under these conditions, a vertically oriented monolayer of surface active agent is generally formed just prior to the c.m.c. [20]. Since poly-(isoprene) rubber constitutes a hydrophobic substrate it was assumed that a monolayer of the anionic surface active ions was formed on this surface at concentrations close to and above the c.m.c.; the most probable situation is illustrated schematically in figure 4a. The surface of glass is composed of mixed siloxane and ionized silicate groupings and it was assumed that the adsorbed layer was less than a complete monolayer on this substrate, as illustrated in figure 4a. On the basis of this model, the thickness of the aqueous core, h, was determined by intensity measurements assuming that this layer also included the hydrated head group. Since the interaction depends on interaction between diffuse double layers, the exact model is not critical because in the relationship between the surface charge and the potential of the diffuse double layer the latter rapidly reaches limiting high values.

Kinetics of film drainage

As the intensity of the light reflected from the aqueous thin film was measured continuously, the thickness, h, was obtained as a function of time. The results obtained using 2×10^{-4} mol dm^{-3} $C_{16}E_3S$ solutions containing zero, 8×10^{-4} mol dm^{-3} and 9.8

$\times\ 10^{-3}$ mol dm^{-3} sodium chloride are illustrated in figure 5. The curves, plotted in the form $1/h^2$ against Δt, where Δt = the time interval from the start of the observations, are linear at short time intervals, then become curved at intermediate values and finally become parallel to the time axis.

The drainage of liquid from between two parallel discs of radius, a, under a net applied pressure P was examined theoretically by Reynolds (21) who derived the equation,

$$\frac{d\,(1/h^2)}{d(\Delta t)} = \frac{4}{3\eta\,a^2}\,P \qquad (8)$$

where η = the viscosity of the liquid flowing between the discs. Reynolds took for the boundary conditions that the velocity of liquid flow was zero at both surfaces, a condition which appears to be fulfilled in the present experiments. Moreover, once a small pressure was applied ($\sim 1.5 \times 10^3$ N m^{-2}) the film area formed immediately and remained essentially constant in cross-sectional diameter during the drainage process.

At long distances of surface separation, i.e. $h > 65$ nm, the repulsive pressure, acting normally to the surface and arising from the overlap of the diffuse electrical double layers on both surfaces, is very small. Consequently, P is equal to the actual applied pressure, P_{ap}, and the Reynolds equation is obeyed (9). These conditions give excellent linearity in the plots of $1/h^2$ against t. From each curve in the linear region, the

viscosity of the film was determined from the initial slope (9).

At shorter distances of surface separation, however, the electrostatic repulsive pressure, P_R, which is a function of h, opposes the applied pressure and the applied pressure in the film becomes,

$$P = |P_{ap}| - |P_R| \qquad (9)$$

and hence,

$$|P_R| = |P_{ap}| - \frac{3\eta\,a^2}{4}\,\frac{d(1/h^2)}{d(\Delta t)}\,. \qquad (10)$$

In this region the net pressure varies continuously with h and hence the plot of $1/h^2$ against Δt is curved. Since all the quantities on the right hand side of equation (10) are known, then P_R can be calculated for a particular value of h. The value of η used in equation (10) was that obtained from the initial region of the $1/h^2$ against Δt curve. Points obtained by this method of analysis are henceforth described as kinetic points.

Finally, at even shorter distances, a situation is reached where the applied pressure is exactly balanced by the electrostatic repulsive pressure, so that,

$$|P_{ap}| = |P_R| \qquad (11)$$

and the plot of $1/h^2$ against Δt becomes parallel to the abscissa indicating that the thin film had attained a constant thickness. Results obtained at this position are henceforth described as equilibrium points. It should be mentioned that following the normal sign convention P_R, the repulsive pressure is taken as positive since it prevents thinning of the film, whereas P_{ap} is essentially negative since it acts to thin the film, i. e. it acts in the same direction as an attractive force.

The values of the film viscosity obtained from the analysis of the initial regions of the $1/h^2$ against Δt curves are listed in table 1. There appears to be a very

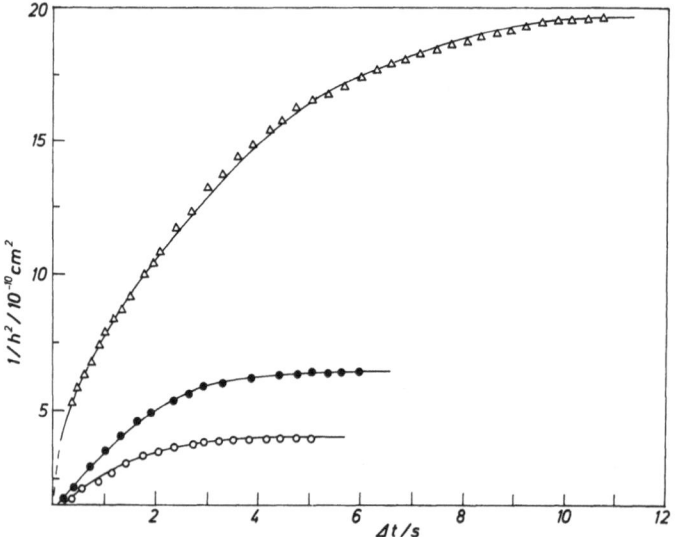

Fig. 5. Curves of $1/h^2$ against Δt illustrating film drainage under applied pressure, $\bigcirc$, 2×10^{-4} mol dm^{-3} C$_{16}$E$_3$S; $\bullet$, 2×10^{-4} mol dm^{-3} C$_{16}$E$_3$S + 8×10^{-4} mol dm^{-3} sodium chloride; $\triangle$, 2×10^{-4} mol dm^{-3} C$_{16}$E$_3$S + 9.8×10^{-3} mol dm^{-3} sodium chloride

Table 1. Thin film viscosities. Temperature 22° ± 1°

Surface active Agents	Concentration /mol dm^{-3}	Sodium chloride concentration /mol dm^{-3}	Ionic strength	Viscosity η/Pa s
C$_{16}$E$_3$S	2×10^{-4}	–	2×10^{-4}	4.01×10^{-3}
	2×10^{-4}	8×10^{-4}	1×10^{-3}	3.26×10^{-3}
	2×10^{-4}	9.8×10^{-4}	1×10^{-2}	2.87×10^{-3}
SDS	6×10^{-3}	–	6×10^{-3}	2.83×10^{-3}
	6×10^{-3}	1.4×10^{-2}	2×10^{-2}	2.51×10^{-3}
	6×10^{-3}	4.4×10^{-2}	5×10^{-2}	1.79×10^{-3}
	6×10^{-3}	9.4×10^{-2}	1×10^{-1}	1.10×10^{-3}

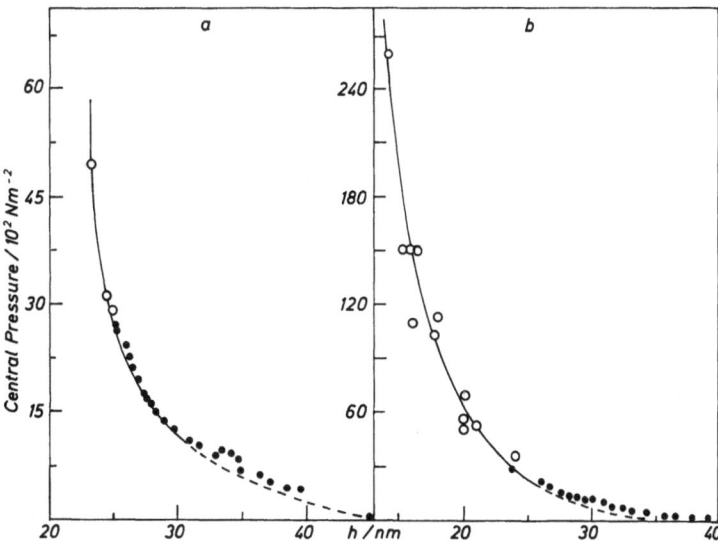

Fig. 6. Central pressure against film thickness, h. a) 4×10^{-3} mol dm^{-3} SDS; b) 6×10^{-3} mol dm^{-3} SDS. Filled points indicate kinetic values. $- - -$, data corrected by the procedure of Hughes and White [19]

significant trend of the results with ionic strength suggesting the possibility of an appreciable electroviscous effect in the drainage results. This would not be surprising in view of the appreciable volume of solution occupied by the elctrical double layer as compared with that of bulk solution. For rather high values of ionic strength, 0.1, the viscosity of the film liquid approaches that expected for the bulk liquid.

Pressure–film thickness results

In figure 6 the results are presented for interaction between glass and rubber surfaces in 4×10^{-3} and 6×10^{-3} mol dm^{-3} SDS in the form of central pressure against film thickness. The equilibrium values of the central pressure are given as open points and those obtained by the kinetic method as filled points. Overall there is excellent concordance between the two experimental methods. The agreement between the two sets of data suggests that the rearrangement of the counter-ions in the overlapping double layers occurs on a time scale which is short compared with the rate of the film thinning process. Also shown in figure 6 are the central pressure curves corrected using the procedure suggested by Hughes and White [19]. As can be seen the correction under the conditions used is small and almost within the experimental error of the measurements even when it is most pronounced at long distances.

Figure 7 gives the results obtained in 10^{-2} and 10^{-1} mol dm^{-1} SDS. It is clear from these that as the concentration of surface active agent is increased the

steeply rising portion of the curve moves to shorter distances of separation. The results obtained on the addition of salt to a 6×10^{-3} mol dm^{-3} SDS solution are illustrated in figure 8. With the lowest concentration of SDS used, 4×10^{-3} mol dm^{-3}, interaction was first distinctly detectable at a film thickness of ca. 47.5 nm. At the highest total ionic strength used (6×10^{-3} mol dm^{-3} SDS + 9.4×10^{-2} mol dm^{-3} sodium chloride) the first positive central pressure was detected at 7.6 nm. These figures (6, 7 and 8) clearly demonstrate the influence of ionic strength on the electrical interaction forces between two adsorbed layers of ionic surface active agent.

The results obtained with $C_{16}E_3S$ both in the absence and presence of salt are given in figure 9. The lower c.m.c. of this material made it feasible to study even lower ionic strengths than with SDS and at the lowest concentration, 2×10^{-4} mol dm^{-3}, an interaction pressure was detected at a distance of 65 nm.

Discussion

Properties of glass and rubber surfaces

Both the glass and the polyisoprene surfaces were prepared in an optically smooth form rather than in a molecularly smooth state, i.e. as can be obtained with mica (4). Nevertheless, electron microscope examination showed that the rubber surfaces were smooth down to the resolution limit of a carbon-platinum replica (ca. 3nm). The work of Lewis [8], who determined the electrokinetic potential of poly-(iso-

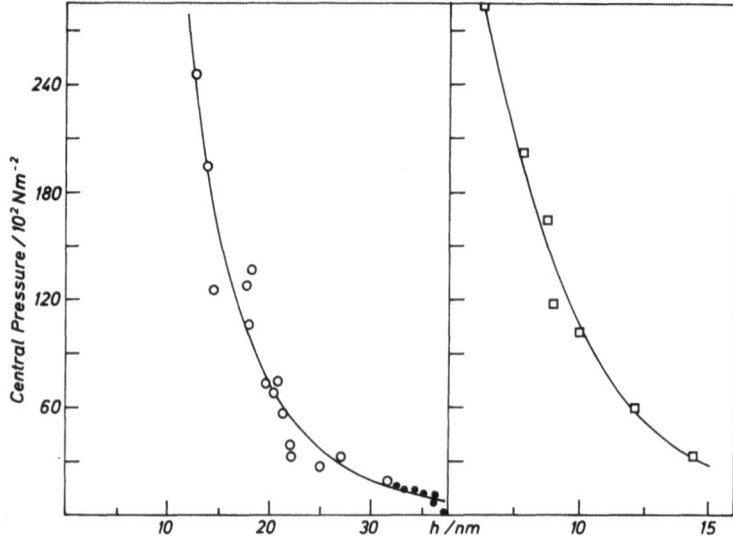

Fig. 7. Central pressure against film thickness, h. $\bigcirc$, 10^{-2} mol dm^{-3} SDS; $\square$, 10^{-1} mol dm^{-3} SDS. Filled points indicate kinetic values

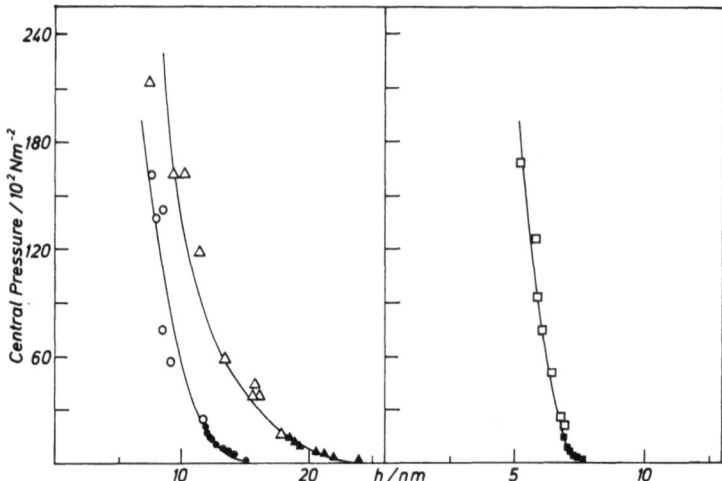

Fig. 8. Central pressure against film thickness h. $\triangle$, 6×10^{-3} mol dm^{-3} SDS + 1.4×10^{-2} mol dm^{-3} sodium chloride; $\bigcirc$, 6×10^{-3} mol dm^{-3} SDS + 4.4×10^{-2} mol dm^{-3} sodium chloride; $\bigcirc$, 6×10^{-3} mol dm^{-3} SDS + 4.4×10^{-2} mol dm^{-3} sodium chloride; $\square$, 6×10^{-3} mol dm^{-3} SDS + 9.4×10^{-2} mol dm^{-3} sodium chloride. Filled points indicate kinetic values

prene) surfaces using moulded capillaries, showed that the poly-(isoprene) surface had a few charged sites present, which presumably arose from the decomposition of the initiator, dicumyl peroxide; consequently the surface was weakly negatively charged. The high contact angle of about 60° for water droplets on the poly-(isoprene) surface indicated that the surface was also very hydrophobic. Glass usually has a negative electrokinetic potential [22] and Lewis [8] found a value of −100 mV for the glass used for the prism in 8 × 10^{-3} mol dm^{-3} SDS solution containing 10^{-3} mol dm^{-3} sodium chloride. In the absence of surface active

agent it was not possible to form a stable thin film between the glass and the rubber, thus inferring the presence of only small electrostatic potentials on the two native surfaces. Experimental evidence on hard glass indicates that the ionized groupings, almost certainly arising from the ionization of silanol groups, are well-spaced and that the surface has a large area of exposed siloxane groups. Hydrophobizing the glass chemically using dichlorodimethylsilane did not change the results obtained in the pressure versus distance measurements suggesting that sufficient adsorption of surface active agent occurred on the

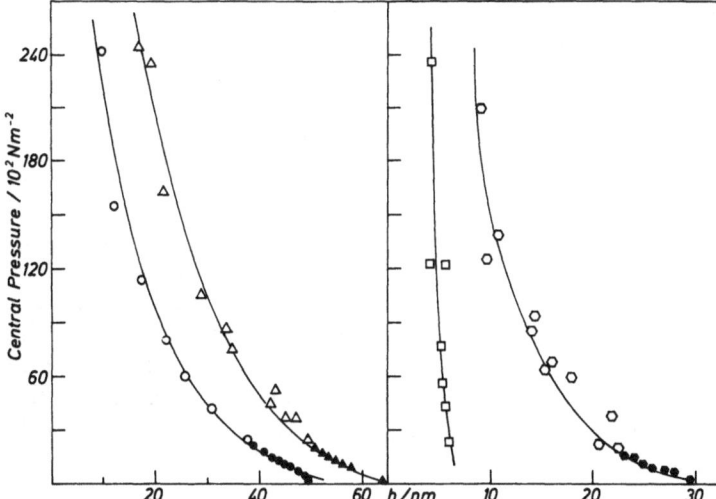

Fig. 9. Central pressure against film thickness h. All solutions contained 2×10^{-4} mol dm^{-3} C$_{16}$E$_3$S. $\triangle$, C$_{16}$E$_3$S alone; $\bigcirc$, + 8 $\times 10^{-4}$ mol dm^{-3} sodium chloride; $\bigcirc$, + 9,8 $\times 10^{-3}$ mol dm^{-3} sodium chloride; $\square$, + 0.1 mol dm^{-3} sodium chloride. Filled points indicate kinetic values

untreated glass to stabilise the film. Thus the model proposed for adsorption in figure 4a seems to be reasonable and the repulsive force set up after adsorption, and the stability of the thin liquid films formed, are in keeping with the suggestion of the presence of adsorbed layers on both surfaces.

Interaction between charged surfaces

According to the theory of Derjaguin and Landau [2] and Verwey and Overbeek [3], the pressure of interaction between two planar surfaces can be written as,

$$P_T = P_R + P_A \qquad (12)$$

where P_T = the total pressure of the interaction, P_R = the pressure of electrostatic interaction and P_A = the attractive pressure arising from van der Waals dispersion interactions.

For the systems used in the present work composed of glass, poly-(isoprene) rubber, water and the hydrocarbon chains of surface active agents with zero distance Hamaker Constants, calculated from Lifshitz theory [23], of 7.28×10^{-20} J, 5.99×10^{-20} J, 3.70×10^{-20} J and 5.12×10^{-20} J respectively, calculations of P_A applying the arguments of Vold [24] for interactions between flat plates showed that the van der Waals contribution was essentially negligible at distances of separation greater than 4 nm. Since attraction effects occurred at distances smaller than those at which pressure measurements were made, we con-

clude that the effects observed were due to electrical double layer repulsion only.

Electrical double layer repulsive pressure

Since, for our measurements, the van der Waals attraction can be neglected then equation (12) can be rewritten as,

$$P_T = P_R. \qquad (13)$$

It was shown by Langmuir [1] that, the repulsive pressure generated by the overlap of the electrical double layers of two approaching flat plates, at identical and constant surface potentials, ψ_s, is given by,

$$P_R = 2 n_o kT [\cosh u - 1] \qquad (14)$$

where n_o = the number of electrolyte ions of each type present per unit volume of the bulk electrolyte phase, k = Boltzmann's constant and T = absolute temperature. For interaction in a 1 : 1 electrolyte the quantity, u, is given by,

$$u = e \, \psi_d / kT \qquad (15)$$

where e = the fundamental unit of charge and ψ_d = the electrostatic potential in a plane mid-way between the two plates. d = the distance from the surface at a potential ψ_s to the mid-plane.

The relationship between d and the surface potential ψ_s, is given by the integral equation (3),

$$- \varkappa d = \int_z^u \frac{dy}{[2(\cosh y - \cosh u)]^{1/2}} . \qquad (16)$$

In this equation,

$$z = e\psi_s/kT \qquad (17)$$

and

$$y = e\psi/kT \qquad (18)$$

where ψ the electrostatic potential expressed as a variable and $\varkappa$, for a 1:1 electrolyte is expressed as,

$$\varkappa^2 = 2 n_o e^2/\varepsilon_r \varepsilon_o kT \qquad (19)$$

with ε_r = the relative permittivity of the medium and ε_o = the permittivity of free space.

As shown by Verwey and Overbeek [3] the solution of equation (14) cannot be obtained explicitly and is therefore expressed in the form given in equation (16). They gave the values of $\varkappa d$ in their monograph for assigned values of u and z and from these, curves can be constructed of $\varkappa d$ against u for constant values of z. These are shown in figure 10 for $z = 4$ and $z = 10$.

In order to compare the experimental data with these curves the experimental curve of central pressure against h obtained in 6×10^{-3} mol dm^{-3} SDS was used (see fig. 6). u was calculated directly from equation (14) using the central pressure and taking n_o for an electrolyte concentration of 6×10^{-3} mol dm^{-3} SDS. As a first approximation, the distance $2d$ was taken as equivalent to the measured distance h. The results are shown in figure 10 assuming that there are experimental errors in both u and d. In the latter case, taking $2d$

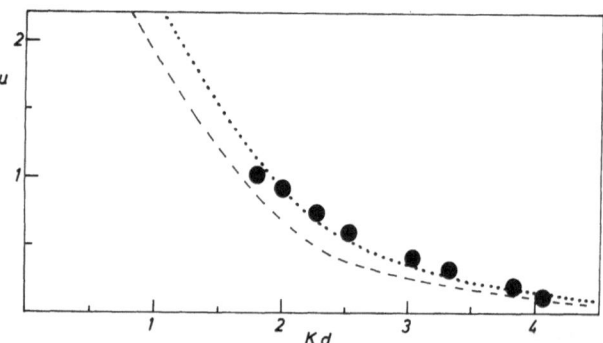

Fig. 10. u against $\varkappa d$.— —, cuve for $z = 4$ (3); - - - -, curve for $z = 10$ (3). ●, experimental values obtained in 6×10^{-3} mol dm^{-3} SDS

$= h$, assumes that the plane of the surface potential, ψ_s, is co-incident with the origin of h, i.e. the value obtained by optical methods. If allowance were made for the presence of the Stern layer then h would probably be an overestimate and d also, and thus the experimental points would be displaced more towards the calculated curve for $z = 4$.

Despite the presence of the experimental errors the points obtained clearly follow the form of the calculated curves and indicate that the value of $|\psi_s|$ is above 100 mV. In the region of high potential the curves become rather insensitive to variation of ψ_s. The inference from figure 10 is that ψ_s remains constant as the two surfaces of adsorbed surface active ions approach.

Weak interaction between electrical double layers

For conditions where the double layer interaction is weak and fulfils the conditions, $\varkappa h > 2$ and $u < 1$, a useful approximation for the electrostatic repulsive pressure is given by (3),

$$P_R = 64 n_o kT \gamma^2 \exp(- \varkappa h) \qquad (20)$$

with $\gamma = (e^{z/2} - 1)/(e^{z/2} + 1)$.

this equation can also be written as

$$\ln P_R = \ln[64 n_o kT \gamma^2] - \varkappa h \qquad (21)$$

from which it is clear that both γ and $\varkappa$ can be obtained if n_o is known. Taking P_R as equivalent to the measured central pressure then the experimental data can be plotted in the form ln |central pressure| against h. The results for 6×10^{-3} mol dm^{-3} SDS, and for 6×10^{-3} mol dm^{-3} SDS in the presence of 1.4×10^{-2} and $4.4. \times 10^{-2}$ mol dm^{-3} sodium chloride are shown in figure 11. Good linear plots are obtained and the increase in the value of ln $|P|$ at $h = 0$ with increasing ionic strength is as anticipated from theory (3). These data confirm the exponential dependence on distance of surface separation of the electrostatic interaction pressure:

All the results obtained for both SDS and $C_{16}E_3S$ gave good linear plots in the range of validity of equation (21). Least squares fits of the experimental data in this form gave values for the intercept at $h = 0$ and the gradient. The experimental values of $\varkappa$ obtained from the gradient were then used to calculate n_o and this value of n_o was used to obtain ψ_s from the intercept value. The results are listed in tables 2 and 3.

The values obtained for $\varkappa$ from the experimental data appear to be in reasonably good agreement with

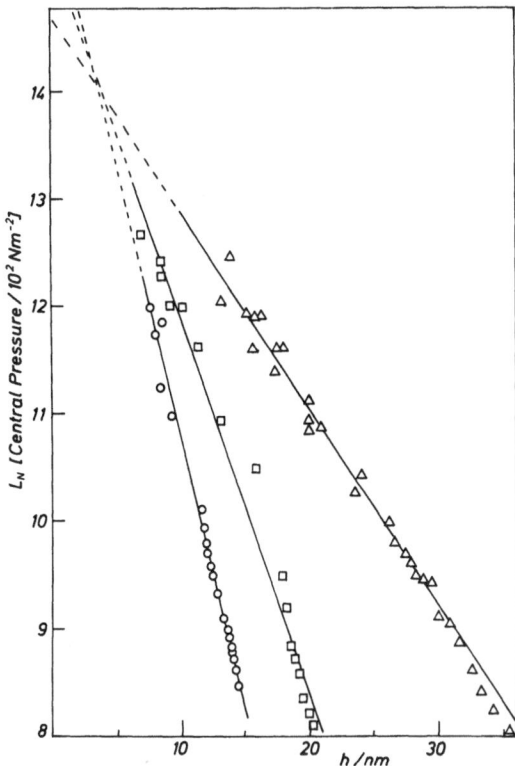

Fig. 11. ln | central pressure | against h. All solutions contained 6×10^{-3} mol dm^{-3} SDS. $\triangle$, SDS alone; $\square$, + 1.4 × 10^{-2} mol dm^{-3} sodium chloride; $\bigcirc$, + 4.4 ×10^{-2} mol dm^{-3} sodium chloride

Table 2. Surface potential and $\varkappa$ values for SDS

SDS concentration / mol dm^{-3}	NaCl concentration / mol dm^{-3}	$\varkappa_{calc}$/ m^{-1} × 10^8	$\varkappa_{exp}$/ m^{-1} × 10^8	$|\psi_s|$/mV
4.0 × 10^{-3}	—	2.08	1.35	64
6.0 × 10^{-3}	—	2.55	1.83	90
1.0 × 10^{-2}	—	2.94^a)	1.28	82
1.0 × 10^{-1}	—	2.94^a)	2.63	43
6.0 × 10^{-3}	1.4 × 10^{-2}	4.65	3.45	60
6.0 × 10^{-3}	4.4 × 10^{-2}	7.35	5.05	50
6.0 × 10^{-3}	9.4 × 10^{-2}	10.40	16.40	—

^{a}) calculated assuming that the c.m.c. was 8 × 10^{-3} mol dm^{-3}

Table 3. Surface potential and $\varkappa$ Values for C$_{16}$E$_3$S

C$_{16}$E$_3$S concentration / mol dm^{-3}	NaCl concentration / mol dm^{-3}	$\varkappa_{calc}$/ m^{-1} × 10^8	$\varkappa_{exp}$/ m^{-1} × 10^8	$|\psi_s|$/mV
2.0 × 10^{-4}	—	0.47	0.74	> 150
2.0 × 10^{-4}	8.0 × 10^{-4}	1.04	0.88	95
2.0 × 10^{-4}	9.8 × 10^{-3}	3.29	1.88	55
2.0 × 10^{-4}	1.0 × 10^{-1}	10.41	8.63	30

those calculated by equation (19). This provides an interesting test of double layer theory since it represents one of the few occasions when a direct measurement of this quantity has been made.

The values of $|\psi_s|$ obtained are clearly sufficiently high to ensure strong electrostatic repulsion between the surfaces, but are nevertheless lower than would be expected from an unscreened monolayer of anionic surface active agent. However, if allowance is made for screening of the anionic head groups by the binding of counter-ions, then the values are those expected for diffuse layer potentials. The results obtained for C$_{16}$E$_3$S show a clear dependence on the ionic strength of the electrolyte solution between the surfaces.

Conclusions

From the results obtained it appears reasonable to conclude that the theory of the diffuse electrical double layer gives a quantitative explanation of the interaction between charged surfaces in an aqueous electrolyte medium. The magnitudes of the potential $|\psi_s|$ obtained indicate that it is the diffuse layer potential which controls the interaction.

No clear evidence has been obtained for desorption of surface active ions during the approach of the surfaces (25, 26, 27) but this could be a consequence of the fact that this only occurs at very close distances and the values of h were too large to observe the effect.

Finally, it should be noted that the film viscosities (table 1) show a dependence on electrolyte concentrations. These appear to indicate an effect of the electric field on the viscosity and would be worthy of further investigation.

Acknowledgements

We wish to express our thanks to the Science and Engineering Research Council for their support of the work described in this paper.

References

1. Langmuir, I., J. Chem. Phys. 6, 873 (1938).
2. Derjaguin, B. V., Landau, L., Acta Physicochim. U.R.S.S. 14, 633 (1941).
3. Verwey, E. J. W., Overbeek, J. Th. G., Theory of Stability of Lyophobic Colloids, Elsevier (Amsterdam 1948).
4. Israelachvili, J. N., Tabor, D., Prog. Surface Membrane Sci. 7, 1 (1973).
5. Derjaguin, B. V., Voropayeva, T. N., Kabanov, B. N., Titiyevskaya, A. S., J. Colloid Interface Sci. 19, 113 (1964).
6. Peschel, G., Aldfinger, R. H., Schwarz, G., Naturwiss. 61, 215 (1974).

7. Roberts, A. D., Tabor, D., Proc. Roy. Soc. **A325**, 323 (1971).
8. Lewis, P., Ph. D. Thesis, University of Bristol (1972).
9. Hough, D. B., Ottewill, R. H., Colloid and Interface Science **IV**, 45. Ed. Kerker, M., Academic Press (New York 1976).
10. Israelachvili, J. N., Adams, G. E. Nature **262**, 774 (1976).
11. Israelachvili, J. N., Adams, G. E., J. Chem. Soc. Faraday, Trans. I. **74**, 975 (1978).
12. Israelachvili, J. N., Faraday Disc. Chem. Soc. **65**, 20 (1978).
13. Pashley, R. M. Israelachvili, J. N., Colloids and Surfaces **2**, 169 (1981).
14. Pashley, R. M., J. Colloid Interface Sci. **80**, 153 (1981).
15. Tamamushi, B., in Colloid Surfactants, Eds. Shinoda, K., Nakagawa, T., Tamamushi, B., Isemura T., Academic Press (New York 1963).
16. Weil, J. K., Bistline, R. G., Stirton, A. J., J. Physical Chem. **62**, 1083 (1958).
17. Cabellero, D., J. Opt. Soc. Amer. **37**, 176 (1947).
18. Hertz, H., Miscellaneous Papers, Macmillan (London 1896).
19. Hughes, B. D., White, L. R., J. Chem. Soc. Faraday Trans. I **76**, 963 (1980).
20. Connor, P., Ottewill, R. H., J. Colloid Interface Sci. **37**, 642 (1971).
21. Reynolds, O., Phil. Trans. Roy. Soc. **177**, 157 (1886).
22. Rutgers, A. J., Trans. Faraday Soc. **36**, 69 (1940).
23. Hough, D. B., White, L. R., Advances Colloid Interface Sci. **14**, 3 (1980).
24. Vold, M. J., J. Colloid Sci. **16**, 1 (1961).
25. Ash, S. G., Everett, D. H., Radke, C., J. Chem. Soc. Faraday Trans. II **69**, 1256 (1973).
26. Hall, D. G., J. Chem. Soc. Faraday Trans II **68**, 2169 (1972).
27. Derjaguin, B. V., Faraday Disc. Chem. Soc. **65**, 306 (1978).

Received March 7, 1983

Authors' addresses:

Dr. D. B. Hough,
Unilever Research,
Port Sunlight Laboratory,
Bebington,
Wirral, L63 3JW.

Professor Dr. R. H. Ottewill,
School of Chemistry,
University of Bristol,
Bristol BS8 ATS.

Progress in Colloid & Polymer Science   Progr. Colloid & Polymer Sci. **68**, 113–121 (1983)

Seitenkettenwechselwirkungen und α-helixinduzierender Effekt bei basischen Poly-[α-aminosäuren]*)

G. Ebert und Y.-H. Kim

Fachbereich Physikalische Chemie – Polymere – der Philipps Universität Marburg/L

Zusammenfassung: Es wird über vergleichende Untersuchungen der α-helixinduzierenden Wirkung von Anionen bei Poly-[N, ε-trimethyl-L-lysin] ($[Me_3Lys]_n$) und Poly-[L-lysin] ($[Lys]_n$) berichtet. Infolge der quartären Ammoniumgruppe in der Seitenkette unterliegt $[Me_3Lys]_n$ im Unterschied zum $[Lys]_n$ keiner pH-induzierten Konformationsumwandlung und kann somit durch pH-Änderung nicht in die α-Helix überführt werden. Dies ist jedoch ähnlich wie bei $[Lys]_n$ und anderen basischen Poly-[α-aminosäuren] möglich, wenn bestimmte Anionen zugesetzt werden. Ihr α-helixinduzierender Effekt nimmt in der Reihenfolge $Cl' < Br' < J' \ll SCN' \approx ClO_4'$ zu. Die erforderliche Konzentration liegt bei $[Me_3Lys]_n$ für ClO_4' bei $c_m = 5{,}0 \cdot 10^{-3}$ Mol/l gegenüber $2{,}1 \cdot 10^{-1}$ Mol/l bei $[Lys]_n$. Verursacht wird der α-helixinduzierende Effekt der Anionen durch eine elektrostatische Abschirmung der positiven Ladungen der Seitengruppen, was eine bestimmte Topologie (Insertion der Anionen) erfordert. Voraussetzung dafür ist offenbar eine spezifische Wechselwirkung der kationischen Seitengruppen mit den Anionen. Diese nimmt mit dem wasserstrukturbrechenden Charakter der Ionen zu, da Hydrathüllen bei Hydratation 1. Art die spezifische Wechselwirkung und damit die α-Helixbildung behindern. Seitengruppenkationen mit Hydratation 2. Art (hydrophobic hydration) begünstigen dagegen die α-Helixbildung, da die hierbei auftretenden hydrophoben Wechselwirkungen der Trialkylgruppen zur Stabilisierung dieser Konformation beitragen. In bestimmten Fällen scheinen Wasserstoffbrücken zwischen Anion und Kation für die spezifische Wechselwirkung und damit für die α-Helixbildung mitverantwortlich zu sein. Dies ergibt sich daraus, daß CH_3SO_4'-Ionen zwar bei $[Lys]_n$ in sehr hohen Konzentrationen (> 3,5 Mol/l) α-Helixbildung induzieren, nicht aber bei $[Me_3Lys]_n$. Auch Tenside wie SOS und SDS bewirken hier keine α-Helix- oder β-Strukturbildung. Dies weist daraufhin, daß in diesen Fällen gleichfalls eine Wechselwirkung zwischen der anionischen Gruppe des Tensids und dem Kation für eine Konformationsumwandlung erforderlich ist.

Abstract: Experiments on the α-helixinducing effect of anions in the case of poly-[N-ε-trimethyl-L-lysin] ($[Me_3Lys]_n$) and poly [L-lysin] ($[Lys]_n$) are reported. As a consequence of the quaternary ammonium group in the side chain $[Me_3Lys]_n$ in contrast to $[Lys]_n$ undergoes no pH-induced conformation change. Therefore it is not possible to get the α-helical conformation by a pH-change. However, like in the case of $[Lys]_n$ and other basic poly-α-aminoacids, α-helix-formation is induced by distinct anions. Their α-helixinducing effect increases according to the lyotropic serie in the order $Cl' < Br' < J' \ll SCN' \approx ClO_4'$. The concentration necessary for α-helix-formation depends on the side-chain. In the case of $[Me_3Lys]_n$ the concentration c_M at the mid-point of transition amounts to $5{,}0 \cdot 10^{-3}$ mol/l for ClO_4' and in that of $[Lys]_n$ it is $2{,}1 \cdot 10^{-1}$ mol/l. Responsible for the α-helixinducing effect of the anions is the shielding of the positive charges of the sidegroups. This requires a distinct topology, i.e. a insertion of the anions between the kationic groups (insertion) as a consequence of a specific interaction. It increases obviously with the water-structure breaking effect of the ions involved because the solvation due to ion-dipole interactions hinders the specific interaction and therefore the α-helix-formation. Sidechain kations mit hydrophobic hydration on the other hand favor the α-helix-formation because of the stabilizing influence of the hydrophobic interactions of the trialkylammonium-groups formed by the conformation transition.

---

*) Dem Gedenken an Herrn Professor Dr. B. Tamamushi, Tokyo in Ehrerbietung und Dankbarkeit gewidmet.

In some cases hydrogen bonds between anion and kation may be responsible for the specific interaction. This yields from the fact that CH$_3$SO$_4$'-anions at very high concentrations with incomplete first hydration shell (> 3,5 mol/l) induce α-helix-formation in the case of [Lys]$_n$ but not in that of [Me$_3$Lys]$_n$. Furthermore, in contrast to [Lys]$_n$ detergents like SOS and SDS do not induce the formation of α-helix or β-structure. This shows that also in these cases primarily interactions between the anion and the kation are necessary for a conformation change.

Schlüsselwörter: Poly-[N,ε-trimethyl-L-lysin], Poly-[L-lysin], α-Helix, Anionenwechsel-wirkung, Hydratation.

Einleitung

Poly-[L-lysin] ([Lys]$_n$) ist eine basische Poly-[α-aminosäure] mit einer Aminoseitengruppe, deren pK-Wert 10,4 beträgt. Es liegt in wäßriger Lösung bei pH-Werten > 10 als α-Helix vor. Beim Erhöhen der H^+-Konzentration findet eine pH-induzierte Konformationsumwandlung statt, die auf die gegenseitige elektrostatische Abstoßung der durch Protonierung entstandenen Ammoniumkationen zurückzuführen ist. Wie gezeigt worden ist, kann diese Umwandlung durch Zugabe von Perchlorat- oder Thiocyanationen in bestimmter Konzentration vollständig unterdrückt werden [1–9]. Ursache dafür ist offensichtlich die elektrostatische Abschirmung der sich abstoßenden kationischen Seitengruppen. Dies wird durch eine spezifische Wechselwirkung der α-helixinduzierenden Anionen mit den Ammoniumgruppen des Polykations bewirkt. Hierdurch soll es zu einer Insertion jener Anionen zwischen die kationischen Seitengruppen und damit zur Ausbildung einer linksgängigen Superhelix um die rechtsgängige α-Helix kommen [5, 6]. Voraussetzung für die zur Insertion der Anionen führende spezifische Wechselwirkung scheint eine „negative" Hydratation (nach Samoilov [10]) der daran beteiligten Ionen zu sein. Dies ergibt sich auch daraus, daß die α-helixinduzierende Wirkung mit dem sog. „wasserstrukturbrechenden Charakter" der Anionen und ebenso der kationischen Seitengruppen zunimmt. Hingegen wird durch die stark hydratisierten, wasserstrukturbildenden Sulfatanionen bei Poly-[L-lysin] und bei Poly-[L-ornithin] bei pH < 10 keine α-Helixbildung hervorgerufen. Die Befunde über die α-helixinduzierende Wirkung von Sulfationen bei Poly-[L-arginin] [11] konnten von Friehmelt [11a] nicht bestätigt werden. Nur das negative einwertige Methylsulfatanion wirkt in sehr hohen Konzentrationen (> 3,5 Mol/l), bei denen die primären Hydrathüllen nicht mehr vollständig ausgebildet werden können, bei [Lys]$_n$ α-helixinduzierend. In diesem Fall ergibt sich die Frage, ob an der Polykation-Anion-Wechselwirkung Wasserstoffbrücken maßgeblich beteiligt sind. Andererseits aber tritt bei [Arg]$_n$ und Poly-[L-homoarginin] ([Har]$_n$) die α-helixinduzierende Wir-

kung z. B. von ClO$_4$' und SCN' bei einer um 2 Zehnerpotenzen niedrigeren Anionenkonzentration auf als bei [Lys]$_n$. Da jene beiden Polymeren in der Seitenkette die stark wasserstrukturbrechende Guanidiniumgruppe enthalten, ist dies wiederum ein Hinweis auf die besondere Rolle die der Hydratation der an der spezifischen Wechselwirkung beteiligten Ionen und damit an der α-Helixbildung zukommt.

Einen weiteren Beitrag zur Beantwortung dieser Fragen sollten Untersuchungen an Poly-[α-aminosäuren] mit einer seitständigen quartären Alkylammoniumgruppe wie des Poly-[N,ε-trimethyl-L-lysins] ([Me$_3$Lys]$_n$) liefern [9]. Hier entfällt infolge der Trimethylierung einmal die Möglichkeit zur Wasserstoffbrückenbindung und ferner liegt im Unterschied zur einfachen Ammonium- und zur Guanidiniumgruppe eine andersartige Hydratation der Seitengruppen (Hydratation 2. Art, hydrophobic hydration) vor.

Experimentelles

Substanzen

Poly-[L-lysin] wurde durch Polymerisation des N-Carboxyanhydrids von N,ε-Carbobenzoxy-L-lysin mit Na-methylat in Dioxan ([M]/[I] = 200) und Abspaltung der Carbobenzoxy-Schutzgruppe von dem erhaltenen Poly-[N,ε-carbobenzoxy-L-lysin] ([CboLys]$_n$) mit HCl/HBr nach Fasman et al. [12] dargestellt. Die viskosimetrisch bestimmten Molekulargewichte (s. u.) betragen 256 000 beim [Cbo-Lys]$_n$ und 118 000 beim [Lys]$_n$(DP ≈ 957).

Poly-[N,ε-trimethyl-L-lysin] wurde durch vollständige Methylierung des [Lys]$_n$ mit Methyljodid bei 25 °C und – mittels pH-Stat konstant gehaltenem – pH 11 erhalten. Das Molekulargewicht wurde viskosimetrisch zu 165 000 ermittelt, (DP ≈ 907).

Die für die optischen Messungen verwendeten Salze und anderen Reagenzien der Fa. Merck, Darmstadt, waren alle von p.A. bzw. Suprapur-Qualität.

Methodik

Die Viskositätsmessungen wurden mit einem automatischen Viskosimeter „Viscomatic" der Fa. Sofica durchgeführt, wobei die Durchflußzeit auf ± 0,01 sec genau bestimmt werden kann. [Cbo-Lys]$_n$ wurde in Dimethylformamid (DMF) im Konzentrationsbereich 0,7 – 22·10^{-3} g/ml gemessen. Aus dem durch Auftragen von η$_{sp}$/c gegen η$_{sp}$ nach Schulz-Blaschke erhaltenen [η] wurde das Molekulargewicht nach Daniel und Katchalski [13] mit Hilfe der

Beziehung $[\eta] = 2{,}24 \cdot 10^{-7} \, M^{1{,}26}$ erhalten. Von $[\text{Lys}]_n$ und $[\text{Me}_3\text{Lys}]_n$ wurde η_{sp} bei 25 °C in 1%iger wäßriger Lösung bestimmt. Unter Verwendung der von Yaron und Berger abgeleiteten Gleichung $\log \text{DP} = 0{,}79 \log \eta_{sp\,(c=1\%)} + 2{,}46$ ist der Polymerisationsgrad DP und damit das Molekulargewicht ermittelt worden [14].

Die CD- und ORD-Messungen wurden mit einem Jasco 20 (Fa. Japan Spec. Co., Tokyo) in Küvetten aus geschmolzenem Quarz (Fa. Hellma) mit Schichtdicken von 0,1 — 1 cm (CD) und 1–5 cm (ORD) unter Thermostatisierung auf ± 0,05 °C aufgenommen. Für die CD-Messungen sind 0,01–0,05%ige, für die ORD-Messungen sind 0,1%ige Polypeptid-Lösungen verwendet worden. Sie wurden aus 0,2 bzw. 1%igen Stammlösungen durch Verdünnen mit den entsprechenden Salzlösungen bzw. Alkoholen bereitet.

Die Reste-Elliptizitäten $[\Theta]_\lambda$ wurden aus den gemessenen Elliptizitäten Θ_λ mittels

$$[\theta]_\lambda = \frac{M}{c'd} \cdot \theta_\lambda = \frac{10}{cd} \cdot \theta_\lambda$$

ermittelt, wobei $[\theta]_\lambda$ die Reste-Elliptizität bei der Wellenlänge λ in Deg cm²/deciMol, θ_λ die gemessene Elliptizität, c' die Polypeptid-Konzentration in g/100 ml, c die Polypeptid-Konzentration in Mol Peptid-Rest/l, M das Molgewicht eines Peptidrestes und d die Schichtdicke in dm ist.

In den Lösungen von unterhalb 230 nm stark absorbierenden Salzlösungen wurden ORD-Spektren von 700 nm – 280 nm aufgenommen und aus den gemessenen optischen Drehungen α_λ bei der Wellenlänge λ die spezifischen und molaren Drehungen $[\alpha]_\lambda$ und $[m]_\lambda$ nach $[\alpha]_\lambda = \frac{100}{c'd} \cdot \alpha_\lambda$ sowie $[m]_\lambda = \frac{M}{c'd} \cdot [\alpha]_\lambda = \frac{10}{cd} \cdot \alpha_\lambda$ bestimmt. Mit Hilfe des Lorentz-Faktors $\frac{3}{n^2+2}$ wurden die $[m]_\lambda$-Werte auf die Werte im Vakuum reduziert und die Brechungsindizes n mit der Sellmeierschen Näherung $n^2 = 1 + \frac{a\lambda^2}{\lambda^2 - \lambda_v^2}$ auf die Wellenlänge λ, bei der die Messung erfolgte, korrigiert.

Ergebnisse

In rein wäßriger Lösung

Das CD-Spektrum des $[\text{Me}_3\text{Lys}]_n$ hat ein Maximum bei 217 nm mit einer Reste-Elliptizität $[\theta]_{217} = +5520$ Deg·cm²/deciMol und ein Minimum bei 198 nm mit $[\theta]_{198} = -31170$. Die Frequenzlagen sind somit identisch mit denen des $[\text{Lys}]_n$ bei pH-Werten mit vollständig protonierter Seitenkette, das nach Tiffany und Krimm hier in Form einer gestreckten Helix vorliegen soll, die der Prolin-II-Helix ähnlich ist, jedoch nur 2,4 statt 3,0 Aminosäurereste pro Windung aufweist [15, 16].

pH-Abhängigkeit

Bei $[\text{Lys}]_n$ wird bei pH-Werten oberhalb von ≈ 9, d. h. mit der Deprotonierung der NH_3^+-Seitenketten zu $-\text{NH}_2$ unter gleichzeitigem Ladungsverlust eine starke Änderung des CD-Spektrums beobachtet, das

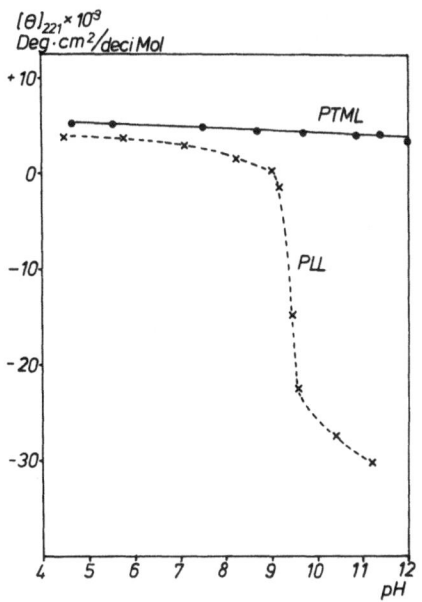

Abb. 1. pH-Abhängigkeit der $[\theta]_{221}$-Werte von $[\text{Lys}]_n$ $(-\times-)$ und $[\text{Me}_3\text{Lys}]_n$ $(-\bullet-)$. Polymerkonzentration $c = 0{,}01\%$, $T=22$ °C, Schichtdicke 1,0 cm

in das der α-Helix übergeht. Bei den $-\text{N}(\text{CH}_3)_3^+$-Seitengruppen des $[\text{Me}_3\text{Lys}]_n$ ist eine solche pH-abhängige Deprotonierung unter Ladungsverlust nicht möglich, so daß die beim $[\text{Lys}]_n$ beobachtete CD- bzw. Konformationsänderung nicht auftreten kann. Diese Erwartung wurde vom Experiment bestätigt, wie Abbildung 1 zeigt, in der $[\theta]_{221}$ gegen den pH-Wert der Lösung aufgetragen ist [9]. Hiernach ist $[\theta]_{221}$ nahezu pH-unabhängig, eine pH-induzierte Konformationsumwandlung tritt nicht ein.

Temperaturabhängigkeit

Tiffany und Krimm sowie Neumann et al. haben gezeigt [16, 18], daß das CD-Spektrum von $[\text{Lys}]_n$ bei 70 °C ein flaches kleines Minimum bei 233 nm mit $[\theta]_{233} = -380$ Deg·cm²/deciMol hat und daß das Maximum bei 217 nm von $[\theta]_{217} = +4850$ Deg·cm²/deciMol auf + 780 abnimmt. Damit ist es dem für ungeordnete (random) Strukturen berechneten CD-Spektren natürlicher Proteine ähnlich. Bei $[\text{Me}_3\text{Lys}]_n$ hingegen wird selbst bei 80 °C das Minimum bei 233 nm auch nicht andeutungsweise beobachtet. Nur das Maximum bei 217 nm nimmt mit der Temperatur ab, allerdings wesentlich weniger als bei $(\text{Lys})_n$, wie die in Tabelle 1 wiedergegebenen Zahlenwerte zeigen. Hier handelt es sich offensichtlich nur um eine quantitative, nicht aber um eine qualitative Änderung des CD-Spektrums.

Tabelle 1. Reste-Elliptizitäten des [Lys]$_n$ und [Me$_3$Lys]$_n$ in Wasser. Konzentration der Polypeptide 0,01%, Schichtdicke 1,0 cm

		Maxima nm	[θ]	Minima nm	[θ] (Deg·cm²/deciMol)
[Me$_3$Lys]$_n$	pH 5,56, 22 °C	217	+5220	198	−31170
	pH 5,56, 80 °C	217	+2610	198	−21500
	pH 11,38, 22 °C	217	+4080		
[Lys]$_n$	pH 5,80, 22 °C	217	+4850	201	−23320
	pH 5,80, 70 °C	217	+ 780	neg. Schulter bei 233 nm, [θ] = −380	
	pH 11,22, 22 °C	191	+68500	221	−30170
				207	−28850

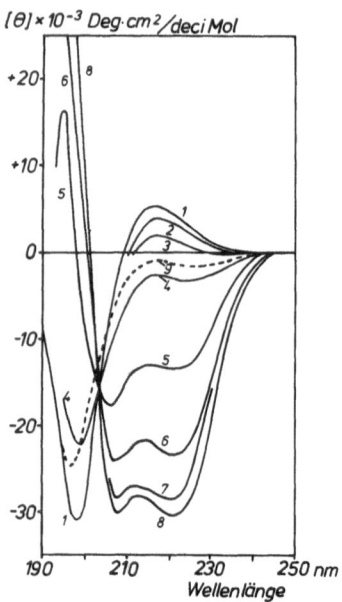

Abb. 2. CD-Spektren von [Me$_3$Lys]$_n$ in NaClO$_4$-Lösungen c=0,01%, T=22 °C, Schichtdicke 1,0 cm, pH 5,5 NaClO$_4$-Konzentration: Kurve 1: 0 Mol/l, 2: 1,0·10^{-3} 3: 2,5·10^{-3}, 4: 4,0·10^{-3}, 5: 5,0·10^{-3}, 6: 8,0·10^{-3}, 7: 1,0·10^{-2}, 8: 3,0·10^{-2}, 9: 3,0·10^{-2} bei 62 °C

Konformation von [Me$_3$Lys]$_n$ in Elektrolytlösungen

CD-Messungen in Perchlorat-Lösungen

Perchlorat-, aber auch einige andere Anionen haben bei protonierten basischen Poly-[α-aminosäuren] wie Poly-[*L*-lysin], aber auch bei Poly-[*L*-ornithin] ([Orn]$_n$), insbesondere aber bei Poly-[*L*-arginin] ([Arg]$_n$) und Poly-[*L*-homoarginin] ([Har]$_n$) eine α-helixinduzierende Wirkung [7, 8]. Wie Ichimura et al. zeigen konnten, bilden die stark basischen [Arg]$_n$ und

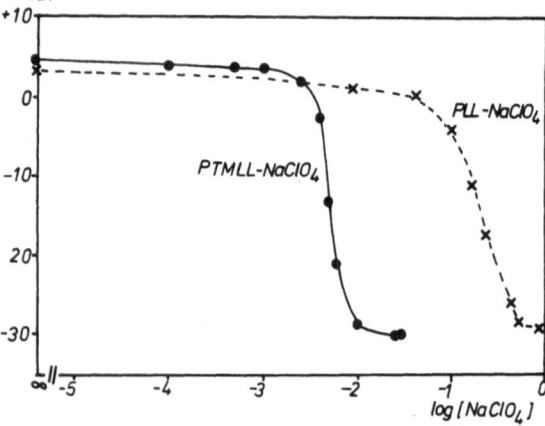

Abb. 3. Abhängigkeit der [θ]$_{221}$-Werte von [Lys]$_n$ (−×−) und [Me$_3$Lys]$_n$ (−○−) vom Logarithmus der NaClO$_4$-Konzentration. Die Kurve von [Lys]$_n$-NaClO$_4$ ist der Arbeit von Peggion et al. [2] entnommen. c=0,01%, c von [Me$_3$Lys]$_n$=4.9·10^{-4} Mol/l, T=22 °C, Schichtdicke: 1,0 cm, pH 5,5

[Har]$_n$ ohne die Anwesenheit solcher Anionen auch bei pH 12 keine α-Helices [11].

Wie aus Abbildung 3 hervorgeht, setzt bereits oberhalb einer ClO$_4'$-Konzentration von 0,4·10^{-2} Mol/l α-Helixbildung ein und bei 1,0·10^{-3} Mol/l beträgt der α-Helixanteil f_H nahezu 100%. Bei 3,0·10^{-2} wurde bei 22 °C für 208 nm ein Minimum der Reste-Elliptizität von [θ] = −30000 Deg·cm²/ deciMol und für 221 nm von −30250 erhalten. Die Umwandlungskurve [θ]$_{221}$-log [ClO$_4$] ist bei [Me$_3$Lys]$_n$ steiler als bei [Lys]$_n$, wie Abbildung 3 zeigt, die Kooperativität also offensichtlich höher. Außerdem aber erfolgt sie im Fall des [Me$_3$Lys]$_n$ bei einer um ca. 2 Größenordnungen niedrigeren ClO$_4$-Konzentration. So beträgt die ClO$_4$-Konzentration am Mittelpunkt der Umwandlung c_m bei [Lys]$_n$ ≈ 2,1·10^{-1} Mol/l, bei [Me$_3$Lys]$_n$ jedoch 5,0·10^{-3}.

Der pH-Wert der Lösungen hatte erwartungsgemäß keinen Einfluß auf die Umwandlungskurven.

Messung der Temperaturabhängigkeit von [θ]$_{221}$ ergaben, daß eine temperaturinduzierte Konformationsumwandlung von der α-Helix in den ungeordneten Zustand auftritt (Abb. 2). Die Temperatur am Mittelpunkt der Umwandlung T_m, wobei definitionsgemäß der α-Helixanteil $f_H = \frac{1}{2}$ ist, zeigt die in Abbildung 4 wiedergegebene Abhängigkeit von der Perchloratkonzentration.

In Abbildung 5 ist die scheinbare oder van't Hoffsche Umwandlungswärme ΔH_{vH}, die bei der Umwandlung einer kooperativen Kettenlänge $\sigma^{-1/2}$ auftritt, als Funktion der Perchloratkonzentration wie-

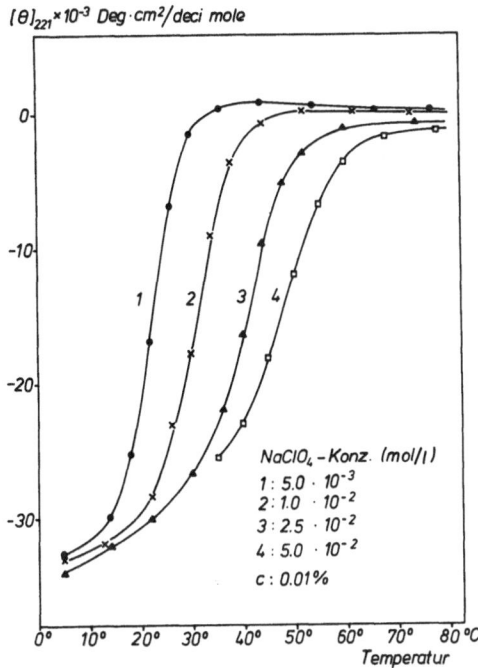

$[\theta]_{221} \times 10^{-3}$ Deg·cm²/deci mole

NaClO₄ - Konz. (mol/l)
1 : 5.0 · 10⁻³
2 : 1.0 · 10⁻²
3 : 2.5 · 10⁻²
4 : 5.0 · 10⁻²
c : 0.01%

Temperatur

Abb. 4. Temperaturabhängigkeit der $[\theta]_{221}$-Werte von $[Me_3Lys]_n$ in NaClO₄-Lösungen. *c*: 0,01%, Schichtdicke: 1,0 cm

dergegeben. Es gilt $\sigma^{1/2} = \Delta H_0 / \Delta H_{vH}$. ΔH_0 ist die bei der Überführung eines Restes z. B. aus dem Helix- in den Knäuelzustand auftretende wahre Umwandlungswärme. σ ist der Kooperativitätsparameter. Geht man davon aus, daß ΔH_0 wie bei anderen basischen Poly-[α-aminosäuren] ≈ 600±50 cal/Rest beträgt [8], so ergibt sich für $[Me_3Lys]_n$ ein $\sigma = 1,2 \cdot 10^{-4}$, für $[Lys]_n$ ein $\sigma = 2,4 \cdot 10^{-4}$ (3, 5, 6).

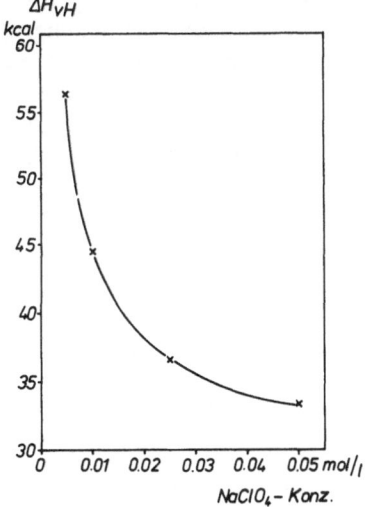

ΔH_{vH}
kcal

NaClO₄ - Konz.

Abb. 5. Abhängigkeit der scheinbaren Umwandlungswärme ΔH_{vH} des $[Me_3Lys]_n$ von der Perchloratkonzentration

Oberhalb einer Perchlorat-Konzentration von $3,0 \cdot 10^{-2}$ Mol/l tritt bei Zimmertemperatur Ausfällung feiner Niederschläge auf. Dabei ändert sich das CD-Spektrum erheblich, wobei ein Minimum bei ≈ 226 nm, ähnlich wie bei der IIβ-Struktur, und eine Schulter bei ≈ 215 nm beobachtet wird (Abb. 6). Wenn die Trübungen durch Erwärmen wieder in Lösung gebracht werden, erhält man erneut das CD-Spektrum der α-Helix. Mit steigender Perchlorat-Konzentration (oberhalb $3,0 \cdot 10^{-2}$ Mol/l) nimmt die zur Auflösung der Trübung bzw. Niederschläge erforderliche Temperatur verständlicherweise zu.

ORD-Messungen in Thiocyanat- und Halogenid-Lösungen

Wegen der starken UV-Absorption von SCN′, J′ und Br′ in dem hier interessierenden Frequenzbereich wurden die ORD-Spektren zwischen 700 und 280 nm aufgenommen und der b_0-Faktor der Moffit-Yang-Gleichung als Maß für den α-Helixanteil ermittelt.

$$[m'] = [m]_\lambda \cdot \frac{3}{n^2 + 2} = a_0 \frac{\lambda_0^2}{\lambda^2 - \lambda_0^2} + b_0 \frac{\lambda_0^4}{(\lambda^2 - \lambda_0^2)^2} .$$

($\lambda_0 = 212$ nm, $b_0 = -630$ für 100% α-Helixanteil)

Die Ergebnisse der Messungen von b_0 an $[Me_3Lys]_n$ in NaCl, NaBr, NaJ und NaSCN-Lösungen zeigt Abbildung 9. Zum Vergleich wurden die von Sugai et al. am System $[Lys]_n$-NaSCN erhaltenen Werte mit aufgenommen, die zeigen, daß beim $[Me_3Lys]_n$ eine um ca. 1 Größenordnung geringere NaSCN-Konzentration für die Konformationsumwandlung erforderlich ist als beim $[Lys]_n$. Weiter erkennt man, daß Chloridionen auch bei hohen Konzentrationen > 1 Mol/l nur eine vernachlässigbare Konformationsänderung zu bewirken scheinen. Tatsächlich erfolgt aber, wie die CD-Spektren in hochkonzentrierten Chlorid-Lösungen (3,5 m CaCl₂) zeigen, bei neutralen pH nur ein Übergang von dem der gestreckten Konformation in das der random-Konformation [7].

CD-Spektren von [Lys]ₙ und [Me₃Lys]ₙ in Sulfat- und Alkylsulfatlösungen

Vom $[Lys]_n$ ist bekannt, daß Sulfationen keine α-helixinduzierende Wirkung haben [5, 6]; Na-methylsulfat hat dagegen in hohen Konzentrationen eine deutliche α-Helixbildung zur Folge [6, 9]. Im Unterschied dazu wird das CD-Spektrum und damit die Konformation des $[Me_3Lys]_n$ durch Na-methylsulfat bis zur Sättigung nicht verändert, wie Abbildung 7

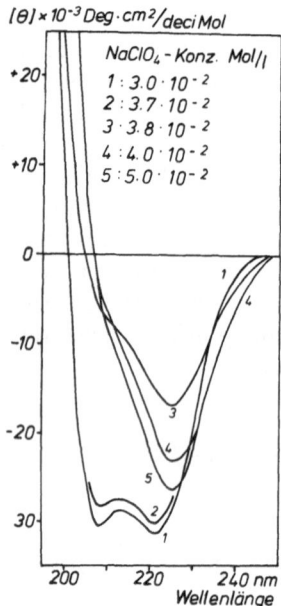

Abb. 6. CD-Spektren von [Me₃Lys]ₙ in NaClO₄-Lösungen 0,03 Mol/l. c: 0,01%, $T = 22$ °C, Schichtdicke: 1,0 cm. 1. keine Trübung; 2. beginnende Trübung; 3.–5. trüb, nach 30 Minuten konstant

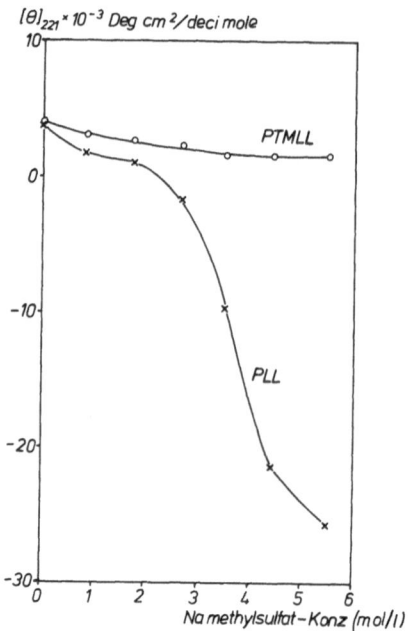

Abb. 7. Abhängigkeit der $[\theta]_{221}$-Werte des [Lys]ₙ ($-\times-$) und des [Me₃Lys]ₙ ($-\bigcirc-$) von der Na-methylsulfat-Konzentration

zeigt. Die temperaturinduzierte α-Helix-Knäuel-Umwandlung in 5,5 molarer Na-methylsulfat-Lösung erfolgt zwischen 0–80 °C über einen sehr breiten Temperaturbereich [9].

In Natriumoctylsufat-Lösungen (NOS) macht sich bei [Me₃Lys]ₙ in 10^{-3} molarer Lösung nur eine geringe Abnahme des CD-Maximums von $[\theta]_{217} = +5590$ Deg·cm²/deciMol auf $+3550$ bemerkbar im Gegensatz zu (Lys)ₙ, das nach Satake und Yang das in Abbildung 8 gezeigte CD-Spektrum zeigt [17]. Oberhalb von $5 \cdot 10^{-3}$ Mol NOS/l bildet es mit [Me₃Lys]ₙ einen Niederschlag, der sich jedoch in 1,0 molarer NOS-Lösung wieder löst. $[\theta]_{217}$ ist mit $+3460$ Deg·cm²/deciMol nahezu unverändert.

Während [Lys]ₙ nach Satake und Yang [17] in Natriumdodecylsufat-Lösungen (SDS) in β-Konformation vorliegt, wurde bei [Me₃Lys]ₙ nur das CD-Spektrum der gestreckten Konformation mit einem $[\theta]_{217} = +2750$ Deg·cm²/deciMol in $5 \cdot 10^{-4}$ molarer SDS-Lösung gegenüber $+5590$ in reinem Wasser beobachtet.

Messungen in Alkoholen

[Me₃Lys]ₙ löst sich recht gut sowohl in 95%igem Methanol als auch in 95%igem n- und iso-Propanol. Wie die CD-Spektren zeigen, liegt in diesen Lösungen keine α-Helix vor, im Unterschied zu [Lys]ₙ, das in >88%igem Methanol und >76%igem iso-Propanol α-Helices bildet.

Diskussion

Im Unterschied zum [Lys]ₙ besitzt [Me₃Lys]ₙ in der Seitenkette eine quartäre Ammoniumgruppe, die also nicht wie die seitständige $-NH_3^+$-Gruppe des [Lys]ₙ bei höheren pH-Werten unter Deprotonierung in eine elektrisch neutrale Form übergehen kann. Die nahezu pH-unabhängigen CD-Spektren (vgl. Abb. 1) zeigen, daß die [Me₃Lys]ₙ-Polykationen im pH-Bereich von 1–12 nicht als α-Helix vorliegen. Welcher Art diese Konformation ist, soll hier nicht diskutiert werden[1]. Dieses Ergebnis bestätigt zunächst die sehr plausible Annahme, daß die elektrische Ladung der Seitengruppen die α-Helixbildung verhindert. Dies wird weiter dadurch erhärtet, daß bei [Me₃Lys]ₙ auch in Methanol und iso-Propanol keine α-Helixbildung stattfindet, während dies bei [Lys]ₙ der Fall ist [18, 19]. Auch die gegenüber [Lys]ₙ beobachtete höhere thermische Stabilität [9] der e.h.-Konformation des [Me₃Lys]ₙ deutet daraufhin, daß die temperaturunabhängige elektro-

---

[1]) Wie bereits erwähnt spricht nach Tiffany und Krimm [15, 16] einiges – wie z. B. das CD-Spektrum – für das Vorliegen einer steilen Helix mit 2,4-Aminosäureresten pro Windung. Hiergegen sind verschiedene Einwände vorgebracht worden. Sicherlich kann man aber zumindest bei niedrigen Ionenstärken von einer stark aufgeweiteten, gestreckten Konformation sprechen, die hier der Kürze wegen als e.h.-Konformation bezeichnet werden soll. Die „random-coil"-Konformation mit einem flachen Minimum bei 233 nm [16, 18] sei hier r-Konformation genannt.

statischen Kräfte hierbei eine dominierende Rolle spielen. Die durch die Trimethylierung der Ammoniumgruppen erhöhte Hydrophobie der Seitenketten reicht erwartungsgemäß auch bei höherer Temperatur[2]) nicht aus, um die α-Helixbildung zu induzieren.

Eine sterische Behinderung der α-Helixbildung durch die drei Methylgruppen ist nicht nur aufgrund von Modellbetrachtungen auszuschließen, sondern vor allem durch die Tatsache, daß bestimmte Anionen wie ClO_4' und SCN', wie bei anderen basischen Poly-[α-aminosäuren] auch, einen starken α-helixinduzierenden Effekt ausüben. Die Ursachen dieser Anionenwirkung sind vielfältiger Natur. Sicherlich ist – wie bereits erwähnt – die Induzierung der α-Helixbildung im wesentlichen auf einen elektrostatischen Abschirmeffekt zurückzuführen. Daneben ist auch eine Änderung der Solvatation der Seitengruppen in Betracht zu ziehen [20]. Die für das Zustandekommen der elektrostatischen Abschirmung erforderliche spezifische Wechselwirkung der kationischen Seitengruppen der Poly-[α-aminosäure] mit den Anionen zeigt sich auch hier in der starken Abhängigkeit der α-Helixbildung von der Art des Anions. Dies ergibt sich aus der Auftragung des den α-Helixanteils f_H charakterisierenden b_0-Parameters gegen die Salzkonzentration in Abbildung 9. Analog der lyotropen Reihe erhält man dabei wie bei $[Lys]_n$ etc. [7, 8] die Anordnung

$$Cl' < Br' < J' \ll SCN \approx ClO_4'.$$

Quantitativ sind diese spezifischen Wechselwirkungen beim $[Lys]_n$ durch Bestimmung der Sedimentationskoeffizienten [21, 22] sowie bei $[Lys]_n$, $[Arg]_n$ und $[Har]_n$ durch Messungen der elektrischen Leitfähigkeit [7, 8] untersucht worden. Wie mehrfach betont [5, 6, 9] ist zweifellos eine der Hauptursachen für die spezifische Wechselwirkung der hier diskutierten Anionen mit den Polykationen ihr Einfluß auf die Wasserstruktur. Je stärker wasserstrukturbrechend ein Anion ist, um so ausgeprägter macht sich die spezifische Wechselwirkung und damit sein α-helixinduzierender Effekt bemerkbar.

In diesem Zusammenhang sind die kürzlich von Zembala und Czarnecki über adsorbierte Schichten von Dodecyltriethylammonium-perchlorat und -chlorid an Quecksilberoberflächen veröffentlichten Untersuchungen von Interesse [23]. Hierbei ergab sich beim Perchlorat eine starke laterale Attraktion der adsor-

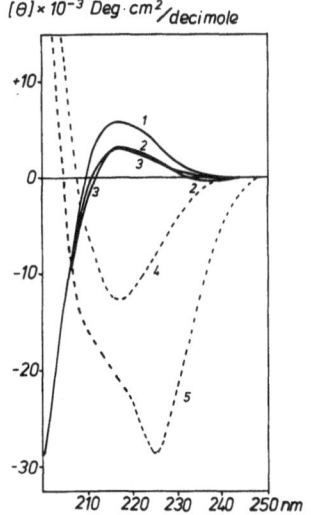

Abb. 8. CD-Spektren von $[Me_3Lys]_n$ und $[Lys]_n$ in Tensid-Lösungen. *c*: 0,01%, *T* = 22 °C, Schichtdicke: 1,0 cm. 1. $[Me_3Lys]_n$ in Wasser; 2. $[Me_3Lys]_n$ $(4,9 \cdot 10^{-4}$ Mol/l) in SOS-Lösung (1,0 Mol/l); 3. $[Me_3Lys]_n$ $(4,9 \cdot 10^{-4}$ Mol/l) in SDS-Lösung $(5,0 \cdot 10^{-4}$ Mol/l); 4. $[Lys]_n$ $(1,2 \cdot 10^{-4}$ Mol/l) in SDS-Lösung $(2,6 \cdot 10^{-2}$ Mol/l); 5. $[Lys]_n$ $(6,2 \cdot 10^{-5}$ Mol/l) in SOS-Lösung (0,7 Mol/l); 4. und 5. sind der Arbeit von Satake und Yang [17] entnommen

bierten Schicht, was auf eine laterale Wechselwirkung der zwischen die Stickstoff-Atome der quartären Ammoniumgruppen eingelagerten ClO_4^- zurückgeführt werden muß [23] und in Übereinstimmung mit der von uns postulierten Insertion der ClO_4' zwischen die Ammonium bzw. Guanidiniumseitengruppen basischer Poly-[α-aminosäuren] ist. Chloridionen hingegen bewirken, wie sich aus der geringen Stabilität der adsorbierten Filme ergibt, keine laterale Stabilisierung. Hier werden die Anionen offensichtlich nicht lateral zwischen den N-Atomen der Ammoniumgruppen angeordnet, sondern befinden sich anscheinend in einer für die Bildung eines stabilen Films weniger günstigen Anordnung. Dabei dürfte die Hydrathülle der Chloridionen auch hier eine entscheidende Rolle spielen. In beiden Fällen sind also offensichtlich die spezifischen Wechselwirkungen der ClO_4'-Anionen mit den Tetraalkylammonium-Ionen für eine durch elektrostatische Abschirmung bedingte stabile molekulare Anordnung verantwortlich. Da die am Hg adsorbierten DTEA-Kationen und die Anionen (Cl', ClO_4') aus räumlichen Gründen sehr nahe benachbart sein müssen, ergibt sich aus diesem Beispiel sehr deutlich die Notwendigkeit einer bestimmten Topologie der Anordnung von Anion und Kation (Insertion) zur Erziehung des o. a. Stabilisierungseffektes [23]. Bei den basischen Poly-[α-aminosäuren] kommt es

---

[2]) Die hydrophoben Wechselwirkungen aliphatischer Gruppen haben bis etwa 58 °C einen positiven Temperaturkoeffizienten.

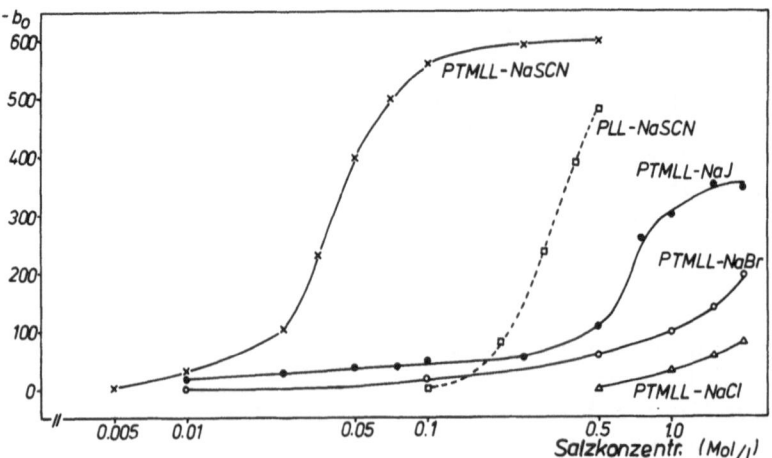

Abb. 9. b_0-Parameter der Moffit-Yang-Gleichung von $[Me_3Lys]_n$ bzw. PTMLL Abhängigkeit von der Salzkonzentration. $c=0,1\%$, T: 22 °C, Schichtdicke: 1,0 cm. Die Kurve PLL bzw. $[Lys]_n$ ist von Sugai et al. [7] übernommen worden

dadurch – wie bereits früher ausführlich dargelegt [5, 6] – zur Ausbildung der linksgängigen Superhelix von Anionen um die rechtsgängige α-Helix.

Daß die spezifischen Wechselwirkungen auch von der Art der kationischen Seitengruppe abhängt, erkennt man an den sehr unterschiedlichen Anionenkonzentrationen, die bei $[Lys]_n$ und $[Me_3Lys]_n$ (Abb. 3) sowie bei $[Lys]_n$ und $[Arg]_n$ bzw. $[Har]_n$ [5] zur α-Helixbildung erforderlich sind.

Geht man nun von dem Befund von Sugai et al. [7] aus, daß die hier diskutierte spezifische Wechselwirkung auch bei den kationischen Seitengruppen mit deren wasserstrukturbrechendem Charakter zunimmt, so sollte man bei dem trimethylierten Produkt eigentlich den entgegengesetzten Effekt erwarten, da in der Umgebung apolarer Gruppen eine Zunahme der Ordnung der Wassermoleküle, also Strukturbildung erfolgt. Bemerkenswerterweise ist aber bei Tetramethylammoniumsalzen im Unterschied zu den höheren Homologen wie Tetraethyl- etc. mittels IR-Untersuchungen ein wasserstrukturbrechender Effekt festgestellt worden [24]. Dieser Befund fügt sich zwar sehr gut in die vorliegenden Betrachtungen ein, jedoch ist er nicht ohne weiteres zu verstehen. Diese Problematik erforderte eine ausführliche, den Rahmen dieses Beitrags sprengende Diskussion, insbesondere auch darüber, ob und inwieweit die zugrundeliegenden Meßmethoden und -resultate die physikalische Realität erfassen. Nimmt man aber in Übereinstimmung mit der üblichen Auffassung an, daß die Trimethylgruppierung einen wasserstrukturbildenden Charakter hat, so ergibt sich anscheinend ein Widerspruch zu dem, was oben über die Bedeutung des wasserstrukturbrechenden Einflusses der Anionen und der Guani-

dinium-Seitengruppen für das Zustandekommen der spezifischen Wechselwirkung und damit der α-Helixstabilisierung gesagt wurde. Die Lösung dieses scheinbaren Widerspruchs ist wahrscheinlich in folgendem zu suchen: Im Unterschied etwa zu der Hydratation 1. Art der einfachen Anionen und Kationen liegt bei den apolaren Alkylseitengruppen der von Hertz als Hydratation 2. Art bezeichnete Typ (hydrophobic hydration) vor. Hier tritt eine erhöhte Wechselwirkung der den apolaren Gruppen benachbarten Wassermoleküle untereinander auf, nicht dagegen mit den „hydratisierten" Gruppen wie im 1. Fall. Durch die gleichzeitig abnehmende Entropie des Wassers in der Umgebung der apolaren Gruppen kommt es bekanntlich zur Ausbildung von hydrophober Wechselwirkung und damit zur Aggregation bzw. zur Mizellbildung bei Tensiden. Dabei wird Entropie gewonnen. In unserem Fall kann man annehmen, daß die α-Helixbildung durch diesen Effekt, der letzthin ein partielles Abstreifen der Hydratwasserhülle bedingt, erleichtert wird. Daß dies offenbar aus sterischen Gründen von großer Bedeutung für α-Helixinduktion ist, zeigen die mit Na-methylsulfat erhaltenen Resultate. Obwohl bei $[Lys]_n > 3,5$ Mol/l Na-Methylsulfat α-Helixbildung stattfindet, ist dies bei $[Me_3Lys]_n$ bis zur Sättigung nicht der Fall. Es ist vorstellbar, daß beim $[Lys]_n$ Wasserstoffbrücken an der Verankerung der CH_3O-SO_3-Ionen an den Lysinseitenketten beteiligt sind. Dann kann man annehmen, daß bei $[Me_3Lys]_n$, bei dem diese Möglichkeit zur H-Brückenbildung nicht besteht, die Hydrathüllen die α-Helixbildung verhindern. Hierfür sprechen auch die Befunde, daß Alkylsulfate mit längerer Seitenkette wie SOS und SDS bei $[Me_3Lys]_n$ keine α-Helix- oder β-Strukturbildung induzieren, obwohl dies beim [Lys]$_n$ der Fall ist. Die hydrophoben Wechselwirkungen der Tensidketten reichen also allein nicht hierfür aus. Auch hier ergibt sich offenkundig, daß für die Bildung geordneter, periodischer Konformation bei den hier diskutierten Polykationen eine die gegenseitige elektrostatische Abschirmung der kationischen Seitengruppen bewirkende spezifische Anion-Kationwechselwirkung erforderlich ist.

Danksagung

Wir danken dem Deutschen Akademischen Austauschdienst (DAAD) für die Gewährung eines Stipendiums für Y.-H. Kim.

Literatur

1. Rifkind, J. M., Biopolymers **8**, 685 (1969).
2. Peggion, E., Cosani, A., Terbojevich, M., Borin, G., Biopolymers **11**, 633 (1972).
3. Ebert, Ch., Ebert G., Werner W., Kolloid Z. Z. Polymere **251**, 504 (1973).
4. Conio, G., Patrone, E., Rialdi, G., Ciferri A., Macromolecules **7**, 654 (1974).
5. Ebert, Ch., Ebert, G., Progr. Colloid & Polymer Sci. **57**, 100 (1975).
6. Ebert, Ch., Ebert, G., Colloid & Polymer Sci. **255**, 1041 (1977).
7. Murai, N., Miyazaki, M., Sugai, Sh., Nihon kagaku kaishi 659 (1976).
8. Miyazaki, M., Yoneyama, M., Sugai, Sh., Polymer **19**, 995 (1978).
9. Kim, Y.-H., Dissertation Marburg/L. 1978.
10. Samoilov, O. J., Die Struktur von wäßrigen Elektrolytlösungen, VEB Verlag der Wissenschaften Leipzig 1961, S. 63 ff.
11. Ichimura, S., Mita, K., Zama, M., Biopolymers **17**, 2769 (1978).
11a. Friehmelt, V., Dissertation Marburg/L. (1981).
12. Fasman, G. D., Idelson M., Blout, E. R., J. Am. Chem. Soc. **83**, 709 (1961).
13. Daniel, E., Katchalski, E., in: „Polyaminoacids, Polypeptides and Proteins" (Hrsg. Stahmann, M. A.) Univ. Wisconsin Press. Wisconsin 1962, S. 183.
14. Yaron, A., Berger, A., Biochim. Biophys. Acta **69**, 397 (1963).
15. Tiffany, M. L., Krimm, S., Biopolymers **8**, 347 (1969).
16. Tiffany, M. L., Krimm, S., Biopolymers **11**, 2309 (1972).
17. Satake, I., Yang, J. T., Biochem. Biophys. Res. Comm. **54**, 930 (1973).
18. Neumann, A. W., Moscarello, M. A., Epand, R. M., Biopolymers **12**, 1945 (1973).
19. Epand, R. M., Scheraga, H. A., Biopolymers **6**, 1383 (1968).
20. Barteri, M., Pispisa, B., Biopolymers **12**, 2309 (1973).
21. Paudjojo, L., Dissertation, Marburg/L. (1979).
22. Ebert, G., Paudjojo, L., in „Dynamic Aspects of Biopolyelectrolytes and Biomembranes" (Hrsg. F. Ooosawa), Kodansha Tokyo 1982, S. 63.
23. Zembala, M., Czarnecki, J., J. Coll. Interf. Sci. **89** 1 (1982).
24. Yamatera, H., Fitzpatrick, B., Gordon, G., J. Mol. Spectr. **14**, 268 (1964).

Eingegangen am 3. Januar 1983

Anschriften der Verfasser:

Prof. Dr. G. Ebert
Fachbereich Physikalische Chemie-Polymere-
der Philipps-Universität
Hans-Meerwein-Straße
D-3550 Marburg/Lahn F.R.G.

Dr. Young-Ha Kim
Korea Advanced Institute for Science
and Technology, Div. of Chem. Engineering
P.O. Box 131, Dongdaemun, Seoul, Korea
MI 48640-2696 USA

Electric-field orientation of charged helical polypeptides in solution*)

K. Yoshioka

Department of Chemistry, College of Arts and Sciences, University of Tokyo (Tokyo), Japan

Abstract: The elctric birefringence of charged polypeptides, including poly(L-lysine hydrobromide), poly(L-ornithine hydrobromide), poly (L-α,γ-diaminobutyric acid hydrochloride) and sodium poly(L-glutamate), in aliphatic alcohol/water mixtures has been measured at various solvent compositions. The field strength dependence of the steady-state birefringence, for both helical and coil conformations, resembled that expected for permanent dipole moment orientation. On the other hand, the transient behaviour indicated a mechanism of induced dipole moment orientation. This conflict can be resolved if saturation of the counterion-induced dipole moment is taken into account. The field strength dependence of the counterion-induced dipole moment and the orientation function is discussed in terms of the one-dimensional lattice model.

Key words: Electric birefringence, electric-field orientation, α-helix, polypeptide, polyelectrolyte.

1. Introduction

The mechanism of the electric-field orientation of uncharged helical polypeptides in solution is well established. When a polypeptide molecule assumes the α-helical conformation, the N–H and C=O groups are directed almost along the helix axis. Hence a large resultant permanent dipole moment is expected for the whole molecule. Dielectric and electro-optical studies on poly(γ-benzyl L-glutamate) and other uncharged polypeptides in helicogenic solvents have demonstrated that the electric-field orientations of these polypeptides is due primarily to the permanent dipole moment [1, 2].

The situation is quite different in the case of charged polypeptides. Electrolytic polypeptides, such as poly(L-glutamic acid) and poly(L-lysine), undergoes a pH-induced helix-coil transition in aqueous media. In some cases, addition of organic solvents to neutral aqueous solutions of ionized polypeptides induces a coil-to-helix transition. Using optical rotatory dispersion and circular dichroism measurements, Epand and Scherage [3] found that poly(L-lysine hydrochloride) undergoes a transition from the random coil to the α-helical conformation between 87 and 90 vol% methanol. Subsequently, Joubert, Lotan and Scheraga [4] showed in a nuclear magnetic resonance study that the ε-amino groups of poly(L-lysine) are charged in 90 vol% methanol at pH 6, even though the conforma-

tion is helical. Thus, poly(L-lysine) in methanol solution seems to be a good system for the study of the transition between the charged coil and the charged helix and the mechanism of the electric-field orientation of these two conformations.

We have carried out transient electric birefringence measurements on a series of basic polypeptides with side chain $R = -(CH_2)_nNH_2$, namely poly(L-lysine) ($n = 4$), poly(L-ornithine) ($n = 3$) and poly(L-α,γ-diaminobutyric acid) ($n = 2$) and an acidic polypeptide with side chain $R = -(CH_2)_2COOH$, poly(L-glutamic acid), in aliphatic alcohol/water mixtures [5–12]. In the present paper the experimental results on the field strength dependence of the steady-state electric birefringence in these systems are summarized and the mechanism of the electric-field orientation of charged polypeptides in solution is discussed in connection with these results.

2. Materials and methods

Poly(L-lysine hydrobromide) (Lot No. L-91, molecular weight given as 140 000) was procured from Pilot Chemicals, Inc. Poly(L-ornithine hydrobromide) (Lot No. 25C–5023, molecular weight given as 122 000) was purchased from Sigma Chemical Co. Poly(L-α,γ-diaminobutyric acid hydrochloride) was prepared from poly(N-carbobenzoxy-L-α,γ-diaminobutyric acid) by decarbobenzoxylation and generously supplied by Dr. S. Kubota of the University of California. Its limiting viscosity number measured in 0.2 M NaCl at 25 °C was 105 cm³ g⁻¹. Poly(L-glutamic acid) was prepared from poly(γ-benzyl L-glutamate) by debenzylation with hydrogen bromide in benzene. Its limiting viscosity number

---

*) Dedicated to the memory of Professor Dr. B. Tamamushi.

measured at pH 7.3 and 25 °C was 182 cm³ g⁻¹, from which the molecular weight of the acid form was estimated to be 87 000.

All organic solvents used were of reagent grade. The solvent composition was expressed in terms of the volume percentage of the organic solvent. The polymer concentration was expressed as kg m⁻³ (= g dm⁻³).

The apparatus and procedure for transient electric birefringence measurements have been described previously [13, 14]. The electric field was applied in the form of single rectangular pulses to a solution in the Kerr cell. The pulse width was from 150 to 500 μs and the pulse amplitude was up to 7 kV. The electric birefringence, Δn, of the solution was calculated from the optical retardation in the presence of an electric field. We define the specific Kerr constant, B/c, of the solution by

$$\frac{B}{c} = \lim_{E \to 0} \frac{\Delta n}{\lambda c E^2} \qquad (1)$$

where E is the field strength, λ is the wavelength of light *in vacuo*, and c is the mass concentration of the solute.

Circular dichroism (CD) measurements were carried out with a JASCO J-20 recording spectropolarimeter. The molar ellipticity at 222 nm (one of two CD minima characteristic of α-helix) was used as a measure of the helix content.

3. Experimental results

The polypeptide/alcohol/water systems hitherto studied are as follows:

poly(L-lysine hydrobromide)/methanol/water [5, 6]
poly(L-ornithine hydrobromide)/methanol/water [7]
poly(L-ornithine hydrobromide)/ethanol/water [8]
poly(L-α,γ-diaminobutyric acid hydrochloride)/methanol/water [9, 10], sodium poly(L-glutamate)/methanol/water [11, 12], sodium poly(L-glutamate)/ethylene glycol/water [11, 12].

Circular dichroism measurements on these systems have revealed that the polypeptide assumes the extended "charged coil" form in water, but is transformed into the helical conformation in the region of high alcohol content. This solvent-induced helix-coil transition was accompanied by an abrupt change in the specific Kerr constant in all these systems. However, the variations of the specific Kerr constant in the transition region were rather small. This is taken as characteristic of the transition between the charged coil and the carged helix. The specific Kerr constant of the charged coil was found to be much larger than that of the uncharged random coil.

The double logarithmic plots of $\Delta n/\lambda c$ versus E^2 in these charged polypeptides for various polymer concentrations and solvent compositions could be superimposed on one another by shifting them horizontally along the abscissa and vertically along the ordinate, except the range where anomalous birefringence transients [5] were observed. Two examples are

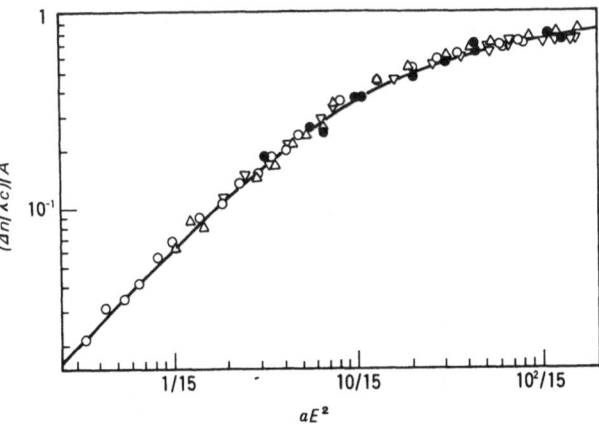

Fig. 1. Double logarithmic plots of $(\Delta n/\lambda c)/A$ versus aE^2 for poly(L-lysine hydrobromide) in 98 vol% methanol at 25 °C. Polymer concentration in 10^{-2} kg m⁻³: ●, 3.1; ▽, 6.2; △ 9.2; ○, 14. The solid curve represents equations (4) or (38)

presented here. Figure 1 is concerned with poly(L-lysine hydrobromide) in 98 vol% methanol at various polymer concentrations, and figure 2 is concerned with sodium poly(L-glutamate) in ethylene glycol/water mixtures at various solvent compositions. In the latter case the polypeptide assumes the helical conformation above 60 vol% ethylene glycol.

This indicates that the electric birefringence of these systems can be expressed by an equation involving two parameters, A and a

$$\frac{1}{A}\frac{\Delta n}{\lambda c} = \Phi(aE^2) \qquad (2)$$

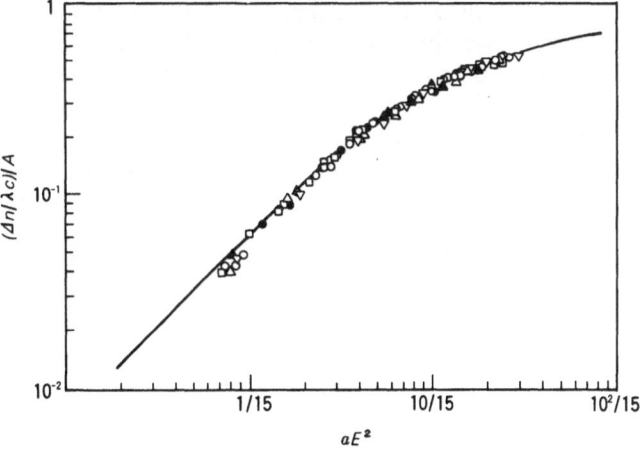

Fig. 2. Double logarithmic plots of $(\Delta n/\lambda c)/A$ versus aE^2 for sodium poly(L-glutamate) in ethylene glycol/water mixtures at 25 °C. Solvent composition in vol% ethylene glycol: ●, 0; ▲, 45; ▽, 60; △, 70; □, 80; ○, 90. Polymer concentration: 0.178 kg m⁻³

where the functional form of the orientation function, Φ, is independent of the polymer concentration and the solvent composition [14]. Note that A is defined by

$$A = \lim_{E \to \infty} (\Delta n/\lambda c) \tag{3}$$

and $a = (B/c)/A$.

The solid curve shown in figures 1 and 2 is a double logarithmic plot of the theoretical equation of the orientation function for the case of permanent dipole moment orientation. This equation has the following form

$$\Phi(E) = 1 - \frac{3\, L(\mu E/kT)}{\mu E/kT} \tag{4}$$

where μ is the apparent permanent dipole moment and $L(x)$ is the Langevin function [15]. Development of equation (4) in a power series of E yields

$$\Phi(E) = aE^2 - \frac{10}{7}\, a^2 E^4 + \frac{15}{7}\, a^3 E^6 - \cdots \tag{5}$$

with $a = \mu^2/15k^2T^2$. Thus, the field strength dependence of the electric birefringence for these systems, in both helical and coil conformations, resembles that expected for permanent dipole moment orientation over a rather wide range of field strengths.

However, it is improbable that the electric-field orientation of the charged coil is due to the permanent dipole moment. Moreover, the apparent permanent dipole moment of the charged helix, calculated by use of equation (4), was sometimes much larger than expected for the backbone permanent dipole moment of the α-helix. For example, in the case of poly(L-lysine hydrobromide) in 98 vol% methanol, the apparent permanent dipole moment thus obtained was 12 D/residue at $c = 3.1 \times 10^{-2}\,\mathrm{kg\,m^{-3}}$ and decreased with increasing polymer concentration [5].

On the contrary, the transient behaviour of the electric birefringence (build-up process and response to the reversing pulse) excludes the mechanism of permanent dipole moment orientation. The build-up and decay processes of the birefringence under the action of rectangular pulses were studied in the case of sodium poly(L-glutamate) in 90 vol% ethylene glycol [11, 12]. The high viscosity of this solvent made it possible to register the transient signals more accurately. The ratio of the area above the rise of the build-up curve and the area below the fall of the decay curve is related to the ratio of the permanent dipole moment contribution to the induced dipole moment contribu-

tion in the expression of the low-field birefringence [16]. This area ratio increased with decreasing field strength and approached unity at low fields in the present case. This fact indicates that the electric-field orientation is due primarily to the induced dipole moment.

The reversing-pulse electric birefringence technique is very powerful for elucidating the mechanism of the electric-field orientation [17, 18]. This technique was applied to poly(L-ornithine hydrobromide) in ethanol/water mixtures by Yamaoka and Ueda [8]. Upon rapid reversal of the pulse field, no transient could be observed. This confirms that the electric field orientation of poly(L-ornithine hydrobromide) results predominantly from the contribution of the counterion-induced dipole moment, regardless of its molecular conformations. Probably the backbone permanent dipole moment of the helical conformation, which amounts to a few Debye units per amino acid residue [1], is largely suppressed by the counterion-induced dipole moment in the charged state, as was suggested in the case of poly(L-glutamic acid) by Kobayasi and Ikegami [19].

4. Discussion

The conflict mentioned in the foregoing section can be resolved if saturation of the counterion-induced dipole moment is taken into account.

Two decades ago, Mandel [20] developed a theory for the induced dipole moment of rod-like polyelectrolytes due to the longitudinal displacement of "bound" counterions in terms of the one-dimensional lattice model. He adopted the two-phase model, in which a certain fraction of counterions are bound to the polyion and the other counterions are free to move in the solution. Subsequently, Neumann and Katchalsky [21] obtained an explicit expression for the field strength dependence of the longitudinal induced dipole moment on the basis of Mandel's theory, as given below

$$m = \frac{nzeL}{2}\left[\coth\frac{zeLE}{2kT} - \frac{1}{N}\coth\frac{zeLE}{2NkT}\right]. \tag{6}$$

Here N is the number of regularly distributed, charged sites per polyion, n the number of "bound" counterions per polyion, L the length of the polyion, e the protonic charge. It is to be noted that the number of "bound" counterions at each site is not restricted and interactions between the counterions are neglected in deriving this equation. If N is very large, equation (6) reduces to

$$m = \frac{nzeL}{2} L\left(\frac{zeLE}{2kT}\right).\tag{7}$$

In the limit of $E \to 0$, we obtain the electric polarizability

$$\alpha = \lim_{E \to 0} \frac{m}{E} = \frac{nz^2e^2L^2}{12kT}.\tag{8}$$

On the other hand, in the limit of $E \to \infty$, we obtain the saturated induced dipole moment

$$m_s = nzeL/2.\tag{9}$$

Using α or m_s, equation (7) can be rewritten as

$$m = \sqrt{3\alpha nkT}\, L\left(\sqrt{\frac{3\alpha}{nkT}}\, E\right)\tag{10}$$

or

$$m = m_s L\left(\frac{m_s E}{nkT}\right).\tag{11}$$

The potential energy of interaction of this induced dipole moment with an electric field which makes an angle θ with the longitudinal axis is given by

$$U = -\frac{nzeL}{2} \int_0^{Eu} L\left(\frac{zeLEu}{2kT}\right) d(Eu)$$

$$= -nkT \ln \frac{\sinh(zeLEu/2kT)}{zeLEu/2kT}\tag{12}$$

where $u = \cos\theta$. Hence, we obtain the following expression for the orientation function [22]

$$\Phi(E) = \frac{3}{2} \frac{\int_0^1 u^2 \exp(-U/kT)\, du}{\int_0^1 \exp(-U/kT)\, du} - \frac{1}{2}$$

$$= \frac{3}{2} \frac{\int_0^1 u^2 \left(\frac{\sinh \chi u}{\chi u}\right)^n du}{\int_0^1 \left(\frac{\sinh \chi u}{\chi u}\right)^n du} - \frac{1}{2}\tag{13}$$

where we have introduced the notation

$$\chi = \frac{zeLE}{2kT} = \sqrt{\frac{3\alpha}{nkT}}\, E = \frac{m_s E}{nkT}.\tag{14}$$

If n is sufficiently small (say $n = 1$), the field strength dependence of this orientation function resembles the behaviour for permanent dipole moment orientation, as described in a previous paper [22]. For large n, equation (13) approaches the classical orientation function (equation (39) referred to later).

Next, we consider the case in which the number of "bound" counterions at each site is restricted to 0 or 1 [23, 24] and interactions between the counterions are neglected. In this case the grand partition function of the rod-like polyelectrolyte is given by

$$\Xi = \prod_{j=-(N-1)/2}^{(N-1)/2} [(1 + \lambda \exp(ja)]\tag{15}$$

where λ is the normalized activity and a is defined by

$$a = \frac{zeLE}{(N-1)kT} \simeq \frac{zeLE}{NkT}.\tag{16}$$

The number of "bound" counterions per polyion is written as

$$n = \lambda \left(\frac{\partial \ln \Xi}{\partial \lambda}\right)_{T,N,E} = \sum_j \frac{1}{\lambda^{-1}\exp(-ja) + 1}\tag{17}$$

$$\simeq \int_{-N/2}^{N/2} \frac{1}{\lambda^{-1}\exp(-ja) + 1}\, dj$$

$$= N - \frac{1}{a} \ln \frac{\lambda^{-1}\exp(Na/2) + 1}{\lambda^{-1}\exp(-Na/2) + 1}.\tag{18}$$

The induced dipole moment is written as

$$m = kT\left(\frac{\partial \ln \Xi}{\partial E}\right)_{T,N,\lambda}$$

$$= \frac{zeL}{N} \sum_j \frac{j}{\lambda^{-1}\exp(-ja) + 1}.\tag{19}$$

If a fixed value is assigned to n, the field strength dependence of the induced dipole moment can be calculated from equations (19) and (18) (λ is eliminated). To the third order in E, the expansion of m is

$$m = \frac{n(N-n)}{N} \frac{z^2e^2L^2}{12\,kT}\, E$$

$$- \frac{n(N-n)\,[N^2 - n(N-n)]}{N^3} \frac{z^4e^4L^4}{720\,k^3T^3}\, E^3$$

$$+ \mathrm{O}(E^5).\tag{20}$$

Hence, the electric polarizability is given by

$$\alpha = \left(1 - \frac{n}{N}\right) \frac{nz^2e^2L^2}{12\,kT}. \tag{21}$$

For large **a**, the limiting form of m becomes

$$m = \frac{n(N-n)zeL}{2} \tanh \frac{n(N-n)zeLE}{2NkT}. \tag{22}$$

The saturated induced dipole moment is given by

$$m_s = \left(1 - \frac{n}{N}\right) \frac{nzeL}{2}. \tag{23}$$

With the help of equation (23), equation (22) can be rewritten as

$$m = m_s \tanh \frac{m_s E}{kT}. \tag{24}$$

As can be seen from equations (21) and (23), the electric polarizability and the saturated induced dipole moment decrease with an increase in n if $n > N/2$.

In general, the induced dipole moment can be written as an old function of E:

$$m = \alpha E + \beta E^3 + O(E^5) \tag{25}$$

where β is the third-order non-linear polarizability. The corresponding orientation function is calculated to be

$$\Phi(E) = \frac{\alpha}{15kT} E^2$$

$$+ \frac{1}{35} \left(\frac{\alpha^2}{9k^2T^2} + \frac{\beta}{kT}\right) E^4 + O(E^6) \tag{26}$$

as reported earlier [25]. The deviation from the Kerr law is expressed by

$$\frac{\Delta n/E^2}{(\Delta n/E^2)_{E\to 0}} = \frac{\Phi/E^2}{(\Phi/E^2)_{E\to 0}} = 1 + \frac{3}{7} \left(\frac{\alpha}{9kT}\right.$$

$$\left. + \frac{\beta}{\alpha}\right) E^2 + O(E^4). \tag{27}$$

Therefore, the sign of the initial slope of $\Delta n/E^2$ *versus* E^2 plot depends on the magnitude of $\beta kT/\alpha^2$. It is positive or negative according as $-\beta kT/\alpha^2 < 1/9$ or $-\beta kT/\alpha^2 > 1/9$.

Using equation (26), we obtain the following orientation function from equation (20)

$$\Phi(E) = \frac{n(N-n)}{N} \frac{z^2e^2L^2}{180\,k^2T^2} E^2 + \frac{n(N-n)[n(N-n\,(5N+9)-9N^2]}{N^3} \frac{z^4e^4L^4}{226800\,k^4T^4} E^4 + O(E^6). \tag{28}$$

The deviation from the Kerr law is given by

$$\frac{\Phi/E^2}{(\Phi/E^2)_{E\to 0}} = 1 + \frac{n(N-n)(5N+9)-9N^2}{N^2} \frac{z^2e^2L^2}{1260\,k^2T^2} E^2 + O(E^4). \tag{29}$$

In the limit of $n = 1$ and $n = N-1$, the orientation function reduces to equation (13) with $n = 1$.

Furthermore, we consider the extreme case in which the number of "bound" counterions at each site is restricted to 0 or 1 and the repulsive interactions between the counterions are large enough that two counterions cannot occupy adjacent sites. In this case the grand partition function is given by

$$\Xi = \begin{bmatrix} 1 & 1 \end{bmatrix} \begin{bmatrix} 0 & \lambda\exp[(N-1)a/2] \\ 1 & 1 \end{bmatrix} \cdots \begin{bmatrix} 0 & \lambda\exp(ja) \\ 1 & 1 \end{bmatrix} \cdots$$

$$\times \begin{bmatrix} 0 & \lambda\exp[-(N-3)a/2] \\ 1 & 1 \end{bmatrix} \begin{bmatrix} \lambda\exp[-(N-1)a/2] \\ 1 \end{bmatrix} \tag{30}$$

where we followed the matrix representation introduced by McTague and Gibbs [23]. If a fixed value is assigned to n, the corresponding canonical ensemble partition function, Q_n, is obtained from $\Xi = \sum_n Q_n \lambda^n$.

Then, the induced dipole moment can be calculated as a function of E by

$$m = kT \left(\frac{\partial \ln Q}{\partial E}\right)_{T,N,n}. \tag{31}$$

In the limit of $n = (N-1)/2$, Q becomes

$$Q = \exp \frac{N-1}{2} a + \exp \frac{N-3}{2} a$$

$$+ 2 \exp \frac{N-5}{2} a + 2 \exp \frac{N-7}{2} a$$

$$+ \ldots + \frac{N-1}{4} \exp(2a)$$

$$+ \frac{N-1}{4} \exp(a) + \frac{N+3}{4}$$

$$+ \frac{N-1}{4} \exp(-a) + \frac{N-1}{4} \exp(-2a)$$

$$+ \ldots + 2 \exp \frac{-(N-7)}{2} a$$

$$+ 2 \exp \frac{-(N-5)}{2} a + \exp \frac{-(N-3)}{2} a$$

$$+ \exp \frac{-(N-1)}{2} a = \frac{\sinh \dfrac{N+1}{2} a}{2 \sinh a}$$

$$+ \frac{\cosh \dfrac{N+1}{2} a - 1}{2 (\cosh a - 1)}. \tag{32}$$

To the third order in E, the expansion of m is

$$m = \frac{(N-1)(N+5)}{N^2} \frac{z^2 e^2 L^2}{24 kT} E - \frac{(N-1)(N+5)(N^2+4N+19)}{N^4} \frac{z^4 e^4 L^4}{5760\, k^3 T^3} E^3 + \mathrm{O}(E^5). \tag{33}$$

For large a, the limiting form of m becomes

$$m = \frac{zeL}{2} \tanh \frac{zeLE}{2kT}. \tag{34}$$

For $N = 1001$ and $n = 500$, we calculated the field strength dependence of the induced dipole moment in the three cases mentioned above. The results are shown in the form of $m/\sqrt{akT}$ *versus* $(\sqrt{\alpha/kT})E$ plots in figure 3. In the last extreme case, saturation of the induced dipole moment is very much pronounced and

the saturated induced dipole moment is greatly depressed. Numerical reduced values of α, β, m_s and $\frac{3}{7}\left(\frac{\alpha}{9kT} + \frac{\beta}{\alpha}\right)$ (the coefficient of E^2 in equation (27)) are listed in table 1. As can be seen from this table, the intial slope of $(\Phi/E^2)/(\Phi/E^2)_{E\to0}$ *versus* E^2 plot is greatly reduced, but still positive in the last extreme case (case 3). Thus, introduction of the counterion repulsion is not sufficient to lead to the negative initial slope of $\Delta n/E^2$ *versus* E^2 plot (see equation (5)) on the basis of the present model.

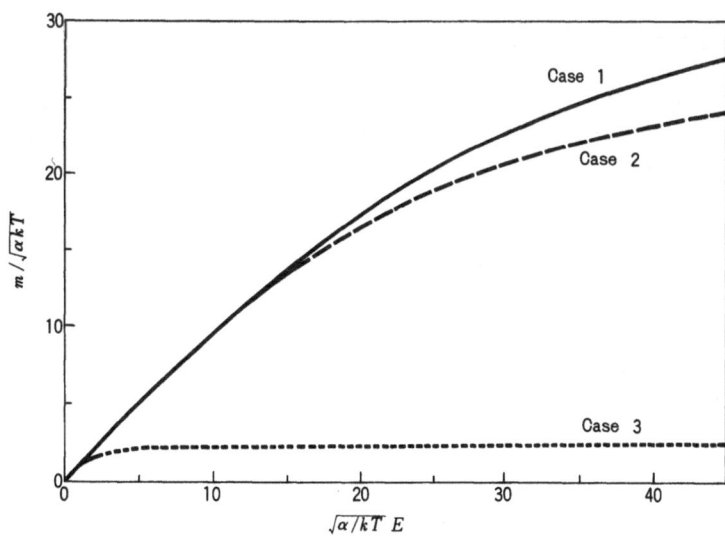

Fig. 3. Calculated dependence of $m/\sqrt{akT}$ on $(\sqrt{\alpha/kT})E$ for $N = 1001$ and $n = 500$ on the basis of the one-dimensional lattice model. Case 1: number of counterions per site is not restricted and counterion interactions are neglected. Case 2: number of counterions per site is restricted to 0 or 1 and counterion interactions are neglected. Case 3: number of counterions per site is restricted to 0 or 1 and two counterions cannot occupy adjacent sites

Table 1. Numerical reduced values of α, β, m_s and $\frac{3}{7}\left(\frac{\alpha}{9kT}+\frac{\beta}{\alpha}\right)$ for $N = 1001$ and $n = 500$ in the three cases on the basis of the one-dimensional lattice model

	Case 1	Case 2	Case 3
$\alpha \left/ \dfrac{z^2e^2L^2}{N^2kT} \right.$	4.18×10^7	2.09×10^7	4.19×10^4
$\beta \left/ \dfrac{z^4e^4L^4}{N^4k^2T^2} \right.$	-6.98×10^{11}	-2.62×10^{11}	-1.76×10^8
$m_s \left/ \dfrac{zeL}{N} \right.$	2.5×10^5	1.25×10^5	500
$\dfrac{3}{7}\left(\dfrac{\alpha}{9kT}+\dfrac{\beta}{\alpha}\right) \left/ \dfrac{z^2e^2L^2}{N^2k^2T^2} \right.$	1.98×10^6	0.99×10^6	2.00×10^2

The induced dipole moment given by equation (24) deserves special attention. If this equation holds over the whole range of field strengths, the potential energy of interaction of this induced dipole moment with an electric field becomes

$$U = -m_s \int_0^{Eu} \tanh \frac{m_s Eu}{kT} \, d(Eu)$$

$$= -kT \cosh \frac{m_s Eu}{kT}. \tag{35}$$

Hence, we obtain the following expression for the orientation function

$$\Phi(E) = \frac{3}{2} \frac{\displaystyle\int_0^1 u^2 \cosh \frac{m_s Eu}{kT} \, du}{\displaystyle\int_0^1 \cosh \frac{m_s Eu}{kT} \, du} - \frac{1}{2}$$

$$= 1 - \frac{3\, L(m_s E/kT)}{m_s E/kT} \tag{36}$$

This equation has the same form as equation (4), the only difference being that μ is replaced by m_s. Development of equation (24) in a power series of E shows that the electric polarizability is related to m_s by

$$\alpha = m_s^2/kT. \tag{37}$$

With the help of equation (37), equation (36) can be rewritten as

$$\Phi(E) = 1 - \frac{3\, L(\sqrt{\alpha/kT}\,E)}{\sqrt{\alpha/kT}\,E}. \tag{38}$$

As is well known, if $m = \alpha E$ holds over the whole range of field strengths, the orientation function is given by the classical equation

$$\Phi(E) = \frac{3}{4}\left[\frac{\exp(\gamma)}{\sqrt{\gamma}\displaystyle\int_0^{\sqrt{\gamma}} \exp(t^2)\,dt} - \frac{1}{\gamma} \right] - \frac{1}{2} \tag{39}$$

where $\gamma = \alpha E^2/2kT$ [15].

A comparison is made among these orientation functions for induced dipole moment orientation in the forms of Φ *versus* $(\alpha/kT)E^2$ and $(\Phi/E^2)/(\Phi/E^2)_{E\to 0}$ *versus* $(\alpha/kT)E^2$ plots, as shown in figures 4 and 5. The solid, dashed and dotted curves in these figures represent equation (39), equation (13) with $n = 1$ and equation (38), respectively. It is to be noted that the initial slope of Φ/E^2 *versus* E^2 plot is negative in the last two cases. The field strength dependence of the corresponding induced dipole moment in these cases is shown in the form of $m/\sqrt{\alpha kT}$ *versus* $(\sqrt{\alpha/kT})E$ plots in figure 6.

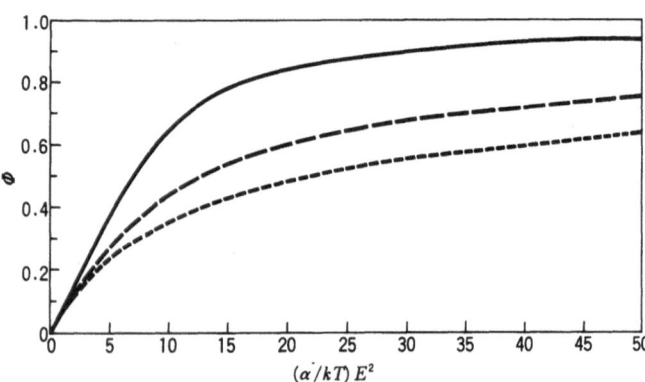

Fig. 4. Plots of Φ *versus* $(\alpha/kT)E^2$. Solid curve: equation (39). Dashed curve: equation (13) with $n = 1$. Dotted curve: equation (38)

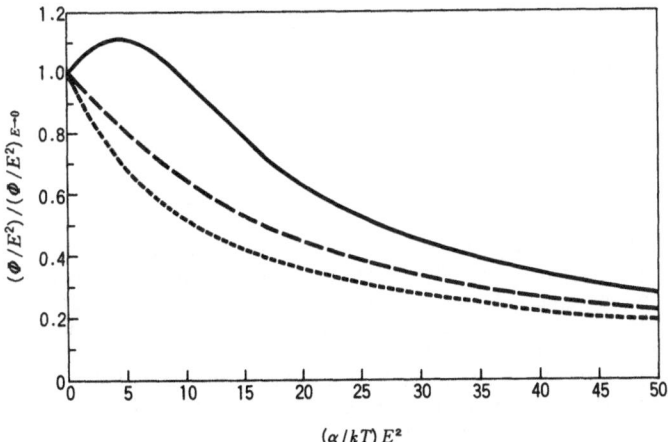

Fig. 5. Plots of $(\Phi/E^2)/(\Phi/E^2)_{E\to0}$ versus $(\alpha/kT)E^2$ *(vide supra)*

The present experimental results regarding charged polypeptides seem to be in favour of equation (38). However, this equation should be regarded as an empirical equation containing α or m_s as one adjustable parameter, because equation (24) is not yet justified theoretically at arbitrary field strengths. Equation (38) can also be fitted to the experimental data on the field strength dependence of the electric birefringence or dichroism over a limited range of field strengths in the cases of DNA (sonicated fragments [26] and enzyme-restricted fragments [27, 28]) and linear flexible polyelectrolytes, such as sodium poly-(ethylenesulfonate) [15, 29], potassium poly(styrenesulfonate) [14] and poly(N-butyl-4-vinylpyridinium bromide) [30], in aqueous solutions.

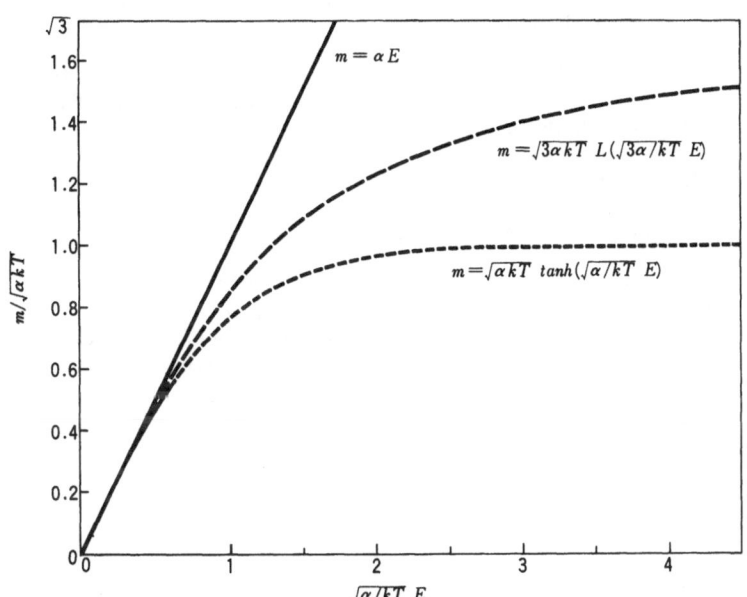

Fig. 6. Plots of $m/\sqrt{\alpha kT}$ versus $(\sqrt{\alpha kT})E$. Solid curve: $m = \alpha E$. Dashed curve: equation (10) with $n = 1$. Dotted curve: equation (24)

5. Concluding remarks

Although the electric-field orientation of the uncharged helix is due primarily to the backbone permanent dipole moment, that of the charged helix is ascribed to the counterion-induced dipole moment. In the latter case, saturation of the induced dipole moment seems to play an important role in determining the field strength dependence of the electric birefringence. However, the exact nature of the counterion-induced dipole moment must await further investigations. Especially, the assumption that the number of bound, but mobile counterions per polyion

is not affected by the field strength should be reexamined. Very recently, Diekmann et al. [31] have proposed a new orientation mechanism in which the induced dipole moment is saturated above a critical field strength. Theoretical calculations of the orientation function on the basis of this mechanism will be published in due course [32].

Note added in proof:

K. Kikuchi [J. Chem. Phys., submitted] have recently found that the field strength dependence of the birefringence of poly(L-lysine hydrobromide) in 98 vol% methanol is given by equation (39) in the range of very low field strengths.

9

Acknowledgments

This work was partly supported by the Grand-in-Aid from the Ministry of Education, Science and Culture of Japan. The experiments were performed by K. Kikuchi, K. Takakusaki and M. Fujimori.

References

1. Wada, A., The α-helix as an electric macro-dipole – Advances in Biophysics (Kotani, M., Ed.) Vol. 9, pp. 1–63 (Tokyo 1976).
2. Yoshioka, K., Electro-optics of polypeptides and proteins – Molecular Electro-optics (O'Konski, C. T., Ed.) Part 2, pp. 601–643 (New York 1978).
3. Epand, R. F., Scheraga, H. A., Biopolymers 6, 1383 (1968).
4. Joubert, F., Lotan, N., Scheraga, H. A., Physiol. Chem. Phys 1, 348 (1969).
5. Kikuchi, K., Yoshioka, K., Biopolymers 12, 2667 (1973).
6. Kikuchi, K., Yoshioka, K., Biopolymers 15, 1669 (1976).
7. Yoshioka, K., Fujimori, M., Kikuchi, K., Int. J. Biol. Macromol. 2, 213 (1980).
8. Yoshioka, K., Fujimori M., Yamaoka, K., Ueda, K., Int. J. Biol. Macromol. 4, 55 (1982).
9. Fujimori, M., Kikuchi, K., Yoshioka, K., Kubota S., Biopolymers 18, 2005 (1979).
10. Yoshioka, K., Kikuchi, K., Fujimori, M., Biophys. Chem. 11, 369 (1980).
11. Takakusaki, K., Kikuchi, K., Yoshioka, K., Polymer 18, 969 (1977).
12. Yoshioka, K., Takakusaki, K., Int. J. Biol. Macromol. 4, 123 (1982).
13. Ikeda, K., Watanabe, H., Shirai, M., Yoshioka, K., Sci. Pap. Coll. Gen. Educ., Univ. Tokyo 15, 139 (1965).
14. Kikuchi, K., Yoshioka, K., J. Phys. Chem. 77, 2101 (1973).
15. O'Konski, C. T., Yoshioka, K., Orttung, W. H., J. Phys. Chem. 63, 1558 (1959).
16. Yoshioka, K., Watanabe, H., Nippon Kagaku Zasshi 84, 626 (1963).
17. O'Konski, C. T., Pytkowicz, R. M., J. Am. Chem. Soc. 79, 4815 (1957).
18. Tinoco, I., Jr., Yamaoka, K., J. Phys. Chem. 63, 423 (1959).
19. Kobayasi, S., Ikegami, A., Biopolymers 14, 543 (1975).
20. Mandel, M., Mol. Phys. 4, 489 (1961).
21. Neumann, E., Katchalsky, A., Proc. Natl. Acad. Sci. U.S. 69, 993 (1972).
22. Kikuchi, K., Yoshioka, K., Biopolymers 15, 583 (1976).
23. McTague, J. P., Gibbs, J. H., J. Chem. Phys. 44, 4295 (1966).
24. Warashina, A., Minakata A., J. Chem. Phys. 58, 4743 (1973).
25. Yoshioka, K., Takakusaki, K., Sci. Pap. Coll. Gen. Educ., Univ. Tokyo 31, 111 (1981).
26. Yamaoka, K., Matsuda, K., Macromolecules 14, 595 (1981).
27. Hogan, M., Dattagupta, N., Crothers, D. M., Proc. Natl. Acad. Sci. U.S. 75, 195 (1978).
28. Stellwagen, N. C., Biopolymers 20, 399 (1981).
29. Yoshioka, K., O'Konski, C. T., J. Polym. Sci., Part A-2 6, 421 (1968).
30. Tricot, M., Houssier, C., Desreux, V., Eur. Polym. J. 12, 575 (1976).
31. Diekmann, S., Hillen, W., Jung, M., Wells, R. D., Pörschke, D., Biophys. Chem. 15, 157 (1982).
32. Yoshioka, K., J. Chem. Phys. 79, 3482 (1983).

Received January 7, 1983;
accepted February 8, 1983

Authors' address:

Prof. Koshiro Yoshioka
Department of Chemistry, College of Arts and Sciences,
University of Tokyo, Komaba, Meguro-ku,
Tokyo 153, Japan

Progress in Colloid & Polymer Science   Progr. Colloid & Polymer Sci. **68**, 131–132 (1983)

A note on the effect of Ca²⁺ on the viscous and elastic moduli of surface layers of fibrinogen*)

A. L. Copley[1]) and R. G. King[2])

Laboratory of Biorheology[1]), Polytechnic Institute of New York, Brooklyn, NY 11201 and Division of Circulatory Physiology and Biophysics[2]), Department of Physiology, College of Physicians and Surgeons, Columbia University, New York, NY 10027, USA.

Key words: Calcium ions, elastic modulus, fibrinogen surface layers, modified Weissenberg Rheogoniometer, viscous modulus.

Ca^{2+} has long been known to play an important role in the integrity of the endothelium [1]. The endothelial junction, which has a significant role in vascular or capillary permeability, is bound together by an interendothelial cement substance, which Chambers and Zweifach named 'calcium proteinate' [2]. In this connection the role of calcium has been investigated, since the beginning of the century, in perfusion studies of the microcirculatory vascular beds [1, 2]. Perfusion fluids without the addition of Ca^{2+} can have a deleterious effect by increasing markedly capillary permeability. One of us (ALC) postulated that the proteinate of calcium is fibrin and he referred to it as 'calcium fibrinate' [1].

Calcium has been found to strengthen the rigidity of fibrin gels, both with plasma [3] and upon the activation of fibrinogen by thrombin [4]. Ca^{2+} may also have an effect on fibrinogen gels formed without thrombin participation. Weisel et al. [5] emphasized that the most common aggregates of fibrinogen which they investigated exhibited a fibrin-like structure. They concluded that their findings provided additional evidence that fibrinogen does not differ greatly from fibrin, although the latter does not contain fibrinopeptides *A* and *B*. It was, therefore, of interest to find out whether the addition of $CaCl_2$ to solutions of fibrinogen would affect the surface viscosity (η_s') and surface elasticity (G_s') of fibrinogen surface gels.

We used, in this study, highly purified human fibrinogen of 98–100 per cent clottability, manufactured according to the method of Blombäck and Blombäck [6] by IMCO Corporation, S-113 30 Stockholm, Sweden. A 0.2 per cent fibrinogen concentration was made in 0.9 per cent NaCl solution containing $CaCl_2$ in concentrations of 5 to 100 mM. The viscous and elastic moduli were measured in a Weissenberg Rheogoniometer, modified by Copley and King [7] and used in the dynamic mode. The pH was controlled at 7.4 by Tris and the temperature was maintained at 37 ± 0.5 °C. Ca^{2+} in concentrations of 5, 20, 50 and 100 mM was added to the preparations.

Figure 1 shows the values of η_s' and G_s' of the 0.2 per cent fibrinogen control preparation. The addition of 5 mM Ca^{2+} causes an increase in the values of η_s' and G_s' of 30 and 40 per cent, respectively. The addition of higher amounts, namely 20, 50 and

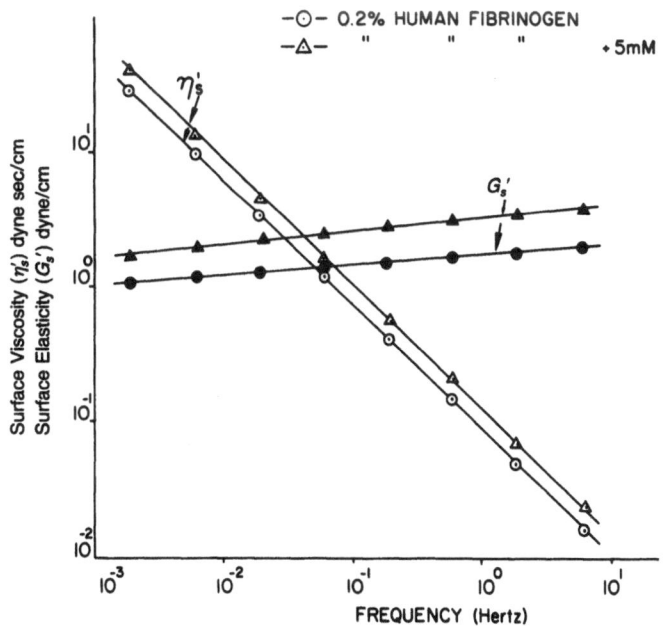

Values η_s' (- O -) and G_s' (- ● -) of a 0.2% fibrinogen preparation as control, compared with values of η_s' (- △ -) and G_s' (- ▲ -) upon the addition of 5 mM Ca^{2+}

---

*) Dedicated to the memory of Professor Dr. B. Tamamushi.

100 mM Ca^{2+} did not cause further increases, and the values for these concentrations are therefore not shown.

Our preliminary findings indicate that Ca^{2+} can markedly affect the surface viscous and elastic moduli by increasing their values. These findings suggest that calcium may affect the interendothelial cement substance which may well contain fibrin(ogen) and thus play an important role in vascular integrity.

Acknowledgement

One of the investigators (R. G. K.) was supported by U. S. P. H. S. Research Grant HL 16851 made to Columbia University College of Physicians and Surgeons.

References

1. Copley, A. L., The endo-endothelial fibrin lining. A historical account. In: The Endoendothelial Fibrin Lining. Copley, A. L., (Ed.), New York-Oxford, Pergamon Press, 1983; Thrombosis Res., 30, Suppl. V, 1–26 (1983).
2. Chambers, R., and Zweifach, B. W., Capillary endothelial cement in relation to permeability. J. Celluar Compar. Physiol. 15, 255–272, (1940).
3. Copley, A. L., Some problems in hemorheology. Proc. 5. Internat. Congr. Rheology, Onogi S. (Ed.) Tokyo, Japan, Baltimore, MD, USA and Manchester, England, Univ. of Tokyo Press and University Park Press, 1970, vol. 2, pp. 3–25.
4. Shen, L. L., Hermans, J., McDonagh, J., McDonagh, R. P. and Carr, M., Effects of calcium ion and covalent cross-linking of formation and elasticity of fibrin gels. Thrombosis Res., 6, 225–265, (1975).
5. Weisel, J. W., Phillips, G. N. and Cohen, C., The structure of fibrinogen and fibrin: II. Architecture of the fibrin clot. Ann. N. Y. Ac. Sci., 408, 371–379 (1983).
6. Blombäck, B. and Blombäck, M., Purification of human and bovine fibrinogen. Arkiv Kemi. , 10, 415–443, (1956).
7. Copley, A. L. and King, R. G., Polymolecular layers of fibrinogen systems and the genesis of thrombosis. In: Hemorheology and Thrombosis, Copley, A. L., and Okamoto, S. (Eds.), New York-Oxford: Pergamon Press, 1976, pp. 393–409; Thrombosis Res. 8 Suppl. II, 393–409, (1976).

Received May 27, 1983

Authors' address:

A. L. Copley
Laboratory of Biorheology
Polytechnic Institute of New York
Brooklyn, NY 11201

Disintegration of poly(N^α, N^ε-L-lysindiylterephthaloyl) microcapsules by fibrinogen*)

S. Sumida, S. Jomura, M. Arakawa, and T. Kondo

Faculty of Pharmaceutical Sciences, Science University of Tokyo, Tokyo, Japan

Abstract: Poly(N^α, N^ε-L-lysinediylterephthaloyl) (abbreviation: PPL) microcapsules were prepared by an interfacial polymerization technique which makes use of the interfacial polycondensation reaction between L-lysine dissolved in water and terephthaloyl dichloride dissolved in an organic solvent. Interaction of PPL microcapsules with fibrinogen was investigated as a function of fibrinogen concentration, pH and ionic strength of the medium, and anion species of added salt.

PPL microcapsules were found to undergo disintegration by the action of fibrinogen molecules at high ionic strengths of the medium, irrespective of pH. The presence of thiocyanate ion, a breaker of water structure, enhanced disintegration of PPL microcapsules by fibrinogen while citrate ion, a maker of water structure, suppressed it. It was shown fluometrically by the use of dansylated fibrinogen that PPL microcapsules adsorb fibrinogen molecules prior to their disintegration.

Disintegration of PPL microcapsules by fibrinogen was suggested to be caused by the hydrophobic interaction of PPL molecules with fibrinogen molecules, leading to intermingling of the polymers involved, based on the experimental findings.

Key words: Interfacial polycondensation reaction; Polyamide microcapsules; Microcapsule disintegration; Adsorption; Fibrinogen; Intermingling.

Introduction

Hemolysate-loaded poly(N^α, N^ε-L-lysinediyl-terephthaloyl) microcapsules have been studied in our laboratory with the hope that they may be used in transfusion as artificial red blood cells [1–4]. So far, their preparation, oxygen absorbability, and rheological properties have been well established. It has then become necessary to investigate the interactions of the microcapsules with various plasma components before putting them into practice because they will experience these interactions when injected into the blood stream.

On the other hand, the interactions between microcapsules and plasma components will provide valuable information about the stability of microcapsules in plasma and the actions of various proteins of synthetic polymers since a microcapsule can be regarded as a matrix of polymers assembled in a very thin spherical membrane. Actually, gelatin molecules were found to cause disintegration of poly(N^α, N^ε-L-lysinediylterephthaloyl) microcapsules in aqueous media of high ionic strengths, irrespective of pH of the

medium, if the molecular weight of gelatin was higher than 5 000 [5]. This suggests a hydrophobic nature of the interaction between the microcapsules and gelatin molecules and intermingling of both polymer components involved. On the contrary, the microcapsules were never disintegrated by either albumin or globulin while they formed aggregates by the action of both of the proteins [6].

In this paper the disintegration phenomenon of poly(N^α, N^ε-L-lysinediylterephthaloyl) microcapsules by fibrinogen, a plasma component, will be described in terms of the protein concentration, pH and ionic strength of the medium, and ionic composition of added salt.

Experimental

Preparation of microcapsules

Poly(N^α, N^ε-L-lysinediylterephthaloyl) microcapsules (abbreviated hereafter to PPL microcapsules) were prepared in a manner quite similar to that in a previous report [2].

Thirty ml of 0.53 M solution of L-lysine dissolved in aqueous 0.67 M sodium carbonate solution was mechanically dispersed in 100 ml of a mixed solvent (cyclohexane-chloroform, 3:1, *v/v*) containing 10% (*v/v*) sorbitan trioleate. After 20 min of stirring

---

*) Dedicated to the memory of Prof. Dr. B. Tamamushi.

100 ml of 0.08 M solution of terephthaloyl dichloride in the mixed organic solvent was added to the emulsion formed and the mixture was stirred for another 20 min. Then, 100 ml of cyclohexane was added to the mixture to stop the interfacial polycondensation reaction between L-lysine and terephthaloyl dichloride. The carboxyl groups of L-lysine molecules remain unchanged after reacting with molecules of the acid dichloride and give negative charges in water to the polymer formed due to their dissociation. PPL microcapsules thus prepared were separated from the organic phase by centrifugation and dispersed into deionized water with the aid of a dispersing agent after washing them three times with cyclohexane on a centrifuge. The dispersed microcapsules were then washed several times with deionized water and dialyzed against deionized water for two weeks. The mean diameter of the microcapsules was about 10 μm.

Observation of microcapsule disintegration

Dialyzed PPL microcapsules were dispersed in each of the veronal-sodium acetate buffers of pH 2.4, 7.5, and 9.4 to give suspensions containing about 2×10^6 capsules in 1 ml. The ionic strengths of the buffers were 0.01, 0.1, and 0.2. To each of the microcapsule suspensions in test tubes was added an equal volume of various concentrations of bovine fibrinogen (95% clottable, Miles Laboratories, Ltd., U.S.A.) solution made up with each of the buffers.

The mixtures were allowed to stand for 1 hr with shaking at 37 °C. At the end of this period, a small portion of each of the mixtures was withdrawn by a pipette and placed on a hemocytometer to observe under an optical microscope if the microcapsules were disintegrated or not. The same procedure was employed to observe the effect of anion species of added salts on microcapsule disintegration by fibrinogen except that solutions of the salts were used as solvents of the protein (ionic strength 0.15).

The degree of microcapsule disintegration was defined by the equation, degree of disintegration $= (N_o - N_f)/N_o \times 100$, where N_o and N_f are the total number of PPL microcapsules in unit volume of the control suspension without fibrinogen and that of the sample suspension remaining unbroken after reacting with fibrinogen, respectively.

Rate of fibrinogen adsorption to macrocapsules

The rate of fibrinogen adsorption to PPL microcapsules was estimated by pursuing change with time in the fluorescence intensity of the supernatant solution of a mixture of PPL microcapsules and fluorescent dansylated bovine fibrinogen. Dansylated fibrinogen was prepared according to the method of Mihalyi and Albert [7]. The number of dansyl radicals introduced to a fibrinogen molecule was less than 6. Equal volumes of PPL microcapsule suspension and fibrinogen (1:1 mixture of fluorescent and non-fluorescent fibrinogen) solution were mixed in an Erlenmeyer flask and the mixture was shaken at 37 °C. A phosphate buffer (pH 7.8) containing 0.1 M KCl was used for both suspension medium of the microcapsules and solvent of the protein. The ionic strength of the buffer was 0.15.

At suitable time intervals, a portion of the mixture was withdrawn and filtered through a Millipore filter to remove the microcapsules. The fluorescence intensity of the filtrate was measured at 450 nm by a Kotaki model UM-2S fluorophotometer using the light of a wavelength of 360 nm for exciting the sample.

Results and discussion

Effect of pH

The effect of pH of the medium on disintegration of PPL microcapsules by bovine fibrinogen is shown in figure 1 where the degree of microcapsule disintegration is plotted against the log concentration of fibrinogen at different pH and at a constant ionic strength of 0.1.

Microcapsule disintegration becomes significant when the fibrinogen concentration reaches a value of 10^{-4} mg/ml at any pH while it is slightly affected by pH at fibrinogen concentrations higher than 10^{-2} mg/ml. This suggests that the interaction of PPL molecules constituting the microcapsules with fibrinogen molecules is of a hydrophobic nature rather than an ionic one inspite of the fact that both PPL and fibrinogen bear electrical charges under the experimental conditions. Actually, fibrinogen molecules are reported to be preferentially adsorbed to the hydrophobic moieties of polymer surfaces [8].

Effect of ionic strength

Figure 2 shows the effect of the ionic strength of the medium on the disintegration phenomenon of PPL microcapsules by bovine fibrinogen at pH 7.4. The degree of microcapsule disintegration is strongly dependent on the ionic strength of the medium at all fibrinogen concentrations; the higher the ionic strength is, the larger is the degree of disintegration. Although not shown here the degree of microcapsule

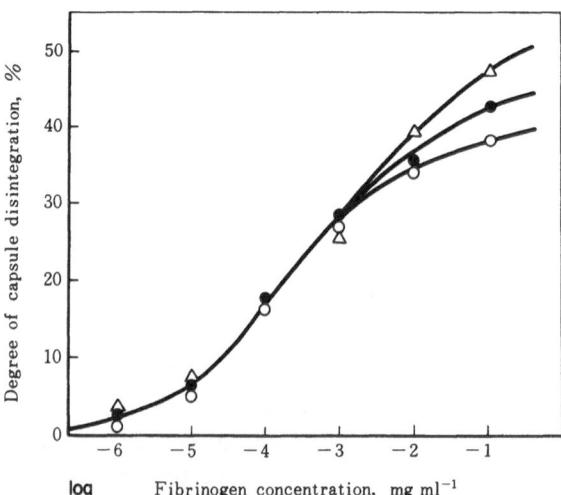

Fig. 1. Effect of pH on disintegration of PPL microcapsules by fibrinogen. pH: ○, 2.4; △, 7.5; ●, 9.4

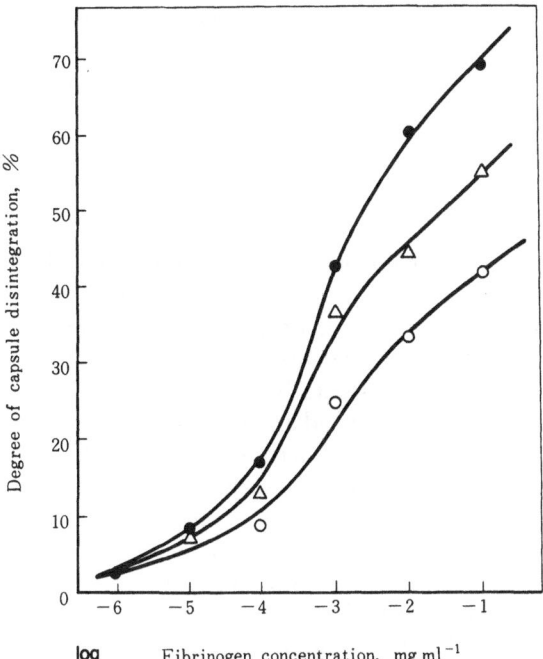

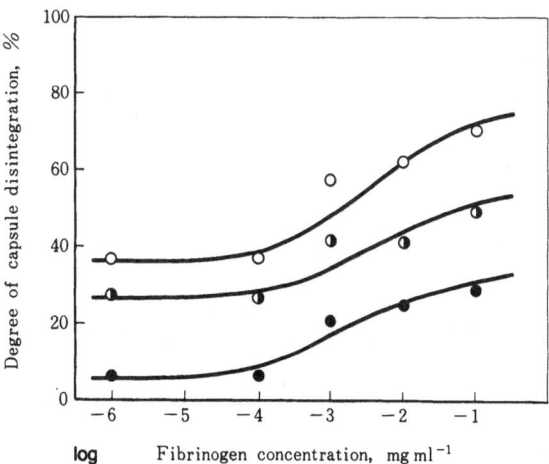

log   Fibrinogen concentration, mg ml^{-1}

Fig. 3. Effect on anion species of added salts on disintegration of PPL microcapsules by fibrinogen. Anion: ○, thiocyanata; ◑, chloride; ●, citrate

Fig. 2. Effect of ionic strength on disintegration of PPL microcapsules by fibrinogen. Ionic strength: ◑, 0.01; △, 0.1; ●, 0,2

disintegration increases with rising ionic strength even at lower pH where fibrinogen and PPL molecules are oppositely charged. This suggests again the importance of the hydrophobic interaction between PPL and fibrinogen molecules in the disintegration phenomenon since it is well known that a high concentration of salt screens the electrostatic interaction between charged particles.

The fact that the degree of capsule disintegration begins to show a steep increase at a fibrinogen concentration of 10^{-4} mg/ml, independent of ionic strength, would indicate the existence of a minimum amount of fibrinogen needed to cause capsule disintegration. A simple calculation gives a value of about 10^{-10} mg fibrinogen per capsule or about 1.4×10^5 fibrinogen molecules per capsule to this minimum amount.

Effect of anion species of added salts

An important role of the hydrophobic interaction of PPL molecules with fibrinogen molecules has so far been suggested. In view of this, it will not be unreasonable to predict that disintegration of PPL microcapsules by fibrinogen is affected by the presence of those ions which cause a great distortion of the water structure around the hydrophobic moieties

of the microcapsules. Figure 3 illustrates how several anions with sodium ion as the common counterion affect the capsule disintegration phenomenon.

By virtue of its linear shape, thiocyanate ion is expected to distort the water structure to a great extent [9]. In fact, the presence of thiocyanate ion significantly enhances disintegration of PPL microcapsules by fibrinogen probably because the anion breaks down extensively the hydrophobic hydration around the microcapsules and fibrinogen molecules to make their direct contact possible. A similar disrupting effect of thiocyanate ion has been reported on the hydrophobic hydration surrounding the hydrocarbon moieties of polypeptides and lipids at the interface [10].

On the contrary, citrate ion, a maker of water structure, seems not to promote but, instead, to suppress capsule disintegration due to its strong tendency to form a Coulombic hydration atmosphere around it. Then, it would be possible that citrate ion strengthens the hydrated structure surrounding the hydrophilic moieties of the microcapsules without affecting the hydrophobic hydration of the hydrophobic moieties. Chloride ion assumes the intermediary position between thiocyanate and citrate ions owing to its moderate water-structure breaking ability.

Adsorption of fibrinogen to microcapsules

Figure 4 shows changes with time in fluorescence intensity for the supernatant solutions of mixtures of the microcapsule suspension and fibrinogen solutions

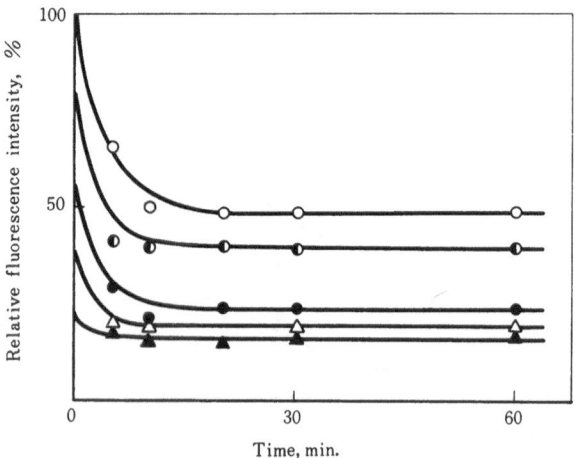

Fig. 4. Rate of fibrinogen adsorption to PPL microcapsules as revealed by change with time in fluorescence intensity of supernatant solution. Fibrinogen conc. (mg/ml): ▲, 0.02; △, 0.04; ●, 0.06; ◑, 0.08; ○, 0.10

of various concentrations at constant pH and ionic strength.

The fluorescence intensity decreases with time very quickly and levels off within 15 min, indicating a rapid adsorption of dansylated fibrinogen molecules to the microcapsules. Since it is reported that unless the number of dansyl radicals bound to a fibrinogen molecule exceeds 6–8 the general physicochemical properties of the protein can be regarded unchanged [7], both dansylated and non-dansylated fibrinogen molecules are expected to be equally adsorbed by the microcapsules and equally contribute to the disintegration phenomenon. Adsorption of fibrinogen molecules to the microcapsules is also evidenced in figure 5 by a photomicrograph a) taken under a fluorescence microscope of PPL microcapsules in the

presence of fluorescent and non-fluorescent fibrinogen. Another photomicrograph b) shows the same microcapsules as in a) taken under a phase contrast microscope for comparison.

Mechanism of microcapsule disintegration

Although it is evident that the phenomenon of PPL microcapsule disintegration starts from the adsorption of fibrinogen to the microcapsules as shown in figures 4 and 5 its mechanism has not yet been fully understood. Earlier papers from our laboratory have suggested that cationic and amphionic polyelectrolyte molecules act on PPL microcapsules at high ionic strengths of the medium in such a way as to cause intermingling of PPL molecules constituting the microcapsules with them [5, 11]. In the present case, fibrinogen molecules are also likely to intermingle with PPL molecules of the microcapsules after being adsorbed by the latter, with liberation of hydrophobically bound water molecules from the hydrophobic moieties of the polymers involved in the interaction. In this sense, disintegration of PPL microcapsules by fibrinogen would be an entropically driven rather than energetically driven phenomenon. However, the detailed mechanism of this disintegration phenomenon still remains inexplainable in molecular terms.

Acknowledgement

This work was supported by a Grant-in-Aid for Scientific Research (Special Research Project for Multiphase Biomedical Materials) from the Ministry of Education, Science, and Culture, Japan, to which our thanks are due.

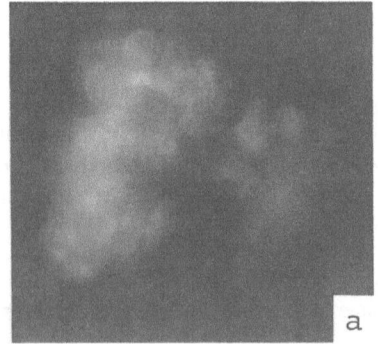
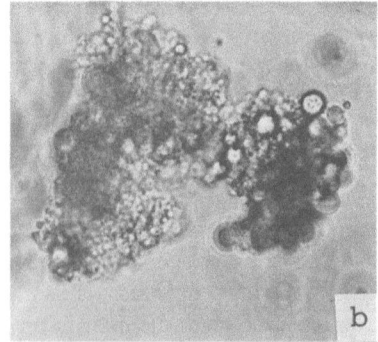

Fig. 5. Photomicrographs of PPL microcapsules in the presence of fibrinogen taken under fluorescence microscope a) and phase contrast microscope b)

References

1. Shigeri, Y., Tomizawa, M., Takahashi, K., Koishi, M., Kondo, T., Can. J. Chem. **49**, 3632 (1971).
2. Arakawa, M., Kondo, T., Can. J. Physiol. Pharmacol. **55**, 1378 (1977).
3. Arakawa, M., Kondo, T., Tamamushi, B., Biorheology **12**, 57 (1975).
4. Arakawa, M., Kondo, T., Can. J. Physiol. Pharmacol. **58**, 183 (1980).
5. Suzuki, S., Kondo, T., J. Colloid Interface Sci. **77**, 280 (1980).
6. Suzuki, S., Kondo, T., Kobunshi Ronbunshu (Polymer rep.) **36**, 197 (1979).
7. Mihalyi, E., Albert, A., Biochemistry **10**, 237 (1971).
8. Okano, T., Nishijima, S., Shinohara, I., Akaike, T., Sakurai, Y., Polymer J. **10**, 239 (1978).
9. Good, W., Biochim. Biophys. Acta **49**, 397 (1961).
10. Shibata, A., Yamashita, S., Yamashita, T., Bull. Chem. Soc. Jpn. **55**, 2814 (1982).
11. Suzuki, S., Kondo, T., J. Colloid Interface Sci. **67**, 441 (1978).

Received December 6, 1982;
accepted February 1, 1983

Author's address:

Prof. T. Kondo
Faculty of Pharmaceutical Sciences
Science University of Tokyo
12, Funagawara-machi
Shinjuku-ku, Tokyo, Japan 162

Progress in Colloid & Polymer Science　　　　　　Progr. Colloid & Polymer Sci. **68**, 138–143 (1983)

Field strength dependence of the electric birefringence of DNA*)

H. Sato and M. Shirai

Department of Chemistry, College of Arts and Sciences, University of Tokyo (Tokyo)

Abstract: The field dependence of electric birefringence for calf thymus DNA and λ DNA at various pH and various ionic strength was measured. The monodisperse and rigid polyions show linear dependence of electric birefringence on the electric field. In order to explain the experimental results the induced moment of a linear polyion is calculated based on the model that a polyion molecule is represented as a line of uniform charge density and the counterions are represented as oppositely charged cloud in a narrow and deep potential valley around a polyion molecule. The orientation factor and the electric birefringence are also calculated. According to the results of these calculations, for large polyions the induced moment is constant, that is, independent of the applied field except in very low electric fields. When the induced moment is constant the orientation factor or the electric birefringence shows linear dependence on the field strength. However, small polyions show quadratic dependence on the field strength. Large polyions usually show linear dependence on the field strength, but they often also show quadratic dependence in very low electric field. The experimental results of electric birefringence for DNA agree with the ones predicted by the theory.

Key words: DNA, electric birefringence, polarizability, induced moment.

1. Introduction

General theories of electric birefringence are in accord with the Kerr law at low field intensities, that is, they predict that the degree of molecular orientation, Φ, and the birefringence are proportional to the square of the field intensity [1, 2]. At sufficiently high electric fields the orientation becomes saturated, and Φ tends toward unity [3]. However, Stellwagen and others found that the square relationship often was not followed in solutions of DNA and polynucleotides, even at quite low degree of orientation [4]. Since then, non-Kerr behavior of this kind also has been reported by several authors [5–7]. They found that the magnitude of electric birefringence frequently depended approximately linearly on the field strength. Recently, Roux et al. [8, 9] found that the solutions of DNA obeyed the Kerr law in very low electric field.

Stellwagen [10], Elias and Eden [11], and Diekman et al. [12] studied the electric birefringence of restriction enzyme fragments of DNA, and found that DNA fragments obeyed the Kerr law over a wide range of field strength except the range where saturation of orientation took place. From the results of studies up to the present we see that DNA of high molecular weight show non-Kerr behavior except in the very low electric fields, whereas small fragments of DNA obey the Kerr law. Since this kind of behavior is found only in polyelectrolyte solutions, the complex polarization effects due to small counterions may be its origin. Therefore, we measured the electric birefringence of various DNA, and investigated the mechanism of orientation in an electric field. Houssier et al. [13] and Hogan et al. [14] tried to explain theoretically the non-Kerr behavior stated above, but their explanations are not always successful [15].

2. Experimental

2.1 Birefringence apparatus

The details of the electric birefringence instrument used in this study is described elsewhere [16]. A Nippon Electric Co. model GLG5311 1 mW He-Ne laser was used as the light source. The light was polarized by the Glan-Thompson prism in a plane at 45° with respect to the direction of the applied field. The cell was a 1 cm path-length quartz spectrophotometer cell. Electrodes were platinum plates supported by a teflon spacer, and the electrode spacing was 2.5 mm. The photodetector employed was Hamamatsu TV Co. model S1223 photodiode, and the preamplifier produced the output voltage. The output signal was converted to the digital signal, and stored in memories by using a Sony-Tektronix Digitizer

---

*) Dedicated to the memory of Professor Dr. B. Tamamushi.

model 390AD. The converter of this model has 10 bit resolution, and has a dual-channel unit, which allowed the signal and pulse to be recorded simultaneously. A Denkenseiki Co. model B401 high voltage pulser and amplifier covered the voltage range from 10 to 2000 V. The rise and fall time of the pulse was less than 0.1 μs. The pulse width could be varied from 1 μs to 20 ms. A micro-cmputer controlled the recorder via GPIB (General Purpose Interface Bus), and triggered the pulser. Average method [11] was effective in improving the signal-to-noise ratio.

2.2 Materials

In this research two kinds of DNA were used. Calf thymus DNA was a Sigma Chemical Company sample, lot no. 59–983. λ DNA was a Miles Chemical Company sample, lot no. 14. The molecular weight of the former was 6×10^6, and that of the latter was 3.2×10^7. All DNA samples were stored in a refrigerator at −20 °C. The DNA samples were dissolved in aqueous NaCl solutions. The concentration of stock solutions prepared were approximately 1 g/1 DNA in 10^{-3} M NaCl solutions. MgDNA samples were prepared from NaDNA by dialyzing NaDNA solution against MgCl₂ solution. Linear λ DNA was prepared by heating its solution of 10^{-3} M NaCl to 50 °C for ten minutes and cooling the tube in ice water to prevent aggregation. Circular λ DNA was prepared by heating similarly, but cooling slowly in the heating bath with the heater disconnected [17, 18].

3. Results

Figure 1 shows field dependence of the electric birefringence for calf thymus NaDNA on a logarithmic scale. The slope is about 1.2 except the saturation region of orientation. From this figure is is seen that the electric birefringence of this DNA is proportional to the 1.2th power of the electric field. In figure 2 field dependence curves of the electric birefringence for circular and linear λ NaDNA are indicated. Except the electric field region where saturation of orientation

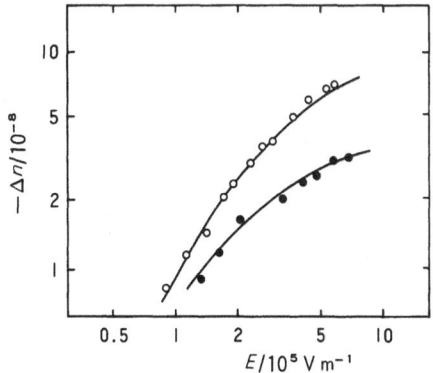

Fig. 2. The electric birefringence of circular and linear λ NaDNA. conc. of NaDNA = 2.5×10^{-3} g/l, conc. of NaCL = 1.0×10^{-3} M. ○, circular; ●, linear

takes place, the birefringence of circular and linear λ DNA is approximately linearly dependent on the field strength. This λ DNA sample is relatively monodisperse, whereas calf thymus DNA is widely polydisperse. In figure 3 field dependence of the electric birefringence at various pH for circular λ DNA is shown The birefringence is approximately linearly dependent on the field. The electric birefringence of circular λ DNA in solutions of various concentrations of added NaCl are given in figure 4. The higher the concentration of added NaCl the smaller the magnitude of electric birefringence. Figure 5 exhibits field dependence of the electric birefringence of calf thymus MgDNA in solutions of various concentrations of MgCl₂. All curves in this figure have a bend at about 7×10^4 V/m. Beyond this field strength the slope is 1.2, and below this field strength the slope is 2.0

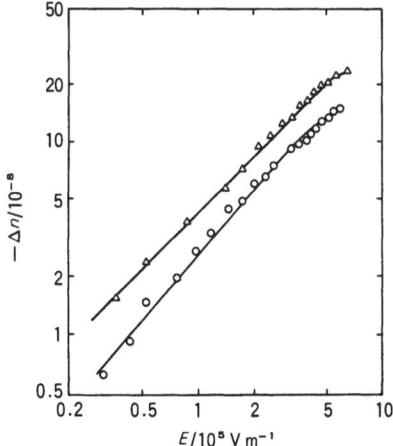

Fig. 1. Field dependence of the electric birefringence of calf thymus NaDNA. conc. of NaDNA = 6.5×10^{-3} g/l. ○, conc. of NaCl = 1.0×10^{-3} M; △, 1.0×10^{-4}

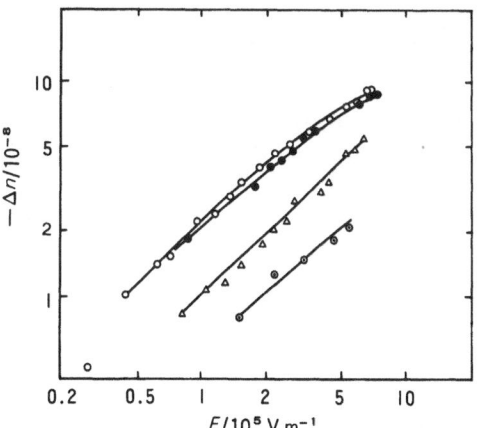

Fig. 3. Field dependence of the electric birefringence of circular λ NaDNA. conc. of NaDNA = 4.6×10^{-3} g/l, conc. of NaCL = 1.0×10^{-3} M. ○, pH = 6; ●, 4.3; △, 3.7; ⊙, 3.4

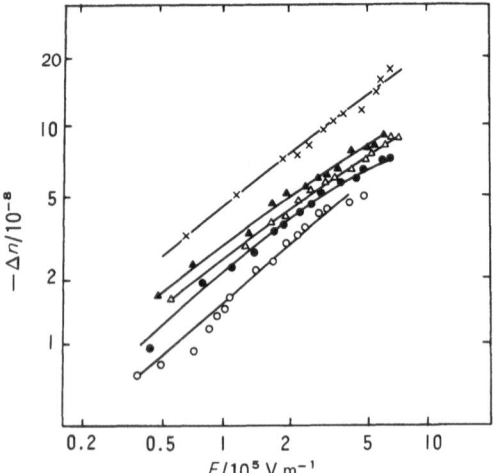

Fig. 4. The ionic strength dependence of electric birefringence of circular λ NaDNA. conc. of NaDNA = 3.9×10^{-3} g/l. $\bigcirc$, conc. of NaCl = 3.75×10^{-3} M; $\bullet$, 1.00×10^{-3}; $\triangle$, 0.75×10^{-3}; $\blacktriangle$, 0.50×10^{-3}; $\times$, 0.25×10^{-3}

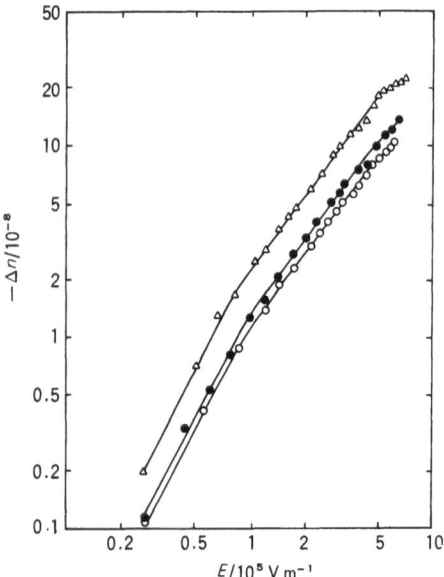

Fig. 5. Field dependence of electric birefringence of calf thymus MgDNA. conc. of MgDNA = 6.5×10^{-3} g/l. $\bigcirc$, conc. of $MgCl_2$ = 5×10^{-4} M; $\bullet$, 3.3×10^{-4}; $\triangle$, 0.5×10^{-4}

4. Discussion

4.1 Field dependence of the electric birefringence

As shown in figures 2, 3, and 4, the electric birefringence of calf thymus DNA and circular and linear λ DNA show approximately linear rather than quadratic dependence on the field strength. This λ

DNA is relatively monodisperse, but calf thymus DNA is considerably polydisperse. For circular λ DNA the segmental orientation should be small in the range of low applied field employed in this work, whereas linear λ DNA is more flexible than the circular one. These facts give rise to a little different field dependence. In view of results, DNA of high molecular weight seems to show approximately linear field dependence, and small molecules show quadratic field dependence. The appearance of bend at about 7×10^4 V/m in the curves of field dependence of the electric birefringence of MgDNA seems to be due to the fact that Mg ions are bound more than Na ions are, the details of which are described in the next section.

4.2 Binding of counterions

Let us consider a rodlike polyion. Many discrete charges are distributed on a polyion. Along the rodlike polyion the potential takes the form of a trough with holes. For polyions such as DNA charged groups are uniformly distributed on a rod, so that, we assume that a polyion can be represented as a line of uniform charge density as assumed by Manning [19]. This polyion is considered to be surrounded by a sharp and deep potential valley. Some counterions are bound in the potential valley around polyions, and they belong to the same kinetic unit as that of the central polyion. When an external electric field is applied, the bound counterions move along the polyion until a steady state is reached, which gives rise to an induced dipole moment.

The difference between the chemical potential, $\bar{\mu}$, of the bound ion and that of the free ion, $\Delta\bar{\mu}$, was calculated by Manning [20, 21]. The equation obtained is as follows,

$$\begin{aligned}\Delta\bar{\mu} &= \bar{\mu}_{\text{bound}} - \bar{\mu}_{\text{free}} \\ &= RT + RT \ln[(10^3/V_p) |z|^{-1} v^{-1} c_s^{-1} r (1 \\ &\quad - e^{-\varkappa b})^{2|z|\xi(1-r)}]\end{aligned} \tag{1}$$

where V_p is the volume of bound region per mole of polyelectrolyte, z the valence of the counterion, v, the number of counterions in a salt molecule, c_s the molarity of simple salt, r the number of moles of bound counterions per mole of ionized group, $\varkappa$ the Debye screening parameter in the simple salt solution without polyelectrolyte, b the distance between the charged groups on the polyion. ξ is given by the following equation [19].

$$\xi = e^2/\varepsilon kTb, \tag{2}$$

where e is the electronic charge, ε the dielectric constant of the medium, k the Boltzmann constant, T the absolute temperature. The method of determining V_p is described by Manning [21].

If the external field is applied, both $\tilde{\mu}_{\text{bound}}$ and $\tilde{\mu}_{\text{free}}$ must be supplemented by an electric potential term $-zeE_l(x-l)N_A$, where 2l is the length of a polyion, and E_l the component of applied field along the polyion axis, N_A Avogadro number. The coordinate along the polyion is denoted by x, and $x = l$, $x = -l$ are the ends of a polyion. The energy is taken to be 0 at $x = l$. In an electric field r is the function of the coordinate and $\Delta\tilde{\mu}$ is the function of r, so that r and $\Delta\tilde{\mu}$ are denoted by r_x and $\Delta\tilde{\mu}(r_x)$, respectively. If $\tilde{\mu}_{\text{free}} = 0$ at $x = l$, then $\tilde{\mu}_{\text{free}} = 2zeE_l l N_A$ at $x = -l$, and

$$\tilde{\mu}_{\text{free}} = -zeE_l(x-l)N_A \quad x < -l \text{ or } x > l \quad (3)$$

$$\tilde{\mu}_{\text{bound}} = \Delta\tilde{\mu}(r_x) - zeE_l(x-l)N_A \quad -l \leq x \leq l. \quad (4)$$

According to Manning's theory [21], in the absence of applied field,

$$\Delta\tilde{\mu}(r) = 0 \quad \text{for } r = 1 - |z|^{-1}\xi^{-1} \quad -l \leq x \leq l. \quad (5)$$

In the presence of applied field,

$$\Delta\tilde{\mu}(r_x) - zeE_l(x-l)N_A = 0 \quad -l \leq x \leq l \quad (6)$$

because the bound and the free counterions are in the equilibrium state. r_x in an electric field can be calculated by equations (1) and (6). From equation (6)

$$\Delta\tilde{\mu}(r_{-l}) + 2zeE_l l N_A = 0. \quad (7)$$

The bound counterions are released by the applied electric field, as is seen from equations (6) and (7). It seems that they overflow from the inclined box. The decrease of bound counterions in weak fields is nearly proportional to r_l, and it is reasonable to assume that certain fraction of counterions, r_c, is easily released without any change of conformation of the polyion molecule. In this way, when the energy is supplied by the electric field, there is a critical field, E_{lc}, below which the bound counterions are easily released by the electric field. Above E_{lc} the decrease of bound counterions is accompanied by some changes of conformation, and needs certain amount of energy. For DNA it is easy to remove bound counterions until $r = 0.75$ in view of the experimental results stated below. The excess electric field beyond E_{lc} only accelerate flow of

free ions, and the charge distribution on a polyion remains unchanged, that is, the induced moment is constant. Much higher electric field will remove further the bound counterions, and will deform the polyion itself.

We assume at first, $r_c = 0.75$, then the calculated values agree with the experimental ones for MgDNA and NaDNA, although this assumption has no theoretical basis. According to equation (5), if $r = r_e = 1 - |z|^{-1}\xi^{-1}$, then $\Delta\tilde{\mu} = 0$, where r_e is equilibrium value of r. Since $\xi = 4.2$ for DNA, $r_e = 0.88$ for MgDNA, and $r_e = 0.76$ for NaDNA. $\Delta\tilde{\mu}$ required to reduce r from 0.88 to 0.75 is 6.5 kJ/mol for MgDNA, which may be reasonable value [22, 23, 24]. This fact means by equation (7) that the electric field which easily remove 1 mole of Mg ions from the polyion is up to 6.8×10^4 V/m, assuming the molecular length is 500 nm [9]. $\Delta\tilde{\mu}$ to reduce r from 0.76 to 0.75 is 0.30 kJ/mol for NaDNA by equation (1), which similarly implies that the electric field which easily remove 1 mole of Na ions from the polyion is up to 6.2×10^3 V/m. In this way the value of E_{lc} for MgDNA and NaDNA are 6.8×10^4 V/m and 6.2×10^3 V/m, respectively. For MgDNA this value of E_{lc} agrees with that shown in figure 5, and for NaDNA the calculated value of E_{lc} nearly agrees with the experimental value 3×10^3 V/m which was found by Roux et al. [8]. According to equation (7) and the followed discussion, E_{lc} is inversely proportional to l. This is the reason why small molecules obey the Kerr law and large molecules do not. Infact the sample DNA under investigation is not rigid, but linear dependence of the electric birefringence on the field strength appears only for long molecules. Therefore, the present calculation is important only for long and rigid hypothetical rod polyions.

4.3 Calculation of induced dipole moment

According to Boltzmann distribution law, the charge distribution ϱ, is given by the following equation,

$$\varrho(x) = A\, e^{-U/kT}, \quad (8)$$

where A is the normalization constant, which is determined by, in weak fields,

$$\int_{-l}^{l} \varrho = nze \quad (9)$$

where n is the number of bound counterions per one polyion, and nz is equal to the product of r and charge of ionized groups in a polyion. In very low applied fields $\Delta\bar{\mu}(r_x)$ is almost independent of the field strength. So that U is given as follows, by using equation (6),

$$U(x) = - zeE_l(x - l) - C \tag{10}$$

where C is a constant. From equations (8) and (10), we have the following equation.

$$\varrho = A \exp\left[- \frac{zeE_l(x - l)}{kT} + C' \right],$$

where C' is a constant. Thus, in weak electric fields,

$$\int_{-l}^{l} A \exp\left[- \frac{zeE_l(x - l)}{kT} + C' \right] dx = nze$$

$$A = \frac{nz^2e^2E_l}{-kT} e^{\left(\frac{zeE_l l}{kT} - C'\right)} \left(e^{\frac{zeE_l l}{kT}} - e^{-\frac{zeE_l l}{kT}} \right)^{-1}. \tag{11}$$

The induced dipole moment μ_i in the electric field $E_l \lesssim E_{lc}$ is given by the following equation,

$$\mu_i = \int_{-l}^{l} \varrho x \, dx = nzel \cdot L\left(\frac{zeE_l l}{kT} \right), \tag{12}$$

where L is Langevin function. In weak fields,

$$\mu_i = \frac{nz^2e^2l^2E_l}{3kT}. \tag{13}$$

Accordingly, the polarizability, α, is given by,

$$\alpha = \frac{nz^2e^2l^2}{3kT}. \tag{14}$$

Since n is proportional to l, α is proportional to l^3 in weak fields. In strong fields, even if $E_l > E_{lc}$, ϱ is constant and is equal to that for E_{lc} as described at the end of section 4.2. So that, μ_i is just like saturated at E_{lc} and μ_i at E_{lc}, μ_{ic} is given by the following equation,

$$\mu_{ic} = nzel \cdot L\left(\frac{zeE_l c_l}{kT} \right). \tag{15}$$

For MgDNA $E_{lc} = 6.8 \times 10^4$ V/m, and $\mu_{ic} = 9.4 \times 10^{-27}$ Cm $= 2.8 \times 10^3$ D.

4.4 Calculation of the orientation factor and the electric birefringence

If the angle between the molecular axis and the direction of the applied electric field is denoted by θ, then $E_l = E\cos\theta$, where E is the strength of applied electric field. The orientation factor is given by the following equation [3].

$$\Phi = \int_0^\pi \frac{3\cos^2\theta - 1}{2} \frac{e^{-U'/kT}}{\int_0^\pi e^{-U'/kT} 2\pi\sin\theta d\theta} 2\pi\sin\theta d\theta \tag{16}$$

$$U' = - \mu E \cos\theta \tag{17}$$

where μ is the permanent dipole moment. The result of calculation for equation (16) in the weak field is as follows;

$$\Phi = \frac{\mu^2}{15k^2T^2} E^2. \tag{18}$$

This is the well-known equation for the permanent dipole moment.

Since the polarity of the induced dipole is reversed by reversing the applied electric field, $\cos\theta$ is equivalent to $\cos(\pi - \theta)$ for the saturated induced dipole, μ_{ic}. Therefore, in the case of saturated induced dipole, the integration over $0 \sim \pi$ in equation (16) must be replaced by the integration over $0 \sim \pi/2$ [25]. Thus,

$$\Phi = \int_0^{\pi/2} \frac{3\cos^2\theta - 1}{2} \frac{e^{-U''/kT}}{\int_0^{\pi/2} e^{-U''/kT} 2\pi\sin\theta d\theta} 2\pi\sin\varrho d\varrho \tag{19}$$

$$U'' = - \mu_{ic} E\cos\theta \tag{20}$$

μ_{ic} is given in equation (15). Even for long molecules μ_{ic} is small, and $|U''| << kT$, so that we expand $e^{-U''/kT}$ in power series and neglect higher order terms. Thus, Φ is calculated as follows;

$$\Phi = \frac{\mu_{ic}}{8kT} E. \tag{21}$$

Therefore, the electric birefringence, Δn is also proportional to E.

$$\Delta n \propto E. \tag{22}$$

According to equations (13) and (14), in weak fields $\mu_i = \alpha E \cos\theta$, then $U'' = -\frac{\alpha}{2} E^2 \cos^2\theta$. In this case the orientation factor obtained is given by the following equation

$$\Phi = \frac{\alpha}{15kT} E^2 . \qquad (23)$$

Equation (23) is the familiar equation for the ordinary induced dipole moment which is proportional to the electric field strength. In equations (18) and (23) Kerr law is valid, but in equations (21) and (22) Φ and Δn depend linearly on the field strength.

If the polyions are polydisperse or flexible, they have the distribution of values of the electric polarizability. The high polarizability terms reach saturation in low fields, and the low polarizability terms reach saturation in high fields. The weighted sum of different values of electric polarizability fits another curve. This seems to be the reason why long DNA molecule except the circular λ DNA molecule does not exhibit linear field dependence of the electric birefringence although it is a long molecule.

4.5 pH dependence of Φ

At NaCl concentration of 10^{-3} M circular λ DNA maintains the same structure in the pH region between 6 and 3.4 [22]. At pH = 6, 4.3, 3.7, 3.4 protonation per one phosphorous is 0, 0.10, 0.22, 0.30, respectively [26]. Therefore, the ratio of the numbers of bound counterions at pH = 6, 4.3, 3.7, 3.4 is 100:90:78:70. The orientation factor, Φ, is proportional to the number of bound counterions according to equations (15) and (21), which agrees with the experimental result given in figure 3.

4.6 Ionic strength dependence

According to equation (1), if the salt concentration c_s increases, the absolute value of $\Delta\bar{\mu}$ decreases through $\ln c_s^{-1}$ and $\varkappa$, and accordingly, r or Δn decreases. So that μ_{ic} decreases by equation (15), which result in decrease of Φ or Δn by equation (21). The results of calculation agree with the experimental ones shown in figure 5.

Acknowledgements

This work was partly supported by Grant-in-Aid for Scientific Research (Project No. 56470003) from the Ministry of Education, Science and Culture of Japan, to which the authors wish to express the gratitude. The authors also wish to thank Prof. K. Yoshioka in this department for useful discussion.

References

1. O'Konski, C. T., Molecular Electro-Optics Part I and II (Marcel Dekker, New York, 1977).
2. Fredericq E. and Houssier, C., Electric Dichroism and Electric Birefringence (Oxford, London, 1973).
3. O'Konski C. T., Yoshioka, K., Orttung, W. H., J. Phys. Chem. **63**, 1558 (1959).
4. Stellwagen, N. C., Ph. D. Thesis, University of California, Berkeley (1967).
5. Houssier, C., Fredericq, E., Biochim. Biophys. Acta. **88**, 450 (1964).
6. Nakayama H., Yoshioka, K., J. Poly. Sci., Part A3, 813 (1965).
7. Kikuchi K., Yoshioka K., J. Phys. Chem. **77**, 2101 (1973).
8. Roux B., Bernengo J. C., Marion C., Hanss, M., J. Colloid and Interface Sci. **66**, 421 (1978).
9. Marion, C., Roux, B., Bernengo, J. C., Hanss, M., Int. J. Biol. Macromol. **2**, 235 (1980).
10. Stellwagen N. C., Biopolymers **20**, 399 (1981).
11. Elias J. G., Eden, D., Macromolecules **14**, 410 (1981).
12. Diekmann S., Hillen, W., Jung, M., Wells, R. D. and Pörschke, D., Biophys. Chem. **15**, 157 (1982).
13. Houssier C., Bontemps J., Edmonds-Alt, Xavier, Fredericq, E., Annals. of N. Y. Ac. Sci. **303**, 107 (1977).
14. Hogan M., Dattagupta, Y., Crothers, D. M., Proc. Natl. Acad. Sci. USA **75**, 195 (1978).
15. Charney E., Yamaoka, K., Manning, G. S., Biophys. Chem. 11, 167 (1980).
16. Kikuchi K., Sato, H., Shirai, M., Yoshioka, K., Sci. Pap. Coll. Gen. Educ., Univ. of Tokyo **32**, 113 (1982).
17. Wang, J. C., Davidson, N., J. Mol. Biol. **15**, 111 (1966).
18. Hershey A. D., Bergi, E., Ingraham, L., Proc. Nat. Acad. Sci. USA **49**, 748 (1963).
19. Manning G. S., J. Chem. Phys. **51**, 924 (1969).
20. Manning G. S., Biophys. Chem. **7**, 95 (1977).
21. Manning G. S., ibid. **9**, 65 (1978).
22. Bloomfield V. A., Crothers, D. M., Tinoco, I., Physical Chemistry of Nucleic Acids (Harper and Row, New York, 1974).
23. Manning, G. S., Biopolymers **11**, 937 (1972).
24. Pörschke, D., ibid. **15**, 1917 (1976).
25. Sokerov, S., Weill, G., Biophys. Chem. **10**, 161 (1979).
26. Jordan, D. O., The Chemistry of Nucleic Acids (Butterworths, Washington, 1960).
27. Vinograd J., Labowitz J., J. Gen. Physiol. **49**, 103 (1966).
28. Cox, R. A., Peacock, A. R., J. Chem. Soc. 4724 (1956).

Received January 12, 1983;
accepted February 7, 1983

Authors' address:

Miss Hisako Sato and Dr. Michio Shirai
Department of Chemistry, College of Arts and Sciences,
University of Tokyo,
Komaba, Meguroku, Tokyo 153, Japan

Mutual adsorption of protein and detergent at the alumina-water interface*)

A. Samanta and D. K. Chattoraj

Department of Food Technology and Biochemical Engineering, Jadavpur University, Calcutta, India

Abstract: Mutual adsorption of bovine serum albumin (BSA) and sodium dodecyl sulphate (SDS) at the alumina-water interface has been determined as functions of protein and detergent concentrations, ionic strength, pH and temperature of the aqueous medium. The extent of adsorption decreases with increase of temperature. The isotherm with respect to the adsorption of protein at a fixed total concentration (C_s^t) of the surfactant is S-shaped in character. The limiting value of the protein adsorption decreases with increase of C_s^t. The study of adsorption at serveral values of pH and ionic strength of the medium indicates that the mutual interactions of BSA and SDS with alumina interface are significantly controlled by the electrostatic effect. At a given value of C_s^t, extent of adsorption (Γ_S) at first decreases with increase of the adsorption (Γ_p) of BSA due to the competitive interaction process whereas Γ_S increases with increase of Γ_p in the relatively higher range of the adsorption of protein as a result of the increase of the mutual interaction of BSA and SDS at the interface. The binding ratio of SDS to BSA in the bulk phase and at the alumina-water interface respectively under identical situation have been compared at various values of the equilibrium bulk concentrations (C_s) of the surfactant in solution. At low values of C_s, these ratios in the two phases are of comparable order in magnitude. However, at high values of C_s, the binding ratio at the interface increases enormously due to the massive co-operative interaction even though the value of this ratio in the bulk remains low.

Key words: Adsorption, at alumina-water interface – adsorption, mutual, of protein and detergent – protein and detergent adsorption – interface, adsorption – interface, alumina-water.

Introduction

Tamamushi and co-workers [1, 2] published two original papers on the adsorption of cationic and anionic detergents from solution to the surface of polar solid powders. The extents of adsorption in all these cases are found to depend on the nature of the solid, nature and concentration of the detergent, pH, temperature and other factors. The isotherms in most of the cases are found to be S-shaped in character. At a given temperature, the limiting value for the extent of adsorption is attained when the concentration of the detergent approaches the critical micelle concentration. The role of electrostatic effect and the van der Waals forces in this kind of adsorption process was also critically examined by these workers. Several other papers were subsequently published in which extents of adsorption of different detergents on the surface of powdered solids were investigated in detail below and above cmc [3–7].

Adsorption of proteins from solution to the surfaces of powdered glass, alumina, polystyrene particles etc. have been extensively studied by many workers [8–15]. These studies indicate that the protein becomes extensively surface denatured when the concentration of the biopolymer in solution is extremely low. At higher protein concentration, the folded protein molecules may saturate the solid surface thus forming an adsorbed monolayer. The extent of adsorption of a protein also depends upon various solution parameters.

It is well-known that lipid and protein components are in interaction with each other at the membrane of the living cell. To understand the feature of this complex process, model studies of the mutual interaction of surfactants and proteins in spread monolayers at the air-water interface had been made by many workers [16–19]. The conclusions obtained from such study are not always quantitative in nature. Besides cell membrane, study of protein-detergent interaction at the interface may be useful in food industry, laundry industry etc. Direct study of the mutual

---

*) Dedicated to the memory of Professor Dr. B. Tamamushi.

adsorption of protein and surfactant at the interface will thus be useful in this respect. Until now no such direct investigation has been carried out and only recently [20] some indirect experimental approach has been considered in this important area of research.

In this paper, our results on the mutual adsorption of bovine serum albumin and sodium dodecyl sulphate from aqueous solution to the alumina interface have been presented at various values of pH, ionic strength and temperature of the medium.

Experimental procedure

Crystalline bovine serum albumin (BSA) used was obtained from Pentax Company, USA. Pure sodium dodecyl sulphate (SDS) was supplied as a gift by Dr. K. S. Birdi from Denmark. Value of cmc of the sample dissolved in water obtained from surface tension measurement was in agreement with that found in the literature [21]. The surface tension versus concentration curves of the detergent solutions indicated absence of any minimum. Alumina powder (chromatography grade) had been supplied by BDH. Electrolytes used were all of analytical grades. Double distilled water was used for the preparation of solutions.

For the determination of the mutual adsorption of BSA and SDS by the powdered alumina, three separate experiments were carried out.

a) Binding of SDS to BSA in bulk phase

A stock quantity of buffer solution of definite pH and ionic strength was prepared. This was treated as solvent. Phosphate buffer was used for pH 6.0 and acetate buffer for pH 4.0. Solutions of known concentrations of proteins and detergent were made from these stock solutions by appropriate dilution with the solvent.

In the equilibrium dialysis experiments [22–24], 2 ml of solvent (without SDS or protein) was taken in a dialysis tubing (5/8 inch in diameter, Arthur Thomas & Co., USA). The casing was initially freed from any trace surface-active material by boiling with water and then repeatedly washing with water. The dialysis bags properly knotted and containing the solvent inside were dropped into the pyrex glass bottles (each fitted with a standard-joint mouth) containing 25 ml solutions of BSA-SDS mixture (BSA solutions of gms percent concentration C_p^t) shaken in horizontal thermostatic shakers that were usually maintained at 28 °C. In all cases, equilibrium was reached within 24 hours time. After this period, the dialysates inside the dialysis tube were taken out and the concentrations of the dialysate solution were measured by the dye-partition technique [25, 26]. A cationic dye methylene blue in N/100 HCl solution was used and the partition was carried out between an aqueous surfactant solution and a chloroform layer. The concentration of the surfactant in the dialysate solution could be estimated with 2 to 3 percent error. Thus from the known concentrations of SDS before and after binding with a known amount of protein present in the equilibrium dialysis experiments, moles (Γ_s^{eq}) of SDS bound per mole of BSA were calculated in the usual manner. From the known amounts of SDS and BSA, present in the total amount of solvent outside and inside the casing, the total initial concentrations C_s^t and C_p^t of SDS and BSA have been calculated. It may be pointed out here that in the conventional procedure [24] for the determination of extent of binding SDS to BSA followed in this laboratory previously, BSA solution was taken inside the dialysis bag and SDS solution outside of it in the bottle.

b) Mutual adsorption of BSA and SDS at the alumina-water interface

Definite amounts of BSA and SDS respectively were dissolved in fixed amount of solvent at a given pH and ionic strength so that their respective weight concentrations were known. 25 ml of this solution were taken in a stoppered bottle and to this, weighed amount (nearly 0.5 gm) of dried alumina was added. The bottle placed in a bath maintained at 28 °C was shaken intermittently for 24 hours. After this period, the bottle was opened and to the system inside was added a knotted dialysis tube containing 2 ml of buffer solution. The closed bottle was again placed in the bath at 28 °C and shaken intermittently for another 24 hours for the attainment of the dialysis equilibrium between the solutions inside and outside the dialysing tube. The tube was then taken out and equilibrium concentration (C_s) of SDS in the dialysis solution inside the casing was estimated by the dye-partition technique. The bottle containing the outside solution and powdered alumina was allowed to remain undisturbed for 6 more hours whereby alumina particles were completely settled down. The supernatant solution in the bottle was taken in definite amount and its protein concentration (C_p) was determined after adding definite amount Folin reagent appropriately. The absorbance of the color thus developed in the protein solution by addition of the reagent was measured in a Klett-Summerson colorimeter using red filter. In the experiments with blank, absorbancy of the solutions of known BSA and SDS concentrations respectively in the absence of alumina were measured under identical conditions. Variation of SDS concentration to some extent in the protein solutions after adsorption (or in the blank) had negligible effect on the value of the absorbancy of the colored solution.

From the known amount of BSA in the bulk solution before and after adsorption by a given weight of powdered alumina, one can easily estimate moles of BSA adsorbed per gram of powdered alumina at a given value of C_p. When surface area per gram of the powder is known, one can then calculate moles (Γ_p) of BSA adsorbed per square meter of alumina surface at a given value of C_p.

SDS in the bulk solution after adsorption by alumina powder remained partly free and partly in the bound state with BSA present in the bulk solution. The concentration (C_s) of SDS in the free state was determined from the experiment directly. From the separate binding experiment described in section (a), amount (Γ_s^{eq}) corresponding to this value of C_s, was evaluated. Γ_s^{eq} represents amounts of SDS bound per gram of BSA. If w_p^b is the weight of BSA present in the bulk solution after adsorption, then amount of SDS bound to it is $w_p^b \cdot \Gamma_s^{eq}$ so that this quantity can be calculated also. From the experiment therefore, weights of SDS remaining in the free and bound state with w_p^b grams of protein in the bulk solution were determined. The sum of these weights was subtracted from the total weight of SDS present before adsorption, whereby amount of SDS absorbed by the given weight of alumina powder added to the system was calculated. From the known specific surface area of alumina, one can then calculate moles (Γ_S) of SDS adsorbed per square meter of alumina powder at given values of C_p and C_s respectively.

c) Determination of the specific surface area of the alumina powder

The specific surface area of powdered alumina was calculated from the adsorption of palmitic acid by alumina in the benzene medium [13, 27]. Alumina powder was dried for three hours at 100–110 °C. Benzene was distilled in the presence of calcium chloride. It was then treated with metallic sodium and finally distilled to remove last trace of water. Vacuum dried palmitic acid was then added to known weight of benzene to which was added weighed amount of alumina powder dried for 3 hours at 100 to 110 °C. The bottle was shaken for 24 hours at 28 °C and then kept undisturbed for 4 hours whereby alumina particles settled down at the bottom completely. An aliquot (2 ml) of the supernatant taken in the weighing bottle was dried in an air-oven at 40 °C and the dry weight of palmitic acid was determined. From the difference in weights before and after adsorption of the acid by alumina powder in benzene media, average weight (W) of the acid adsorbed per gram of alumina were determined. The concentrations of the acid in benzene were always relatively high so that W became independent of acid concentration. This amount of palmitic acid may be assumed to form a compact monolayer on the surface of one gram of alumina powder. The number of palmitic acid molecules in the monolayer per gram of alumina can thus be calculated. Assuming that one molecule of palmitic acid covers 20.5 sq. Ångstrom per molecule, specific surface area per gram of powdered alumina was estimated. The average value of the specific surface area of alumina powder was 74.5 ± 1.5 square meter per gram.

Results and discussion

It has already been shown by Tamamushi and coworkers [1, 2] that SDS dissolved in aqueous solvent can be adsorbed at the surface of the powdered alumina, under suitable conditions. In agreement with their observations, we have also noted in the present work that all the isotherms for the adsorption of SDS at the alumina-water interfaces are S-shaped and their exact nature depends on pH, ionic strength and temperature of the medium. At higher values of the surfactant concentration close to cmc, the extent of adsorption of SDS becomes independent of the concentration of SDS in the bulk solution.

Mitra and Chattoraj [13] have also shown that bovine serum albumin initially present in the aqueous solvent may be significantly adsorbed at the alumina-water interface and the extent of adsorption depends upon pH, ionic strength, temperature and the bulk concentration of the biopolymer. They have further shown that at C_p greater than 0.01 percent, Γp increases regularly with increase of C_p and finally at high values of the protein concentration, Γ_p attains its maximum value Γ_p^m which does not alter further with further increase of C_p.

In agreement with the observation of Mitra et al. [13], the isotherm obtained from the plot of Γ_p against C_p at pH 6.0, ionic strength 0.10 and temperature 28 °C is found to be S-shaped in character (vide fig. 2) in the absence of SDS. The range of C_p in these experiments was 0.01 to 0.4 percent. Value of Γ_p^m in this set of experiment is nearly 2.8×10^{-8} moles/m². It has been shown by Norde [15] that if the ellipsoidal-shaped BSA molecules are packed on the solid surface in the side-on or horizontal direction forming a compact monolayer, then the value of Γ_p^m should be 3.8×10^{-8} moles/m². The experimental value of this quantity is considerably less than the theoretical value. This means that the protein molecule at the interface has expanded to cover up the surface and thus undergoes mild surface denaturation process. This is in agreement with the previous observation of Mitra et al. [13].

Mitra et al. [13] have further demonstrated that in the range of C_p between 0.0001 to 0.01 grams percent, BSA at the interface becomes extensively surface-denatured and the adsorption isotherm in this range of concentration exhibits a maximum in agreement with the earlier observation of Bull [8, 9] for the adsorption of proteins on glass surface. Bull has also demonstrated that protein adsorbed on the solid surface at such low value of C_p cannot be desorbed from the surface by repeated washing. In the present series of experiments, measurement is always carried out above 0.01 gram percent protein concentration to avoid extensive irreversible denaturation of the protein at the solid interface.

In the adsorption experiments, mixtures of BSA and SDS solutions were used prior to the addition of alumina powder. BSA in the bulk solution undergoes binding interaction with SDS whereby the biopolymer changes its conformation by denaturation process occurring in the bulk medium. When powdered alumina is added to this solution, protein-SDS complex will be adsorbed at the interface and this may or may not be difficult to desorb. Because of the existence of this uncertainty factor, one should first carefully examine whether the extent of mutual adsorption of BSA and SDS by alumina powder depends upon the procedure of mixing BSA and SDS in the aqueous solution at the time of the adsorption experiments. To settle this point, we also determined Γ_p and Γ_s at given values of C_p and C_s using three different mixing procedures: a) BSA and SDS were mixed in the bulk solution and then powdered alumina was added to the mixture. b) BSA was dissolved in solvent and powdered alumina was added to it. After waiting for twenty hours, SDS was added to the solvent in contact with the powder. c) SDS was dissolved in solvent and powdered alumina was added to it. After waiting for twenty hours BSA was added to the solvent in contact with the powder. Value of C_s^t in all these experiments was kept constant at 5×10^{-4} (M). In figure 1, the isotherms obtained from the plot of Γ_p versus C_p in three types of mixing processes are found to be identical. This means that BSA denatured by the binding interaction either in the bulk phase or at the interface may be adsorbed or desorbed in identical

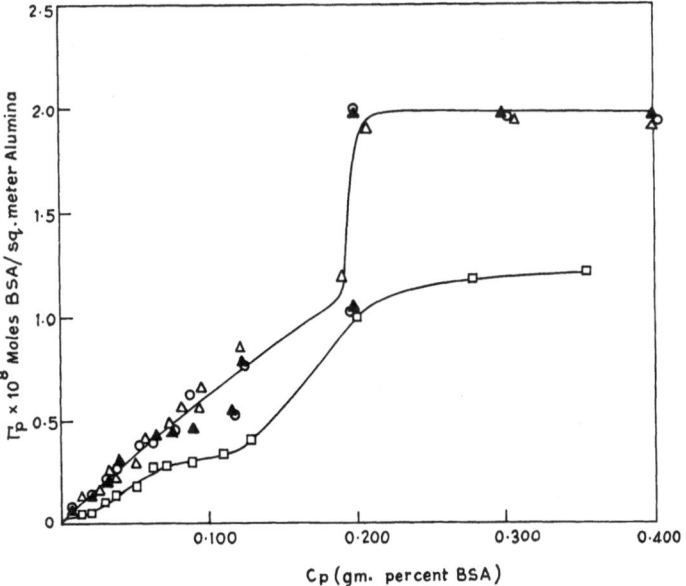

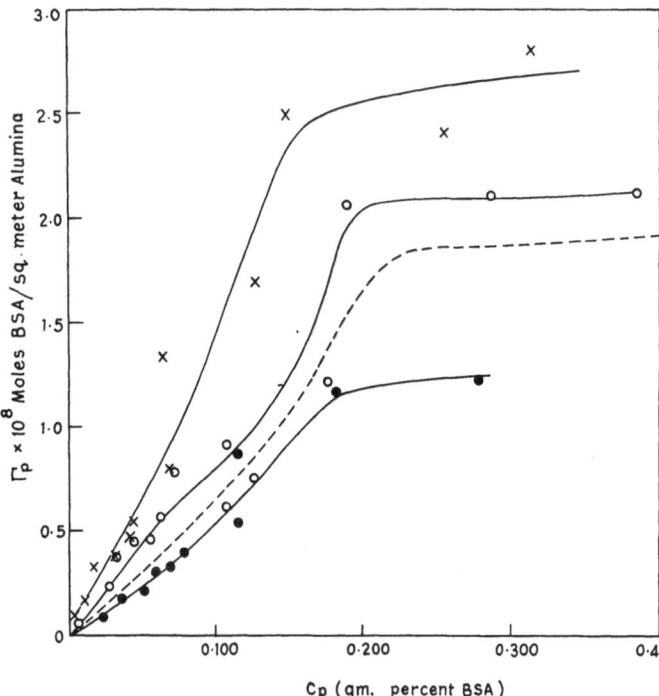

Fig. 1. Plot of moles (Γ_p) of BSA adsorbed per square meter of alumina surface against equilibrium concentration (C_p) of protein in bulk at 28 °C, pH 6.0, ionic strength 0.10 and C_s^t equal to 5×10^{-4} (M). Alumina powder added to a) Mixture of BSA and SDS solution at 28 °C, $\triangle$; b) BSA solution followed by addition of SDS at 28 °C, $\bigcirc$; c) SDS solution followed by addition of BSA at 28 °C, $\blacktriangle$; d) Mixture of BSA and SDS at 45 °C, $\square$

Fig. 2. Plot of moles (Γ_p) of BSA adsorbed per square meter of alumina surface against equilibrium concentration (C_p) of protein in bulk at 28 °C, pH 6.0, ionic strength 0.10. Alumina powder added to a) $C_s^t = 0$: BSA solution, $\times$; b) Mixture of $C_s^t = 1 \times 10^{-4}$ (M): BSA solution, $\bigcirc$; c) Mixture of $C_s^t = 5 \times 10^{-4}$ (M): BSA solution, dotted line; d) Mixture of $C_s^t = 1 \times 10^{-3}$ (M): BSA solution, $\bullet$

manner for the attainment of equilibrium in both the phases and this equilibrium state is not dependent on the nature of the mixing process. In subsequent experiments, the mixing procedure a) was used throughout this work.

In figure 1, the isotherms for protein adsorption at 28 °C and 45 °C have also been compared for C_s^t equal to 5×10^{-4} (M), Γ_p is found to decrease with increase of temperature. This indicates that the adsorption is physical in nature. For the case of physical adsorption, the state of equilibrium is established when rates of adsorption and desorption become equal to each other. Mitra et al. [13] have already shown that adsorption of pure BSA at the alumina-water interface in the absence of SDS is physical in nature.

In figure 2, the isotherms for adsorption of BSA at three different values of C_s^t have been compared with each other. In all these adsorption isotherms, Γ_p at first increases with increase of C_p non-linearly and then it reaches the limiting value Γ_p^m when C_p is high. However, values of Γ_p (at a given value of C_p) or value of Γ_p^m itself decrease with increase in the concentration of C_s^t in the system. Values of Γ_p and Γ_p^m for C_s^t equal to zero are always highest in this respect. With increase of C_s^t in the system, more and more SDS will be adsorbed at the interface as a result of which lesser amount of the protein will be present in the boundary

region. The protein molecules at the interface on interaction with adsorbed SDS may expand laterally by surface denaturation which may be another cause for the decrease of Γ_p^m with increase of C_s^t. However, existence of the limiting value of Γ_p^m for each value of C_s^t indicates that the mutual adsorption of BSA and SDS at interface is possibly monomolecular in nature with respect to protein.

In figure 3, the isotherms for the adsorption of BSA at ionic strengths 0.01, 0.10 and 0.50 have been compared with each other. All these measurements were carried out at pH 6.0, temperature 28 °C and C_s^t equal to 5×10^{-4} (M). Γ_p at a given value of C_p and Γ_p^m respectively are observed to decrease with increase of the ionic strength of the medium. This indicates that the extent of mutual adsorption of BSA in the presence of SDS is controlled by the electrostatic effect. Alumina particles at pH 6.0 are positively charged [28]. BSA having isoelectric point near pH 5 is negatively charged at pH 6.0 and its charge is expected to increase as a result of the binding interaction of the protein with SDS. With decrease of the ionic strength the ion-atmosphere of positively charged alumina and

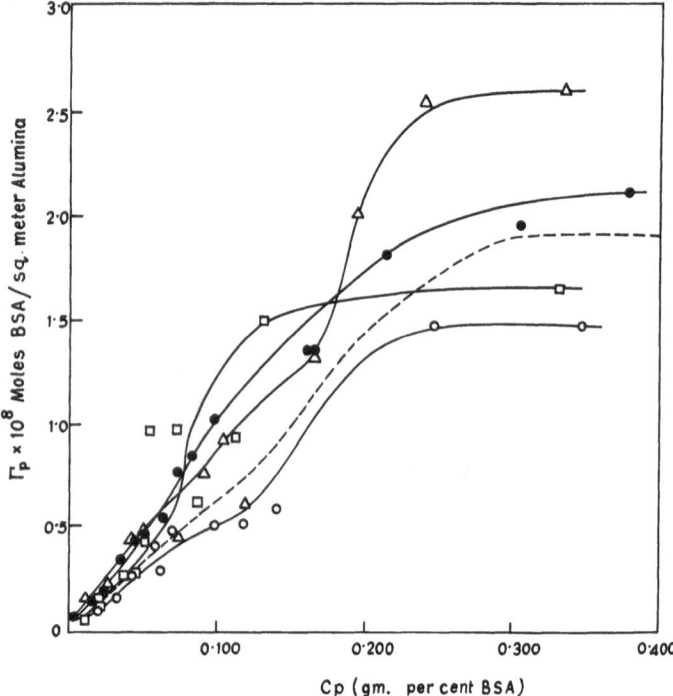

Fig. 3. Plot of moles (Γ_p) of BSA adsorbed per square meter of alumina surface against equilibrium % concentration (C_p) of protein in bulk at 28 °C, pH 6.0 (Phosphate buffer), C_s^t equal to 5×10^{-4} (M): a) ionic strength 0.010, ●; b) ionic strength 0.10, dotted line; c) ionic strength 0.50, O, at pH 4.0 (acetate buffer); d) $C_s^t = 1 \times 10^{-4}$ (M), ionic strength 0.10, △; e) C_s^t 5×10^{-4} (M), ionic strength 0.10, □

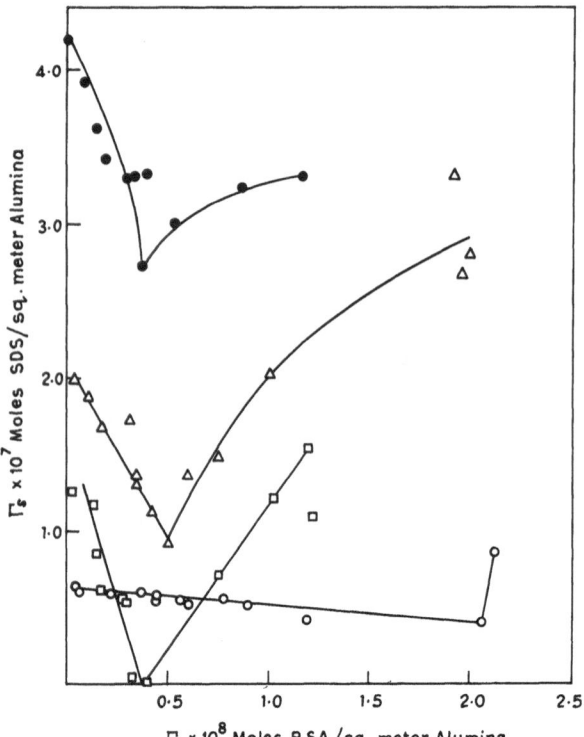

Fig. 4. Plot of Γ_s vs. Γ_p at 28 °C, ionic strength 0.10, pH 6.0 (phosphate buffer), a) $C_s^t = 1 \times 10^{-3}$ (M), ●; b) $C_s^t = 5 \times 10^{-4}$ (M), △; c) $C_s^t = 1 \times 10^{-4}$ (M), O; d) $C_s^t = 5 \times 10^{-4}$ (M) at 45 °C, □

that of negatively charged BSA-SDS complex are extended as a result of which adsorption is enhanced. In the same figure 3, the binding isotherms at pH 4.0 have been included for C_s^t values 1×10^{-4} (M) and 5×10^{-4} (M) respectively. At this pH also, one finds that Γ_p^m decreases with increase in the value of C_s^t in the medium. Comparisons of the isotherms in these figures at C_s^t equal to 5×10^{-4} (M) and ionic strength 0.1 further indicate that Γ_p at a low value of C_p is greater at pH 4 than that at pH 6.0. However, Γ_p^m at pH 6.0 is greater than that at pH 4.0. The anomaly here partly arises due to the decrease of the positive charge of BSA with increased binding of negatively charged dodecyl sulphate ions. When concentration of SDS is high, net charge of BSA decreases and it may even become negative due to its binding in excessive amount to the biopolymer molecules in solution.

In figures 4 and 5, values of the extent of adsorption (Γs) of SDS have been plotted against corresponding values of Γ_p at fixed values of C_s^t, pH, temperature and ionic strength. In all the cases, it is noted with interest that Γ_s decreases at first with increase of Γ_p until a

minimum value is reached. With increase of Γ_p further, Γ_s also increases without reaching any limiting value. From these observations, one can make general conclusion that in the relatively low range of Γ_p, both protein and surfactant compete for the adsorption at the alumina-water interface. Adsorption of more protein means displacement of SDS in this region of C_p under consideration. Beyond the minimum, adsorption of protein enhances the adsorption of SDS so that the net result is the mutual adsorption of both. It may be possible that protein at the interface changes conformation in a manner so that more and more SDS remains strongly bound to adsorbed BSA. The results in this high region of Γ_p may be of considerable interest to the system of biomembrane where considerable amounts of protein remain associated with large amounts of lipids thus forming the bilayer structure. From close analysis, one finds that minima in all the curves in figure 4 are indeed quite sharp. Variation of Γ_s with Γ_p for C_s^t equal to 1×10^{-4} (M) in the lower range of protein adsorption is however considerably slow. In figure 5 also, minima of the curve for C_s^t equal

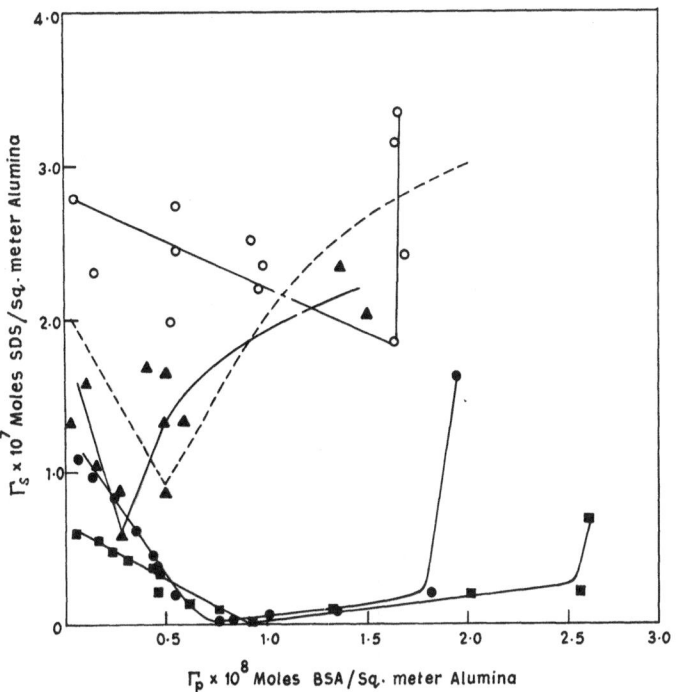

Fig. 5. Plot of Γ_s vs. Γ_p at 28 °C, pH 6.0 (phosphate buffer). a) ionic strength 0.010, $C_s^t = 5 \times 10^{-7}$ (M), ●; b) ionic strength 0.10, $C_s^t = 5 \times 10^{-5}$ (M), dotted line; c) ionic strength 0.50, $C_s^t = 5 \times 10^{-4}$ (M), ▲; and d) ionic strength 0.10, $C_s^t = 1 \times 10^{-4}$ (M), pH 4.0 (acetate buffer), ■; e) ionic strength 0.10, $C_s^t = 5 \times 10^{-4}$ (M), pH 4.0 (acetate buffer), ○

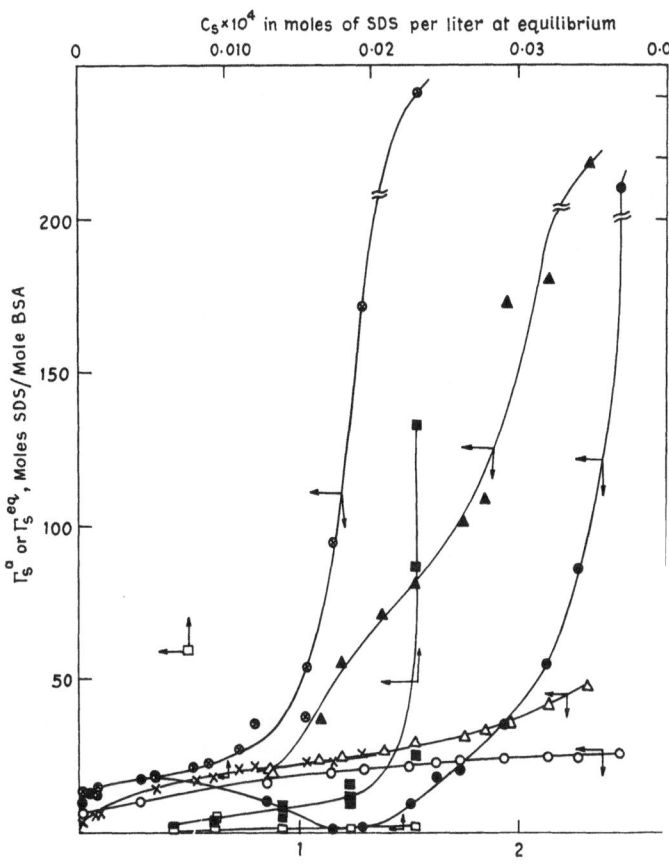

Fig. 6. Plot of Γ_s^a or Γ_s^{eq}, vs. C_s at alumina surface and in bulk at pH 6.0 (phosphate buffer), 28 °C, ionic strength 0.10. Left lower scale, a) $C_s^t = 1 \times 10^{-3}$ (M), in bulk △, at interface, ▲; b) $C_s^t = 5 \times 10^{-4}$ (M), in bulk, ×, at interface, ⊗; c) $C_s^t = 5 \times 10^{-4}$ (M), in bulk, ○, at interface, ●, at 45 °C; d) $C_s^t = 1 \times 10^{-4}$ (M), in bulk, □, at interface, ■; Upper scale on the right

to 5×10^{-4} (M) with respective pH values 6.0 and 4.0 and ionic strengths 0.5 and 0.1 are sharp. In contrast to this, the minimum obtained for the curve at pH 6.0 at ionic strength 0.01 and C_s^t equal to 5×10^{-4} (M) and at pH 4.0, ionic strength 0.1 and C_s^t equal to 1×10^{-4} (M) are not at all sharp but they spread in wide region of Γ_p. In these two cases Γ_s becomes very close to zero in wide region of Γ_p which means that SDS is completely absent in the surface of the solid. In figure 4, Γ_s is also found to be zero at the point of minimum at 45 °C. At 28 °C however, protein is not able to completely displace SDS from the surface at the minimum point possibly because of the strong affinity of SDS to adsorbed BSA. With increase of C_s^t more and more SDS remains bound with protein at the interface at the minimum point. The effect of pH and ionic strength on the minimum (vide fig. 5) are also noted with interest.

In the complex systems under consideration, the protein both in the bulk phase as well as at the interface are in binding interaction with SDS. However, affinities of binding in the two phases for the same system are expected to be different. In the bulk

phase, SDS is partly bound to BSA and partly remaining in the free state depending upon the equilibrium concentration of the surfactant in the solution. The binding isotherms in the bulk phase representing the plot of Γ_s^{eq} against C_s for various systems under consideration at a fixed temperature, pH and ionic strength are presented in figures 6 and 7. Here C_s in all cases are less than 3×10^{-4} (M) and values of Γ_s^{eq} do not exceed 60. The values of the binding depend upon C_s, pH, ionic strength and temperature of the medium. The results included in the binding isotherm are consistent with those earlier obtained by Sen, Mitra and Chattoraj [24] for C_s less than 3×10^{-4} (M).

It may be pointed out here that total number of moles (Γ_s^a) associated per mole of adsorbed BSA at a given value of C_s is equal to the ratio Γ_s/Γ_p. At a given

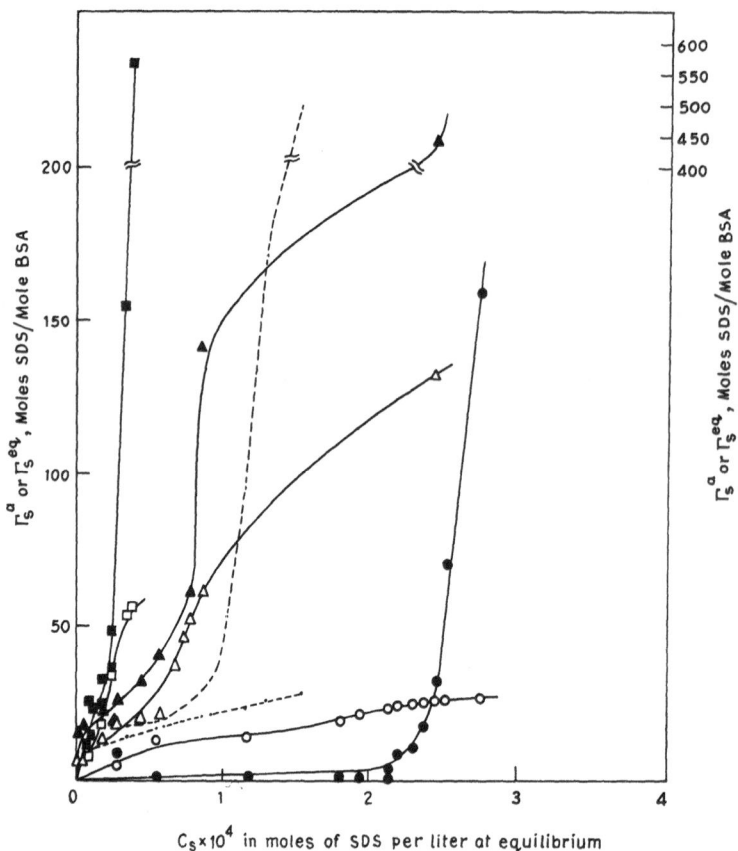

Fig. 7. Plot of Γ_s^a or Γ_s^{eq} vs. C_s at pH 6.0 (phosphate buffer), 28 °C, $C_s^t = 5 \times 10^{-4}$ (M). a) ionic strength 0.010, in bulk, $\bigcirc$, at interface, $\bullet$; b) ionic strength 0.50 in bulk $\triangle$, at interface $\blacktriangle$; c) ionic strength 0.10, in bulk dashed-dotted line and at interface dotted line; d) ionic strength 0.10 at pH 4.0 (acetate buffer), in bulk $\square$, at interface, $\blacksquare$

value of C_s (or C_p), value of Γ_s^a can be computed from the experimental data. Let us also conveniently assume that whole amount of SDS present at the interface is in binding interaction with adsorbed protein. In figures 6 and 7, the binding isotherm in the surface phase thus obtained from the plot of Γ_s^a against C_s is compared with that obtained in the case of bulk phase under identical conditions of temperature, pH and ionic strength of the medium. For C_s equal to or less than 1×10^{-4} (M), Γ_s^a and Γ_s^{eq} in many cases are close to each other and in few other cases Γ_s^a is even significantly less than Γ_s^{eq}. The nature of surfactant – protein interaction therefore at low detergent concentrations in the bulk and surface phases has some similarities from the qualitative point of view. However, for C_s varying between 1×10^{-4} (M) to 3×10^{-4} (M), Γ_s^a sharply increases in all cases without limit due to the massive type of binding interaction occurring in the interfacial phase. A molecule of BSA containing 590 amino-acid residues [29] is able to bind in this region 400 to 600 SDS molecules or possibly even

more. At pH 4.0, such massive binding interaction occurs when C_s is even considerably less than 1×10^{-4} (M). Further from the results given in figures 1 to 3, it may be qualitatively concluded that although BSA undergoes lateral expansion at the alumina-water interface at high values of C_p, the surface denaturation is neither very extensive nor irreversible in nature. It is highly probable that due to the massive and co-operative type of interaction, adsorbed SDS at protein-covered interface from multilayer ordered structure. This observation is quite relevant with the ordered bilayer structure of the lipid in the cell membrane in the presence of membrane bound proteins.

Tanford [30] and also Sen et al. [24] have already pointed out that SDS undergoes massive binding interaction with BSA even in the aqueous phase in the bulk when C_s is close to or above 10×10^{-4} (M). The value of Γ_s^{eq} in these cases may be as high as 400 or more and even then saturation in binding in the bulk phase is not attained [24].

The cmc of SDS in the bulk is 8.0×10^{-3} (M). The massive binding interaction thus occurs in the surface phase when C_s is nearly hundred times less than cmc. Interface thus favours this massive and co-operative binding interaction between protein and the detergent even when C_s is low. In cell membrane, proteins and lipids are in co-operative interaction with each other. The values of cmc and solubilities of these lipids are indeed insignificantly small. Even at this state, protein-lipid interaction at the interface may be massive and co-operative under favourable condition.

Acknowledgement

The authors are grateful to Dr. K. P. Das and Dr. R. Chatterjee for discussion and help. One of the authors (AS) likes to thank the Council of Scientific and Industrial Research, India for financial assistance.

References

1. Tamamushi, B., Tamaki, K., "Proceedings of the Second International Congress of Surface Activity", Butterworths, London, 1957, Vol. III, p. 449.
2. Tamamushi, B., Tamaki, K., Trans. Faraday Soc. **55**, 1007 (1959).
3. Connor, P., Ottewill, R. H., J. Colloid Interface Sci. **37**, 595 (1964).
4. Mukherjee, P., Anavil, A., in: "Adsorption at Interfaces" edited by K. L. Mittal, ACS Symposium Series No. 8, American Chemical Society, Washington D. C., p. 107 (1975).
5. Sommasundaran, P. and Fuerstenau, D. H., J. Phys. Chem. **70**, 90 (1966).
6. Sommasundaran, P., Healy, T. W., Fuerstenau, D. H., J. Phys. Chem. **68**, 3562 (1964).
7. Trogor, F. J., Schechter, R. S., Wade, W. H., J. Colloid Interface Sci. **70**, 293 (1979).
8. Bull, H. B., Biochem. Biophys. Acta **19**, 464 (1956).
9. Bull, H. B., Archs. Biochem. Biophys. **68**, 102 (1957).
10. Chattoraj, D. K. and Bull, H. B., J. Amer. Chem. Soc., **81**, 5128 (1959).
11. Dilman, W. J., Miller, I. R., J. Colloid Interface Sci. **44**, 221 (1974).
12. Morrissey, B. W., Stromberg, R. R., J. Colloid Interface Sci. **44**, 221 (1974).
13. Mitra, S. P., Chattoraj, D. K., Indian J. Biochem. Biophys. **15**, 147 (1978).
14. Norde, W., Lyklema, J., J. Colloid Interface Sci. **66**, 257, 266, 277, 285, 295 (1978).
15. Norde, W., "Proteins at Interfaces", Doctoral Thesis, Agricultural University, Wageningen, (Netherlands, 1976).
16. Schulman, J. H. and Rideal, E. K., Proc. Roy. Soc. Ser. B, **122**, 46 (1937).
17. Eley, D. D., Hedge, D. G., J. Colloid Sci. **11**, 445 (1957).
18. Bull, H. B., J. Amer. Chem. Soc. **67**, 10 (1945).
19. Pearson, J. T., J. Colloid Interface Sci. **27**, 64 (1968).
20. Arnebrant, T., Nylander, T., Hegg, P. O., Larson, Kara, Abstracts, "International Symposium on Surfactants in Solution", June-July, 1982, Lund, Sweden, page 30.
21. Mukherjee, P., Mysels, K. J., "Critical Micelle concentration", National Bureau of Standards, (Washington D. C., 1971).
22. Chatterjee, R., Chattoraj, D. K., Biopolymer **18**, 147 (1979).
23. Sen, M., Mitra, S. P., and Chattoraj, D. K., Colloids and Surfaces **2**, 259 (1981).
24. Sen, M., Mitra, S. P., Chattoraj, D. K., Indian J. Biochem. Biophys. **17**, 370 (1980).
25. Mukherjee, P., Anal. Chem. **28**, 870 (1956).
26. Biswas, H. K., Mondal, B. M., Anal. Chem. **44** (9), 1636 (1972).
27. Orr, C., Dallavalle, J. M., Fine Particle Measurement: Size, Surface and Pore Volume, pp. 207–215, The McMillar Company (New York, 1959).
28. Chattoraj, D. K., Chowrashi, P., Chakravarti, K., Biopolymer **5**, 173 (1967).
29. Spahr, P. F., Edsall, J. T., J. Biol. Chem. **239**, 850 (1964).
30. Tanford, C., "Hydrophobic Effect", Formation of Micelles and Biological Membranes, Wiley, (New York, 1980).

Received January 12, 1983;
accepted January 29, 1983

Authors' address:

Dr. D. K. Chattoraj
Head, Dept. of Food Technology
and Biochemical Engineering,
Jadavpur University,
Calcutta-700032, India.

Progress in Colloid & Polymer Science

Progr. Colloid & Polymer Sci. **68**, 152–157 (1983)

The selective adsorption of NO on synthetic iron(III) oxide hydroxides

T. Ishiwaka and K. Inouye

Department of Chemistry, Chiba University, Chiba, Japan

Abstract: In order to verify the mechanism of selective adsorption of NO from mixed gases on synthetic crystals of α-, β- and γ-FeOOH, the following measurements were carried out at 30 °C: (1) the amount of chemisorption of NO, SO_2 and CO_2 on each FeOOH polymorph which is prechemisorbed by one of the gases, (2) the amount of "exchange" adsorption between NO and SO_2, when a gas is introduced to the surfaces preoccupied by chemisorption of another gas, and (3) the comparison of the sum of chemisorption values measured separately for NO, SO_2 and CO_2 with the total chemisorption from mixed gas systems NO-SO_2 and NO-CO_2. The interaction between adsorbate molecules and FeOOH surface has been discussed to deleate probable positions of chemisorption of each adsorbate on predominant crystal surfaces of FeOOH.

Key words: FeOOH surface, adsorption of NO, SO_2, CO_2 and H_2O, selective adsorption of NO, adsorption sites, chemisorption

Introduction

The theory of adsorption was one of the problems in which Professor B. Tamamushi was interested for nearly half a century since he published several papers on adsorption during his first period of research in the colloid-surface chemistry [1]. There is no need to mention that numerous studies have been carried out since then by many workers on the mechanism of adsorption phenomena and at the same time on the nature of surfaces. In recent years, however, this classical problem seems to have gained a renewed recognition of importance with respect to the environmental control through adsorptive removal of pollutants. The gaseous pollutants such as SO_x and NO_x contained in the industrial and urban atmospheres particularly of developed countries could be removed, if we find more appropriate selective adsorbents, since the pollutants exist in diluted state in a huge amount of mixed gases.

We have tried to find new potential adsorbents suitable to this purpose, which should have some unavoidable characteristics such as easy preparation, thermal stability, high adsorptive capacity and rapid adsorption rate. Our special concern lies in finding such crystalline substances, not amorphous substances like carbons, but ones that the adsorption sites are exposed on crystal surfaces together with necessary characteristics for adsorbent mentioned above. Amongst the materials so far examined, several materials seem to be more or less promising as for the SO_2 adsorbents, synthetic iron(III) oxide hydroxides (α-, β- and γ-FeOOH) [2], synthetic alunite [3], and synthetic hydroxyapatite [4]; as for the NO adsorbents, synthetic FeOOH [5, 6], synthetic jarosite [7], synthetic alunite [3], and synthetic hydroxyapatite [4]. There are some problems to be improved and clarified about these adsorbents; the increase of specific surface area in the case of jarosite and the increase in the adsorption rate of NO for most of the samples described above. Meanwhile, the confirmation of adsorption sites on solid surface has been considered significant in order to obtain further practicability. The adsorption sites on synthetic FeOOH crystal surfaces are particularly interesting, because these crystals have the predominant surfaces that occupy over 90% of the total surface areas, based on their specific layer structures [8] and consequently on their particle morphology [9].

The selectivity of adsorption of the component gas in mixed gases is also to be examined, since it determines the feasibility of the adsorbent for a particular gas. This paper discusses the selectivity of NO adsorption on synthetic FeOOH by examining the adsorptive sites for NO, SO_2, CO_2 and H_2O. The adsorption measurements of a gas on the surfaces preadsorbed by another gas, with reference to the crystallographic surface structure of FeOOH, enabled us to estimate reasonable adsorption sites for the SO_2 adsorption [10]. The similar technique is employed and explained in the present work also for the NO adsorption.

Experimental

Materials

α-FeOOH was prepared by hydrolysing 0.6 mol dm^{-3} Fe$_2$(SO$_4$)$_3$ solution at 50 °C for 30 h with addition of 1.0 mol dm^{-3} NaOH to adjust the pH at 13.6. β-FeOOH was obtained by heating 0.1 mol dm^{-3} FeCl$_3$ solution at 95–99 °C for 5 h in the presence of urea. The oxidation of 0.2 mol dm^{-3} FeCl$_2$ solution with urotropin ((CH$_2$)$_6$N$_4$), NaNO$_2$ and HCl at 60 °C for 45 min gave γ-FeOOH. Each precipitate was washed thoroughly with distilled water until no SO$_4^{2-}$ or Cl^{-1} was detected and dried at 60 °C for α- and β-FeOOH and 80 °C for γ-FeOOH. The X-ray diffraction, by use of Fe *K*α irradiation at 30 kV and 10 mA filtered by Mn foil, indicated that each FeOOH sample is crystallographically pure. The specific surface area determination was done by the BET method from nitrogen adsorption at liquid nitrogen temperature.

Adsorption measurements

Three types of adsorption measurements have been experimented in this work: (1) the static adsorption measurement for obtaining the NO adsorption isotherms at 30 °C, using a quartz spring with a sensitivity of 215 mm g^{-1}. The sample was pretreated at 110 °C and 10^{-3} Pa for 14 h, followed by lowering the temperature to 30 °C and introducing the gas to be adsorbed at different pressures up to 500 Torr (6.67 × 10^4 Pa). The effect of preadsorbed SO$_2$ and CO$_2$ on the adsorption of NO was examined by the adsorption isotherm of NO on the surfaces that were precedingly adsorbed with SO$_2$ or CO$_2$ at 6.67 × 10^4 Pa and desorbed at 10^{-3} Pa and 30 °C for 24 h (called "prechemisorbed" hereafter); (2) the chemisorption values, denoted as A_{NO}, A_{SO_2} and A_{CO_2} for the cases of NO, SO$_2$ and CO$_2$, respectively, measured by evacuating at 10^{-3} Pa and 30 °C for 24 h the sample which was previously treated with the gas at 6.67 × 10^4 Pa. When a gas (X) was adsorbed onto the surfaces which had been prechemisorbed with a different gas (Y), the amount of chemisorption of X gas, denoted as $A_X^{*(Y)}$, was measured after the similar evacuation of Y gas as described above; (3) the direct determination of the amounts of NO and SO$_2$ chemisorbed on surfaces. The sample was heated at 600 °C in a nitrogen stream for 30 min for complete desorption of NO [11] and at 800 °C in air for 30 min for the desorption of SO$_2$ [6], followed by analyses described below. NO was analyzed by colorimetry at 545 nm of NO$_2^-$, produced by complete ozone-oxidation of NO and following reduction with Zn, in the presence of sulfanyl amide and naphthyl ethylene diamine [12]. SO$_2$ was determined by titration of SO$_4^{2-}$ with the standard barium acetate solution in the presence of Arcenazo III in aqueous H$_2$O$_2$ [13].

The pure NO used in this study was further purified by passing through a solid CO$_2$-ethanol trap. SO$_2$ and CO$_2$ were dried over P$_2$O$_5$ and subjected to repeated bulb-to-bulb distillations.

Results

Chemisorption of NO on surface prechemisorbed by SO$_2$ and CO$_2$

In figures 1, 2 and 3, are given the changes of NO adsorption at 30 °C on α-, β- and γ-FeOOH crystals, respectively, caused by the preceding chemisorption of SO$_2$ and CO$_2$. It is evident that the preoccupying molecules on crystal surfaces decrease considerably the adsorption of NO molecules. The decreasing

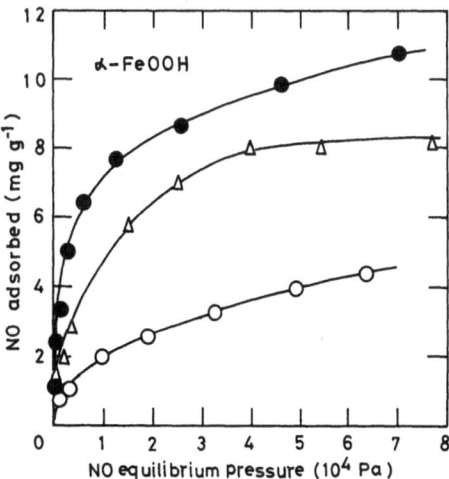

Fig. 1. Decrease of NO adsorption on α-FeOOH at 30 °C by prechemisorbed SO$_2$ and CO$_2$. ●: No adsorption on bare surface; ○: NO adsorption on surface prechemisorbed by SO$_2$; △: NO adsorption on surface prechemisorbed by CO$_2$

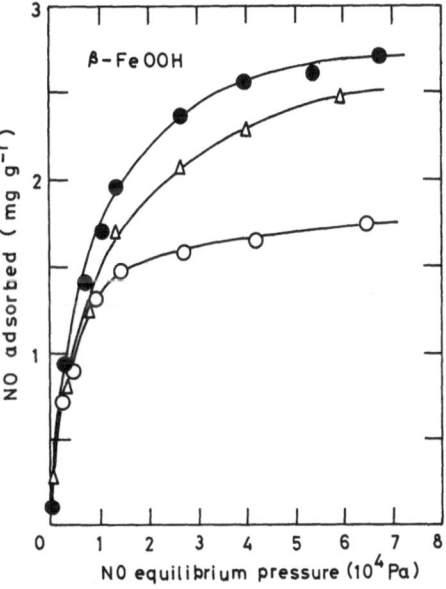

Fig. 2. Decrease of NO adsorption on β-FeOOH at 30 °C by prechemisorbed SO$_2$ and CO$_2$. ●: NO adsorption on bare surface; ○: NO adsorption on surface prechemisorbed by SO$_2$; △: NO adsorption on surface prechemisorbed by CO$_2$

effect of prechemisorbed CO$_2$ molecules on the adsorption of NO is in a less extent than the effect of SO$_2$. The effect of prechemisorbed molecules on the subsequent adsorption varies with the type of crystals, on that we will discuss later.

Table 1 gives the chemisorption values on each FeOOH polymorph and the ratios between the chemisorption of a gas on surfaces preoccupied by

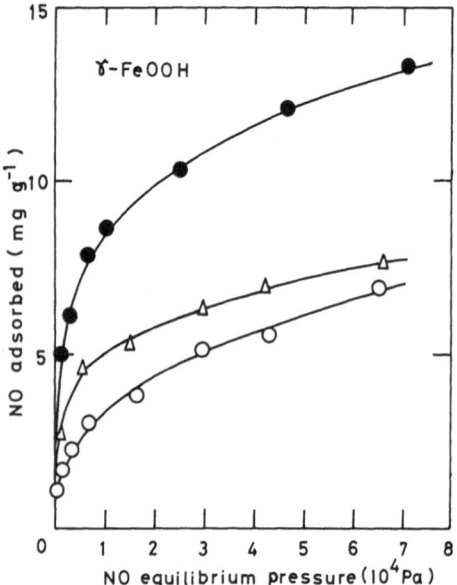

Fig. 3. Decrease of NO adsorption on γ-FeOOH at 30 °C by prechemisorbed SO_2 and CO_2. ●: NO adsorption on bare surface; ○: NO adsorption on surface prechemisorbed by SO_2; △: NO adsorption on surface prechemisorbed by CO_2

Table 1. Changes of chemisorption by prechemisorption of another gas at 30 °C.
A_X denotes the chemisorption of X gas on bare surfaces and $A_X^{*(Y)}$ denotes the chemisorption of X gas on the surfaces prechemisorbed by Y gas. For the numerals in the parentheses, see text.

Adsorbent	α-FeOOH	β-FeOOH	γ-FeOOH
A_{NO} (mg g^{-1})	9.8	2.7	9.9
A_{SO_2} (mg g^{-1})	19.0	8.0	30.1
A_{CO_2} (mg g^{-1})	2.4	1.5	2.8
$A_{NO}^{*(SO_2)}$ (mg g^{-1})	3.1 (9.0)	1.7 (2.8)	4.2 (4.5)
$A_{NO}^{*(CO_2)}$ (mg g^{-1})	8.0	2.5	5.6
$A_{SO_2}^{*(NO)}$ (mg g^{-1})	9.7	5.8	18.4
$A_{CO_2}^{*(NO)}$ (mg g^{-1})	0.05	0.00	0.02
$A_{NO}^{*(SO_2)}/A_{NO}$	0.31 (0.92)	0.64 (1.03)	0.41 (0.45)
$A_{NO}^{*(CO_2)}/A_{NO}$	0.82	0.92	0.57
$A_{SO_2}^{*(NO)}/A_{SO_2}$	0.51	0.72	0.61
$A_{CO_2}^{*(NO)}/A_{CO_2}$	0.02	0.00	0.01

another gas and the chemisorption value on bare crystal surfaces.

When SO_2 is chemisorbed prior to the introduction of NO, a portion of SO_2 comes out of surfaces, as will described in the next section; in table 1, the values of $A_{NO}^{*(SO_2)}$ in the parentheses were obtained by adjusting the amount of NO under an assumption that NO is adsorbed in the amount corresponding to the SO_2 desorbed.

The relationships of chemisorption between the gases on each FeOOH are summarized as follows: (1) The existence of SO_2 and CO_2 molecules prechemisorbed exhibits interfering effects on the subsequent chemisorption of NO molecules in different degrees specific to each FeOOH. The effect, however, seems to be more impressively produced by SO_2 than by CO_2, because $A_{NO}^{*(SO_2)}/A_{NO}$ values are generally smaller than $A_{NO}^{*(CO_2)}/A_{NO}$ values. (2) The existence of NO molecules chemisorbed on each FeOOH reduces the chemisorption of SO_2 later introduced, because the ratios $A_{SO_2}^{*(NO)}/A_{SO_2}$ are in the range 0.5–0.7. (3) The existence of prechemisorbed NO on each FeOOH hinders almost completely the chemisorption of CO_2 molecules later introduced, because the $A_{CO_2}^{*(NO)}/A_{CO_2}$ ratios are nearly null.

The correlation between NO adsorption and SO_2 adsorption

We tried to determine the amounts of NO and SO_2 existing on the crystal surfaces by means of analyzing each gas desorbed during the calcination of the samples at appropriate conditions (cf. Experimental), to reveal the "exchange" adsorption between NO and

Table 2. The "exchange" chemisorption between NO and SO_2 at 30 °C.
$A_{X(Y)}$ denotes the amount of X gas remained on surfaces after Y gas was introduced.

Adsorbent	α-FeOOH	β-FeOOH	γ-FeOOH
$A_{NO(SO_2)}$ (mg g^{-1})	10.0	2.7	10.2
A_{NO} (mg g^{-1})	9.8	2.7	9.9
Expelled NO by SO_2 (%)	0	0	0
$A_{SO_2(NO)}$ (mg g^{-1})	13.1	6.0	29.8
A_{SO_2} (mg g^{-1})	19.0	8.0	30.1
Expelled SO_2 by NO (%)	31	25	1

SO_2. Table 2 gives the analytical values and the degree of exchange of a gas chemisorbed with another gas. It appears that NO molecules reject some SO_2 molecules which have existed on the surfaces, in different degrees characteristic to each FeOOH; it is noticeable that γ-FeOOH shows very low degree of the exchange of SO_2 by NO. The SO_2 molecules, on the other hand, show no exchange action with prechemisorbed NO molecules on each FeOOH.

Table 3. Chemisorption from mixed gases at 30 °C.

Adsorbent	α-FeOOH	β-FeOOH	γ-FeOOH
(1) Total chemisorption from NO–SO$_2$ mixture (mg g^{-1})	23.0	8.3	38.0
(2) $A_{NO} + A_{SO_2}$ (mg g^{-1})	28.8	10.7	40.0
Ratio (1)/(2)	0.80	0.78	0.95
(3) Total chemisorption from NO–CO$_2$ mixture (mg g^{-1})	10.2	2.7	6.0
(4) $A_{NO} + A_{CO_2}$ (mg g^{-1})	12.2	4.2	12.7
Ratio (3)/(4)	0.84	0.64	0.47

Chemisorption from mixed gases NO–SO$_2$ and NO–CO$_2$

The equal molar mixtures of NO and SO$_2$ as well as NO and CO$_2$ were treated with FeOOH samples to determine the total chemisorption to be compared with the sums $A_{NO} + A_{SO_2}$ and $A_{NO} + A_{CO_2}$, respectively. Comparing the values in table 3, we can recognize that the total chemisorption from mixed gases is, without an exception, smaller than the sum of chemisorption for each gas. It is noteworthy that the difference between the total chemisorption and the sum of chemisorption for each gas is remarkable in the cases of adsorption pertinent to CO$_2$; the total chemisorption from NO–CO$_2$ mixture is close to A_{NO} in table 1, excluding perhaps the case of γ-FeOOH.

Discussion

We reported in a previous paper [6] on the adsorption of NO on FeOOH by a flow-gas method at 100 °C from a simulated flue gas composed of 300 ppm NO, 500 ppm SO$_2$, 10% H$_2$O, 1% O$_2$, 10% CO$_2$ and the rest N$_2$. It was revealed that co-existing gases decrease the NO adsorption, the retarding effect on the NO adsorption being most marked for H$_2$O, followed by SO$_2$ and then CO$_2$ which shows almost no effect on NO adsorption. This result seems to be coincident with the present experiments, where the interferences in the NO adsorption by the prechemisorption of different gases have been examined from different viewpoints.

Figures 1, 2 and 3 indicate qualitatively that the preoccupation of adsorptive sites influences on the amount of NO static adsorption at 30 °C in the similar order as observed in the flow-gas method at 100 °C: the NO adsorption is decreased by the prechemisorp-

tion of SO$_2$ considerably and of CO$_2$ in less extent. It must be reminded that these results express the overall adsorption of NO on each FeOOH polymorph, whereas all the values given in tables 1, 2 and 3 concern with only the chemisorption determined by evacuating the adsorbed sample to eliminate the portion of physisorption. From the chemisorption experiment we can clarify the kind of plausible adsorption sites on FeOOH surfaces for each gas and at the same time discuss the interaction between the specific site and each adsorbate.

The chemisorption of NO on FeOOH crystals was reasonably expressed, as in our previous report [5], by the Langmuir saturation adsorption (constant b) in table 4 calculated from the results which are illustrated in figs. 1, 2 and 3. Comparing the $b_{NO}^{*(SO_2)}$ and $b_{NO}^{*(CO_2)}$ in table 4 with A_{NO}, $A_{NO}^{*(SO_2)}$ and $A_{NO}^{*(CO_2)}$ in table 1, respectively, we can recognize that these values are fairly well coincident. It is therefore natural that the ratios $b_{NO}^{*(SO_2)}/b_{NO}$ and $b_{NO}^{*(CO_2)}/b_{NO}$ are close to $A_{NO}^{*(SO_2)}/A_{NO}$ and $A_{NO}^{*(CO_2)}/A_{NO}$, respectively. These coincidences also support the plausibility of the Langmuir saturation adsorption as a measure of chemisorption.

Table 4. Effects of prechemisorbed SO$_2$ and CO$_2$ on the Langmuir constant b of NO adsorption at 30 °C.

Adsorbent	α-FeOOH	β-FeOOH	γ-FeOOH
b_{NO}, NO adsorption on bare surfaces (mg g^{-1})	10.8	2.97	10.7
$b_{NO}^{*(SO_2)}$, NO adsorption on the surfaces prechmisorbed by SO$_2$ (mg g^{-1})	3.00	1.65	4.35
$b_{NO}^{*(CO_2)}$, NO adsorption on the surfaces prechemisorbed by CO$_2$ (mg g^{-1})	8.61	2.65	5.64
$b_{NO}^{*(SO_2)}/b_{NO}$	0.28	0.56	0.41
$b_{NO}^{*(CO_2)}/b_{NO}$	0.80	0.89	0.53

On the other hand, a portion of SO$_2$ prechemisorbed on FeOOH surfaces appears to be desorbed by NO introduced subsequently (table 2); that was known from analyses of SO$_2$ remaining on the sample. If the same number of NO molecules are adsorbed on the sites as the number of desorbed SO$_2$ compensationally, the values $A_{NO}^{*(SO_2)}$ and accordingly $A_{NO}^{*(SO_2)}/A_{NO}$ must become larger, as given in the parentheses in table 1. Experimental results described above show that the exchange between the desorbed SO$_2$ with NO is unbelievable to take place. In a previous experiment on the thermal desorption analysis of chemisorbed SO_2 on FeOOH [16], it was found that SO_2 is desorbed mostly as SO_3 at temperatures characteristic to each FeOOH. Since SO_2 is bonded with O^{2-} on surfaces, the surface structure seems to change after the desorption to one which is unsuitable to the NO adsorption.

The results suggest an overwhelmingly stronger adsorptive bonding of NO with the surface sites than SO_2 and in particular CO_2. The strong adsorption of NO results in several facts as follows: (1) the negligibly low capability of CO_2 adsorption on the surfaces prechemisorbed by NO (table 1, $A_{CO_2}^{*(NO)}/A_{CO_2}$), (2) low interference of prechemisorbed CO_2 on the subsequent chemisorption of NO (table 1, $A_{NO}^{*(CO_2)}/A_{NO}$) for α- and β-FeOOH, and (3) no "exchange" between SO_2 and NO, when NO molecules are prechemisorbed before SO_2 (table 2).

The interpretation of the results in table 3, comparing the total chemisorption from NO–SO_2 and NO–CO_2 mixed gases with the sums of chemisorption of NO, SO_2 and CO_2 on bare surfaces, seems significant with regard to the adsorption sites. If this comparison gives equal amounts of chemisorption, we will be able to assume that the sites for a gas and another are different. As for the chemisorptive sites for NO and SO_2, the ratio of the total chemisorption to the sum of chemisorption ranges from 0.78 to 0.95. This fact does not simply mean that the chemisorptive sites for NO and SO_2 are entirely different or identical. We remind the preceding conclusion that the adsorption sites of SO_2 and H_2O are at different positions; SO_2 on O^{2-} and H_2O on OH^- for each FeOOH polymorph [10]. It is also interesting to recall that the NO adsorption in a flow-gas method is interfered with H_2O and SO_2 as well [6]. We may rather assume that the chemisorption site could not be represented simply as one on a flat crystal surface, but that the steric fitness of surface to molecules is important. The similar consideration will be valid to the correlation between the NO site and CO_2 site (table 3), but lower ratios of the total chemisorption to the sum $A_{NO} + A_{CO_2}$ for β-FeOOH (0.64) and γ-FeOOH (0.47) than α-FeOOH (0.84) need further consideration. The prechemisorption of CO_2 reduces the chemisorption of NO from the chemisorption on bare surfaces in different degrees of 0.81, 0.91 and 0.55 for α-, β- and γ-FeOOH, respectively (tables 1 and 4). These results seem to contain no inconsistency and suggest that even though the amount of CO_2 chemisorption is relatively low due to weak bonding with surface, a certain interfering effect on the NO adsorption is exhibited in different extents characteristic to each FeOOH polymorph, the interfe-

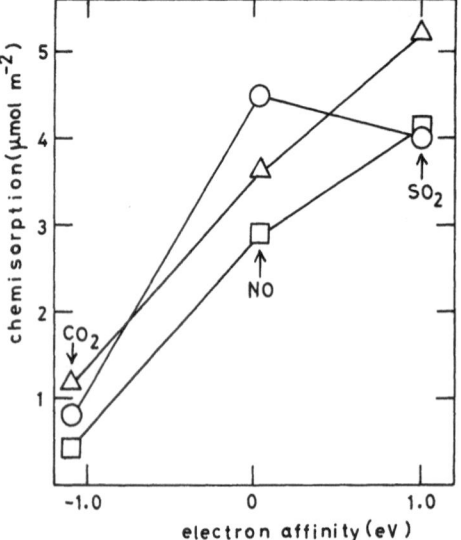

Fig. 4. Correlation between the chemisorption on FeOOH at 30 °C and the electron affinity of adsorbate molecules. ○: α-FeOOH, △: β-FeOOH, □: γ-FeOOH

rence being particularly noticeable for γ-FeOOH. Basically it seems plausible to presume that the same places on surface are occupied by CO_2 and NO. This hindrance varies with the difference in surface structure, as will qualitatively be understood later in figure 5.

There will be at least two viewpoints to discuss in general the causes of chemisorption: (1) the electronic factor controlling the chemisorption and (2) the structural factor based on the surface structure and the molecular morphology of adsorbate. As for the electronic factor, the chemisorption will be pertinent to the electron affinity of adsorbate molecule. Figure 4 shows the relationship between the chemisorption and the electron affinity of NO, SO_2 and CO_2. The chemisorption as adsorptive activity is expressed purposely in the unit 10^{-6} mol m^{-2}, using the BET specific surface area. The chemisorptive activity appears to increase with the electron affinity for each FeOOH, except the case of a high value of NO chemisorption on α-FeOOH, which may be influenced by another particular reason. The relationship in figure 4 makes us presume that the chemisorption is predominantly related to the electronic factor. Concerning with the chemisorption of NO and SO_2, we have revealed previously that the electron transfer occurs from FeOOH surfaces toward adsorbate molecules from the observation of decreasing changes in electrical conductivity of FeOOH with the adsorption of the gases [14, 15]. The higher chemisorptive activity of NO on α-FeOOH may be due to higher density of residual electro-static valence of the NO-adsorption

Progress in Colloid & Polymer Science   Progr. Colloid & Polymer Sci. **68**, 158–162 (1983)

Thermoanalytical investigation on the coagel-gel-liquid crystal transition in some water-amphiphile systems*)

M. Kodama and S. Seki

Department of Chemistry, Faculty of Science, Kwansei Gakuin University, Nishinomiya, 662 Japan

Abstract: Thermoanalytical studies on the binary systems of water and amphiphiles such as surfactant (octadecyltrimethyl ammonium chloride and potassium stearate) and lecithin (dipalmitoyl phosphatidylcholine) revealed the stepwise decreasing T_c transition curves, due to the so-called chain-melting, with increasing water content. Detailed and careful annealing treatment for the samples indicated the existence of a new phase transition corresponding to the so-called coagel-to-gel at a temperature below the T_c curves. For this transition, we proposed the symbol T_{gel} curve shown in the phase diagram. Furthermore, the interrelationship between the T_{gel} transition and the ice-melting phenomena was examined precisely and the predominent roles of 'newly incorporated water (intermediate water)' between the bilayers of the amphiphilic molecules at the T_{gel} transition was pointed out.

Key words: water-amphiphile system, coagel, gel, liquid crystal, phase transition, T_c-curve, T_{gel}-curve, octadecyltrimethylammonium chloride, potassium stearate, dipalmitoyl phosphatidylcholine.

Introduction

In the so-called amphiphilic compound which has long hydrocarbon chain as hydrophobic group and polar hydrophilic head group in its constituents, the hydrocarbon chains are known to undergo a kind of order-disorder configurational transition, i. e., the so-called 'chain-melting transition', depending on both temperature and water content, before reaching the isotropic liquid state. Aggregation states of amphiphile molecules at a temperature above this transition have been studied by many investigators, mainly from the viewpoint of lyotropic and/or thermotropic liquid crystals [1–4]. In contrast to this trend, phase studies at a temperature below the chain-melting transition temperature have been behind in their progress and several confused results have been reported.

In order to make some contributions in this low temperature as well as at the lower water content regions, the present work reports our thermoanalytical investigations on the behaviour of chain-melting transition and on the nature of phases existing below its transition temperature in the binary system of water and amphiphiles. Here, we should like to

illustrate the case of two kinds of typical surfactant and a lecithin, focusing our special attention on the roles of water on the transition phenomena.

1) Water-octadecyltrimethylammonium chloride system

As the first example, we refer to our result on the binary system of water and octadecyltrimethylammonium chloride ($C_{18}Cl$) as a typical cationic surfactant [5].

Figure 1 shows the phase diagram of the water-$C_{18}Cl$ system obtained from all the DSC (differential scanning calorimetry) thermograms of a number of different samples. The T_c curve in this figure represents the change of the transition temperature caused by the chain-melting with the variation of water content. The curve goes down stepwise to the limiting temperature of 15.5 °C, corresponding to the so-called Krafft point, and reflects a total of three endothermic peaks which appear at successively lower temperatures with increasing water content. Such a successive phase transition behavior obtained here is apparently distinguished from the smooth T_c curves for amphiphile-water systems reported by many other investigators [6, 7, 8].

At a temperature below the T_c curve, we discovered a new phase transition corresponding to the so-called

---

*) Dedicated to the memory of Professor Dr. B. Tamamushi.

sites on the *a–c* surfaces, on that we discussed in an earlier paper [14]. In table 2, the expelled SO_2 by the subsequently adsorbed NO amounts 31, 25 and 1% for α-, β- and γ-FeOOH, respectively. This result seems not to be due to the difference in electron affinity between SO_2 and NO, but to the steric fitness of each molecule to the space surrounding chemisorptive sites.

The adsorbate molecules chemisorbed at the most appropriate positions on surface projections of each FeOOH are illustrated in figure 5. Each molecule is drawn in the size relative to ionic radii of composing atoms. The molecules chemisorbed are situated at the positions supposed from the above considerations and in accordance with our previous results [2, 5, 10], namely NO, SO_2 and CO_2 are bonded with O^{2-}, whereas H_2O with OH^-. The FeOOH crystal structure is characteristically constructed by stacking twins composed of two Fe-centered octahedra sharing edges. In α-FeOOH, the twins are connected each other at corners [17]. The O^{2-} with the strongest bonding with NO, SO_2 and CO_2 exists at the specific positions shown in figure 5, where the residual electrostatic valence gives the highest value [14]. H_2O molecules,

on the other hand, bind primarily with OH^- at the bottom of shallow groove, and in addition with neighboring O^{2-} ions by hydrogen-bondings [18]. As for β-FeOOH, the crystallographic data available [8, 19] suggest possible adsorption positions as accordingly delineated in figure 5 for each adsorbate. The parallel grooves in the *c*-direction of γ-FeOOH, where both OH^- and O^{2-} stand in row [20], serve the places for the chemisorption of NO, SO_2, CO_2 and H_2O as well, causing the noticeable hindrance of chemisorption of a subsequently introduced molecules by initially chemisorbed molecules, even if the bonding with surface is weak as in the case of CO_2.

Acknowledgement

The authors wish to acknowledge the financial support from the Environmental Sciences Project of the Ministry of Education, Science and Culture, Japanese Government.

References

1. Tamamushi, B., e. g., Kolloid-Z. **4**, 58 (1929).
2. Ishiwaka, T., Inouye, K., Nippon Kagaku Zasshi (J. Chem. Soc. Japan, Pure Chem. Sect.) **91**, 935 (1970).
3. Kurata, M., Kaneko, K., Inouye, K., J. Phys. Chem., in press.
4. Inouye, K., Environmental Control and Chemistry for Prevention, Ed. Saegusa, T., Sumitomo, H., Takahashi, H., Yoshikawa, S., Chapt. 1, Tkoyo Univ. Press, Tokyo (1981).
5. Hattori, T., Kaneko, K., Ishiwaka, T., Inouye, K., Nippon Kagaku Kaishi (J. Chem. Soc. Japan) 1979, 423.
6. Ishiwaka, T., Inouye, K., Nippon Kagaku Kaishi (J. Chem. Soc. Japan) 1979, 697.
7. Inouye, K., Nagumo, I., Kaneko, K., Ishikawa, T., Z. Phys. Chem., Neue Folge, **131**, 199 (1982).
8. Mackay, A. L., Mineral. Mag. **32**, 545 (1960).
9. Inouye, K., Ishikawa, T., Hyomen (Surface) **16**, 129 (1978).
10. Ishikawa, T., Inouye, K., Nippon Kagaku Kaishi (J. Chem. Soc. Japan) 1980, 681.
11. Hattori, T., Kaneko, K., Ishikawa, T., Inouye, K., Nippon Kagaku Kaishi (J. Chem. Soc. Japan) 1979, 911.
12. Japanese Ind. Standard, K-0104 (1974).
13. Seidman, E. B., Anal. Chem. **30**, 1680 (1958).
14. Kaneko, K., Ishikawa, T., Inouye, K., Nippon Kagaku Kaishi (J. Chem. Soc. Japan) 1977, 162.
15. Kaneko, K., Inouye, K., Corrosion Sci. **27**, 639 (1981).
16. Ishikawa, T., Inouye, K., Nippon Kagaku Kaishi (J. Chem. Soc. Japan) 1981, 1840.
17. Ewing, F. J., J. Chem. Phys. **3**, 203 (1935).
18. Kaneko, K., Serizawa, M., Ishikawa, T., Inouye, K., Bull. Chem. Soc. Japan **48**, 1764 (1975).
19. Watson, J. H., Cardell, R. R. Jr., Heller, W., J. Phys. Chem. **66**, 1757 (1962).
20. Ewing, F. J., J. Chem. Phys. **3**, 420 (1935).

Received January 25, 1983;
accepted February 3, 1983

Authors' address:

Prof. Katsuya Inouye
Department of Chemistry,
Faculty of Science, Chiba University,
Chiba 260, Japan

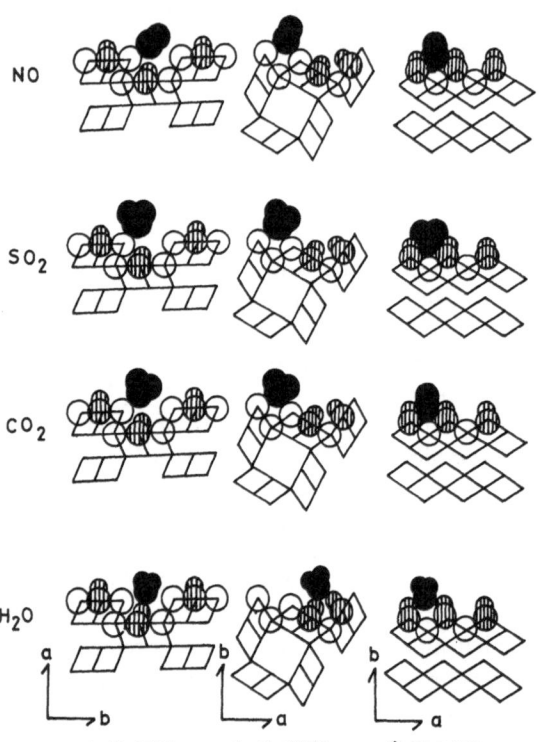

Fig. 5. Chemisorption of NO, SO_2, CO_2 and H_2O on α-, β- and γ-FeOOH viewed from the *c*-axis direction. Open and hatched circles on surfaces are O^{2-} and OH^-, respectively. Adsorbate molecules are represented by filled symbols.

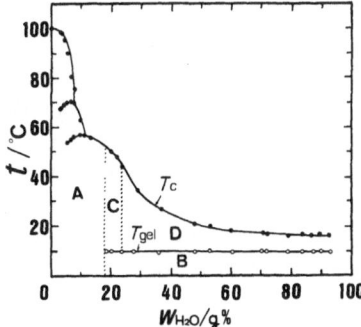

Fig. 1. Phase diagram of the water-octadecyltrimethylammonium chloride system. Phases at a temperature below the T_c curves are divided into four regions of A, B, C and D, depending on both water content and temperature

coagel-to-gel, to which we give the symbol T_{gel} described in figure 1. Before proceeding to a discussion of the character of the T_{gel} transition, we explain the nature of the regions of A, B, C and D at a temperature below the T_c curve, shown in the phase diagram, depending on both temperature and water content.

In the region A at a water content below 17 g%, all the water added to the samples does not crystallize on cooling down to -100 °C, indicating that all the water in this region exists as a "bound water" incorporated between the bilayers of the $C_{18}Cl$ molecules. The phase of this region is constructed with an anhydrous crystal or a hydrated one, but the amount of the bound water does not reach the limiting value ($4H_2O$) for the coagel formation which is achieved at the water content of 17 g%. The structure of water added beyond this limiting value depends on a temperature, above or below the T_{gel} transition temperature. Thus, at a temperature below the T_{gel} curve corresponding to the region B in the phase diagram, more water is present as a bulk free water coexisting with the coagel phase having the limiting amount of the bound water. While, at a temperature above the T_{gel} curve corresponding to the region C, this excess water is additionally interposed between the bilayers, but exists as the weak-bound water, i. e., the so-called "intermediate water" [9] which causes a partly ionized fused state of the polar head groups of the $C_{18}Cl$ molecules. Thus, the gel phase comes into existence in the region C. This gel phase starts to coexist with the bulk free water at a water content above 23 g%, corresponding to the region D in the phase diagram.

Now we should like to proceed to the discussion on the nature of T_c curve. The gel phase of the water-$C_{18}Cl$ system mentioned above is allowed to exist as a

metastable, supercooled phase, down to -100 °C, but by a suitable annealing treatment this supercooled gel phase is transformed into the stable state, the coagel phase. This means that by the annealing treatment the intermediate water as the members of the gel phase is released outside the bilayers and is converted into the bulk free water coexisting with the resulting new phase, the coagel phase. This free water is again intercalated between the bilayers at the elevated T_{gel} transition temperature, giving rise to the phase change from the coagel to the gel phase.

We want to stress again that the coagel-gel transition originates from a change in the water structure from the free to the intermediate water and the existence of the bulk free water coexisting with the coagel phase is a requirement for the appearance of the coagel-gel transition [9].

2) Water-potassium stearate system

As the second example we explain the result of the thermoanalytical investigation on the water-potassium stearate which is a typical anionic surfactant and has been already studied by McBain et al. [6] and Skoulios et al. [10, 11, 12]. In particular, the diagram obtained by McBain has been frequently cited in many texts as a typical example of the surfactant-water systems.

The phase diagram obtained by the present authors is shown in figure 2. As to the anhydrous crystal, before reaching the transition at about 260 °C where it is transformed into the liquid crystal, we find out a total of seven, successive transitions which persist up

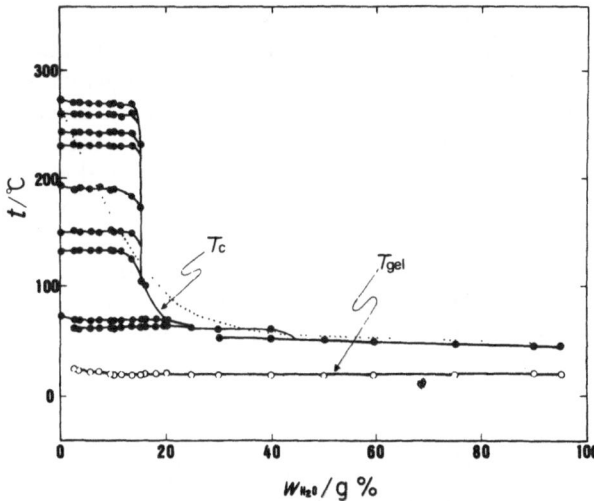

Fig. 2. Phase diagram of the water-potassium stearate system. The T_c curves obtained by McBain et al. [6] are indicated by dotted line

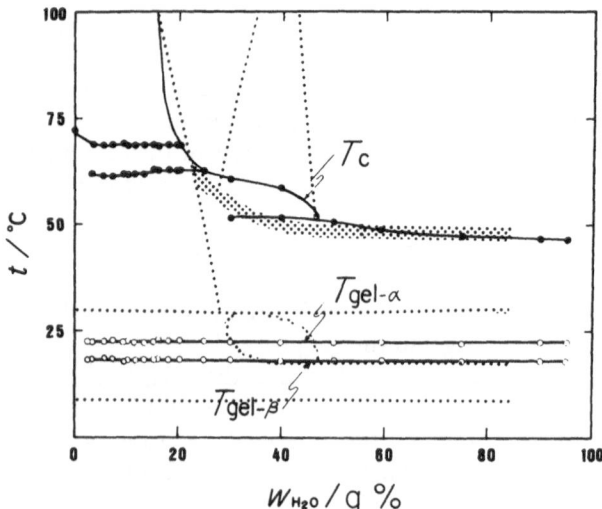

Fig. 3. Detailed phase diagram of the water-potassium stearate system around the T_{gel} transition temperature. The phase diagram reported by Skoulios et al. [12] are indicated by dotted lines

to about 15 g% water content. The first transition at about 70 °C is not observed by Skoulios et al. [10] from the X-ray diffraction methods. These findings obtained here indicates the existence of a successive seven stages of semi-crystalline mesophases caused by stepwise partial meltings of the hydrocarbon chains. The T_c curves obtained here is quite different from the smooth curve proposed by McBain et al. [6] from the turbidity measurements (see fig. 2).

When the water-potassium stearate mixtures are cooled down to a temperature below the T_c curves, the gel phase comes into existence. This gel phase is also allowed to exist as the supercooled one down to −20 °C, similarly to the water-$C_{18}Cl$ gel phase mentioned above. Detailed and careful investigations on the annealing condition for this supercooled gel phase confirmed the existence of two kinds of coagel phase, stable and metastable states, each of which is refered to, by us, as α- and β-phases, respectively. Figure 3 shows the enlarged phase diagram around the T_{gel} transition temperature, where the $T_{gel-\alpha}$ curve corresponds to the phase transition from the coagel α to the gel phase, while the $T_{gel-\beta}$ curve to the coagel β-gel transition. These behaviors around T_{gel} are greatly different from the result obtained from the X-ray method by Skoulios et al. [12] who revealed the existence of composite, metastable gel phases, as shown in figure 3. Furthermore, we notice here that the T_{gel} transitions come to appear at a extremely low water content, quite differently from the water-$C_{18}Cl$ system. Consistently with this finding, we can observe the ice-melting peak at 0 °C at the same low water

content. The estimation of the enthalpy change (ΔH) due to the melting of ice at 0 °C reveals that at a temperature below the T_{gel} curve all the water added to the samples is present as the free water coexisting with the coagel phase, indicating that in the coagel phase of this system the bound water is not interposed between the bilayers. Here, it is emphasized again that the conditions necessary for the appearance of the coagel-gel transition is the existence of free water coexisting with the coagel phase.

3) Water-dipalmitoyl phosphatidylcholine system

Finally, we should like to illustrate the case of dipalmitoyl phosphatidylcholine (DPPC), one of the main constituents of cell membrances [13, 14].

Figure 4 shows the phase diagram of the water-DPPC system obtained from all the DSC thermograms. The so-called main transition temperatures (T_m curve) go down stepwise to the limiting temperature of 42.6 °C with increasing water content, reflecting a total of five endothermic peaks, which is quite different from the smooth direct drift of the T_m curves as depicted by Chapman et al. [7]. The magnified phase diagram below 10 g% water content more evidently indicates the successive phase transition behavior. Similarly to our results, the X-ray observations of Luzzati et al. [15] on egg phosphatidylcholine-water system had revealed also the existence of the composite phases (H_α, O_α and L_α) at the high-temperature high-lipid region, but their transition temperatures are not consistent with our results obtained on the synthetic DPPC employed in the present study (see fig. 5).

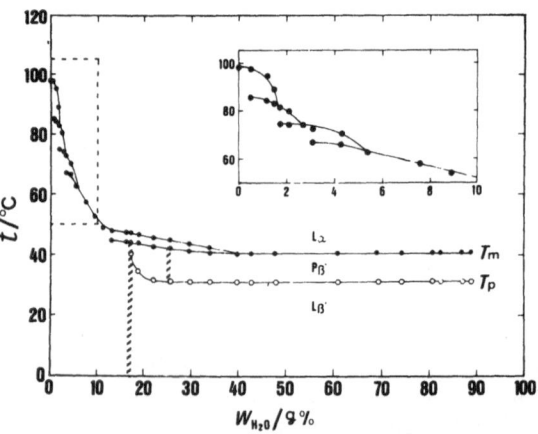

Fig. 4. Phase diagram of the water-dipalmitoyl phosphatidylcholine system. Water contents for phase separations are indicated by hatched lines

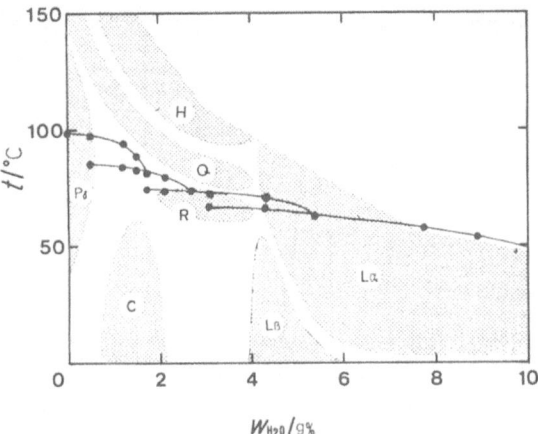

Fig. 5. Comparison of our data with the results by Luzzati et al. [15] given by the hatched areas in the region of low water contents of the water-DPPC system

As to the appearance of the so-called pretransition (T_p curve) phenomenon from L'_β to P'_β phase, we have performed detailed experiments on several specimens with different water contents from 17 g% up to 25 g%. Figure 6 shows the comparison of our data with the curves by Doniach [16] based on the results of Chapman et al. [7] and Janiak et al. [17]. As is shown in the figure, the pretransition temperature which was discovered at 17 g% water content for the first time shifts to lower temperatures with increasing water content up to ca. 25 g% water content and then

reach the nearly constant limiting value of 31.2 °C. Comparing the effect of water content on the pretransition phenomenon with the endothermic peak due to the ice-melting, the sharp endothermic peak at 0 °C makes its appearance for the first time at the same water content of about 17 g%. This finding indicates the necessity of the free water coexisting with the L'_β phase for the appearance of the pretransition phenomenon. This free water is interposed between the bilayers of the DPPC molecules at the elevated pretransition temperature, which causes the sudden increase in the thickness of the water layer at that temperature found by Mitsui et al. [18]. Thus, the resulting new phase corresponds to the so-called P'_β phase of periodic rippled lamellar structure proposed by Tardieu et al. [19] and Janiak et al. [17]. This ripple structure may be caused by the co-operational interaction between the polar head groups of the DPPC molecules and the newly incorporated water (discovered by us) at the pretransition temperature, instead of the structurally bound water proposed by Janiak et al. [17]. Consistently with the observations in two examples mentioned above, it is strongly presumed that the pretransition (T_p) in the water-DPPC system corresponds to the coagel-gel transition (T_{gel}) in the water-surfactant system, while the main transition (T_m) to the gel-liquid crystal one (T_c).

References

1. Luzzati, V., Tardieu, A., Ann. Rev. Phys. Chem. **25**, 79 (1974).
2. Charvolin, J., Tardieu, A., Solid State Physics, Supplement 14 (1978) p. 209.
3. Tiddy, G. J. T., Physics Reports 57, No. 1, 1 (1980).
4. Kelker, H., Hatz, R., Handbook of Liquid Crystals, Verlag Chemie, Weinhein (1980) p. 512.
5. Kodama, M., Kuwabara, M., Seki, S., Thermal analysis, Vol. 2, Proceeding of the 7th ICTA, Heyden & Son. Inc. (1982) p. 822.
6. McBain, J. W., Sierichs, W. C., J. Am. Chem. Soc. **25**, 221 (1948).
7. Chapman, D., Williams, R. M., Iadbrooke, B. D., Chem. Phys. Lipids 1, 445 (1967).
8. Fontell, K., Mol. Cryst. Liq. Cryst. **63**, (1/4), 59 (1981).
9. Kodama, M., Kuwabara, M., Seki, S., Thermochimica Acta, **50**, 8 (1981).
10. Gallot, B., Skoulios, A., Acta Cryst. 15, 826 (1962).
11. Luzzati, V., Mustacchi, H., Skoulios, A., Discuss Faraday Soc. **25**, 43 (1958).
12. Vincent, J. M., Skoulios, A., Acta Cryst. 20, 432 (1966).
13. Kodama, M., Kuwabara, M., Seki, S., Biochim. Biophys. Acta **689**, 567 (1982).
14. Kodama, M., Ogawa, Y., Kuwabara, M., Seki, S., Ions and Molecules in Solution, Studies in Physical and Theoretical Chemistry, Vol. 27 (1982) p. 449 Elsevier Sci. Pub. Co. (1982).
15. Luzzati, V., Gulik-Krzywicki, T., Tardieu, A., Nature 218, 1031 (1968).
16. Doniach, S., J. Chem. Phys. 70, 4587 (1979).

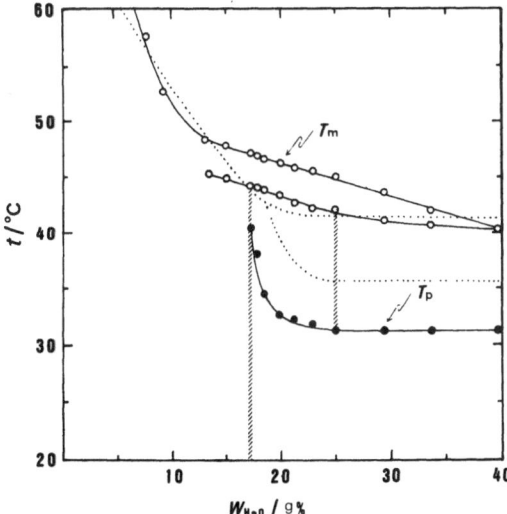

Fig. 6. Detailed phase diagram of the water-DPPC system around the pretransition temperature. The dotted line is taken from the fig. 8 by Doniach [16].

Progress in Colloid & Polymer Science, Vol. 68 (1983)

17. Janiak, M. J., Small, D. M., Shipley, G. G., Biochemistry **15**, 4575 (1976).
18. Inoko, Y., Mitsui, T., J. Phys. Soc. Japan **44**, 1918 (1978).
19. Tardieu, A., Luzzati, V., Reman, F. C., Mol. Biol. **75**, 711 (1973).

Received January 19, 1983;
accepted February 28, 1983

Authors' address:

M. Kodama
Department of Chemistry
Faculty of Science
Kwansei Gakuin University
Nishinomiya, 662 Japan

Subject Index